Pre-Calculus
Know-It-ALL

About the Author

Stan Gibilisco is an electronics engineer, researcher, and mathematician who has authored a number of titles for the McGraw-Hill *Demystified* series, along with more than 30 other books and dozens of magazine articles. His work has been published in several languages.

Pre-Calculus
Know-It-ALL

Stan Gibilisco

New York   Chicago   San Francisco   Lisbon   London   Madrid
Mexico City   Milan   New Delhi   San Juan   Seoul
Singapore   Sydney   Toronto

Cataloging-in-Publication Data is on file with the Library of Congress.

McGraw-Hill books are available at special quantity discounts to use as premiums and sales promotions, or for use in corporate training programs. To contact a representative please e-mail us at bulksales@mcgraw-hill.com.

Pre-Calculus Know-It-ALL

1 2 3 4 5 6 7 8 9 0   DOC/DOC   0 1 4 3 2 1 0 9

ISBN 978-0-07-162702-3
MHID 0-07-162702-2

The pages within this book were printed on acid-free paper.

Sponsoring Editor
  Judy Bass

Acquisitions Coordinator
  Michael Mulcahy

Editorial Supervisor
  David E. Fogarty

Project Manager
  Vasundhara Sawhney

Copy Editor
  Priyanka Sinha

Proofreader
  Bhavna Gupta

Production Supervisor
  Richard C. Ruzycka

Composition
  Glyph International

Art Director, Cover
  Jeff Weeks

To Emma, Samuel, Tony, and Tim

Contents

Preface

This book is intended to complement standard pre-calculus texts at the high-school, trade-school, and college undergraduate levels. It can also serve as a self-teaching or home-schooling supplement. Prerequisites include beginning and intermediate algebra, geometry, and trigonometry. *Pre-Calculus Know-It-ALL* forms an ideal "bridge" between *Algebra Know-It-ALL* and *Calculus Know-It-ALL*.

This course is split into two major sections. Part 1 (Chapters 1 through 10) deals with coordinate systems and vectors. Part 2 (Chapters 11 through 20) is devoted to analytic geometry. Chapters 1 through 9 and 11 through 19 end with practice exercises. They're "open-book" quizzes. You may (and should) refer to the text as you work out your answers. Detailed solutions appear in Appendices A and B. In many cases, these solutions don't represent the only way a problem can be figured out. Feel free to try alternatives!

Chapters 10 and 20 contain question-and-answer sets that finish up Parts 1 and 2, respectively. These chapters aren't tests. They're designed to help you review the material, and to strengthen your grasp of the concepts.

A multiple-choice Final Exam concludes the course. It's a "closed-book" test. Don't look back at the chapters, or use any other external references, while taking it. You'll find these questions more general (and easier) than the practice exercises at the ends of the chapters. The exam is meant to gauge your overall understanding of the concepts, not to measure how fast you can perform calculations or how well you can memorize formulas. The correct answers are listed in Appendix C.

I've tried to introduce "mathematicalese" as the book proceeds. That way, you'll get used to the jargon as you work your way through the examples and problems. If you complete one chapter a week, you'll get through this course in a school year with time to spare, but don't hurry. Proceed at your own pace.

Stan Gibilisco

Acknowledgments

I extend thanks to my nephew Tony Boutelle, a technical writer based in Minneapolis, Minnesota, who offered insights and suggestions from the viewpoint of the intended audience, and found a few arithmetic errors before they got into print!

I'm also grateful to Andrew A. Fedor, M.B.A., P.Eng (afedor@look.ca), a freelance consultant from Hampton, Ontario, Canada, for his proofreading help. Andrew has often provided suggestions for my existing publications and ideas for new ones.

Pre-Calculus
Know-It-ALL

PART
1

Coordinates and Vectors

1

Cartesian Two-Space

If you've taken a course in algebra or geometry, you've learned about the graphing system called *Cartesian* (pronounced "car-TEE-zhun") *two-space*, also known as *Cartesian coordinates* or the *Cartesian plane*. Let's review the basics of this system, and then we'll learn how to calculate distances in it.

How It's Assembled

We can put together a Cartesian plane by positioning two identical *real-number lines* so they intersect at their zero points and are perpendicular to each other. The point of intersection is called the *origin*. Each number line forms an *axis* that can represent the values of a mathematical *variable*.

The variables

Figure 1-1 shows a simple set of Cartesian coordinates. One variable is portrayed along a horizontal line, and the other variable is portrayed along a vertical line. The number-line scales are graduated in increments of the same size.

Figure 1-2 shows how several *ordered pairs* of the form (x,y) are plotted as points on the Cartesian plane. Here, x represents the *independent variable* (the "input"), and y represents the *dependent variable* (the "output"). Technically, when we work in the Cartesian plane, the numbers in an ordered pair represent the *coordinates* of a point on the plane. People sometimes say or write things as if the ordered pair *actually is* the point, but technically the ordered pair is the *name* of the point.

Interval notation

In pre-calculus and calculus, we'll often want to express a continuous span of values that a variable can attain. Such a span is called an *interval*. An interval always has a certain minimum value and a certain maximum value. These are the *extremes* of the interval. Let's be sure that

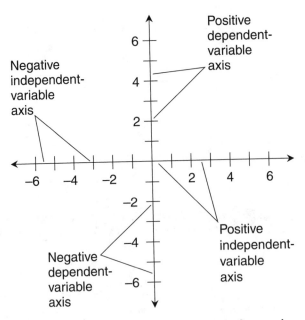

Figure 1-1   The Cartesian plane consists of two real-number lines intersecting at a right angle, forming axes for the variables.

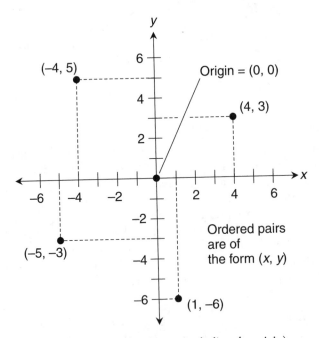

Figure 1-2   Five ordered pairs (including the origin) plotted as points on the Cartesian plane. The dashed lines are for axis location reference.

you're familiar with standard interval terminology and notation, so it won't confuse you later on. Consider these four situations:

$$0 < x < 2$$
$$-1 \leq y < 0$$
$$4 < z \leq 8$$
$$-\pi \leq \theta \leq \pi$$

These expressions have the following meanings, in order:

- The value of x is larger than 0, but smaller than 2.
- The value of y is larger than or equal to -1, but smaller than 0.
- The value of z is larger than 4, but smaller than or equal to 8.
- The value of θ is larger than or equal to $-\pi$, but smaller than or equal to π.

The first case is an example of an *open interval*, which we can write as

$$x \in (0,2)$$

which translates to "x is an element of the open interval (0,2)." Don't mistake this open interval for an ordered pair! The notations look the same, but the meanings are completely different. The second and third cases are examples of *half-open intervals*. We denote this type of interval with a square bracket on the side of the included value and a rounded parenthesis on the side of the non-included value. We can write

$$y \in [-1,0)$$

which means "y is an element of the half-open interval $[-1,0)$," and

$$z \in (4,8]$$

which means "z is an element of the half-open interval $(4,8]$." The fourth case is an example of a *closed interval*. We use square brackets on both sides to show that both extremes are included. We can write this as

$$\theta \in [-\pi,\pi]$$

which translates to "θ is an element of the closed interval $[-\pi,\pi]$."

Relations and functions

Do you remember the definitions of the terms *relation* and *function* from your algebra courses? (If you read *Algebra Know-It-All*, you should!) These terms are used often in pre-calculus, so it's important that you be familiar with them. A *relation* is an operation that transforms, or *maps*, values of a variable into values of another variable. A *function* is a relation in which there is never more than one value of the dependent variable for any value of the independent variable. In other words, there can't be more than one output for any input. (If a particular input produces no output, that's okay.) The Cartesian plane gives us an excellent way to illustrate relations and functions.

The axes

In a Cartesian plane, both axes are *linear*, and both axes are graduated in *increments* of the same size. On either axis, the change in value is always directly proportional to the physical displacement. For example, if we travel 5 millimeters along an axis and the value changes by 1 unit, then that fact is true everywhere along that axis, and it's also true everywhere along the other axis.

The quadrants

Any pair of intersecting lines divides a plane into four parts. In the Cartesian system, these parts are called *quadrants*, as shown in Fig. 1-3:

- In the *first quadrant*, both variables are positive.
- In the *second quadrant*, the independent variable is negative and the dependent variable is positive.
- In the *third quadrant*, both variables are negative.
- In the *fourth quadrant*, the independent variable is positive and the dependent variable is negative.

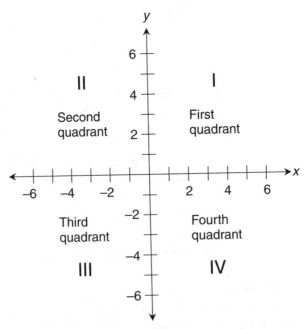

Figure 1-3   The Cartesian plane is divided into quadrants. The first, second, third, and fourth quadrants are sometimes labeled I, II, III, and IV, respectively.

The quadrants are sometimes labeled with Roman numerals, so that

- Quadrant I is at the upper right
- Quadrant II is at the upper left
- Quadrant III is at the lower left
- Quadrant IV is at the lower right

If a point lies on one of the axes or at the origin, then it is not in any quadrant.

Are you confused?

Why do we insist that the increments be the same size on both axes in a Cartesian two-space graph? The answer is simple: That's how the Cartesian plane is defined! But there are other types of coordinate systems in which this exactness is not required. In a more generalized system called *rectangular coordinates* or the *rectangular coordinate plane*, the two axes can be graduated in divisions of different size. For example, the value on one axis might change by 1 unit for every 5 millimeters, while the value on the other axis changes by 1 unit for every 10 millimeters.

Here's a challenge!

Imagine an ordered pair (x,y), where both variables are nonzero real numbers. Suppose that you've plotted a point (call it P) on the Cartesian plane. Because $x \neq 0$ and $y \neq 0$, the point P does not lie on either axis. What will happen to the location of P if you multiply x by -1 and leave y the same? If you multiply y by -1 and leave x the same? If you multiply both x and y by -1?

Solution

If you multiply x by -1 and do not change the value of y, P will move to the opposite side of the y axis, but will stay the same distance away from that axis. The point will, in effect, be "reflected" by the y axis, moving to the left if x is positive to begin with, and to the right if x is negative to begin with.

- If P starts out in the first quadrant, it will move to the second.
- If P starts out in the second quadrant, it will move to the first.
- If P starts out in the third quadrant, it will move to the fourth.
- If P starts out in the fourth quadrant, it will move to the third.

If you multiply y by -1 and leave x unchanged, P will move to the opposite side of the x axis, but will stay the same distance away from that axis. In a sense, P will be "reflected" by the x axis, moving straight downward if y is initially positive and straight upward if y is initially negative.

- If P starts out in the first quadrant, it will move to the fourth.
- If P starts out in the second quadrant, it will move to the third.
- If P starts out in the third quadrant, it will move to the second.
- If P starts out in the fourth quadrant, it will move to the first.

If you multiply both *x* and *y* by −1, *P* will move diagonally to the opposite quadrant. It will, in effect, be "reflected" by both axes.

- If *P* starts out in the first quadrant, it will move to the third.
- If *P* starts out in the second quadrant, it will move to the fourth.
- If *P* starts out in the third quadrant, it will move to the first.
- If *P* starts out in the fourth quadrant, it will move to the second.

If you have trouble envisioning these point maneuvers, draw a Cartesian plane on a piece of graph paper. Then plot a point or two in each quadrant. Calculate how the *x* and *y* values change when you multiply either or both of them by −1, and then plot the new points.

Distance of a Point from Origin

On a straight number line, the distance of any point from the origin is equal to the absolute value of the number corresponding to the point. In the Cartesian plane, the distance of a point from the origin depends on both of the numbers in the point's ordered pair.

An example

Figure 1-4 shows the point (4,3) plotted in the Cartesian plane. Suppose that we want to find the distance *d* of (4,3) from the origin (0,0). How can this be done?

We can calculate *d* using the *Pythagorean theorem* from geometry. In case you've forgotten that principle, here's a refresher. Suppose we have a right triangle defined by points *P*, *Q*, and *R*. Suppose the sides of the triangle have lengths *b*, *h*, and *d* as shown in Fig. 1-5. Then

$$b^2 + h^2 = d^2$$

We can rewrite this as

$$d = (b^2 + h^2)^{1/2}$$

where the 1/2 power represents the nonnegative square root. Now let's make the following point assignments between the situations of Figs. 1-4 and 1-5:

- The origin in Fig. 1-4 corresponds to the point *Q* in Fig. 1-5.
- The point (4,0) in Fig. 1-4 corresponds to the point *R* in Fig. 1-5.
- The point (4,3) in Fig. 1-4 corresponds to the point *P* in Fig. 1-5.

Continuing with this analogy, we can see the following facts:

- The line segment connecting the origin and (4,0) has length *b* = 4.
- The line segment connecting (4,0) and (4,3) has height *h* = 3.
- The line segment connecting the origin and (4,3) has length *d* (unknown).

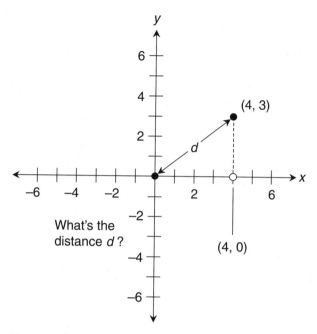

Figure 1-4   We can use the Pythagorean theorem to find the distance *d* of the point (4,3) from the origin (0,0) in the Cartesian plane.

The side of the right triangle having length *d* is the longest side, called the *hypotenuse*. Using the Pythagorean formula, we can calculate

$$d = (b^2 + h^2)^{1/2} = (4^2 + 3^2)^{1/2} = (16 + 9)^{1/2} = 25^{1/2} = 5$$

We've determined that the point (4,3) is 5 units distant from the origin in Cartesian coordinates, as measured along a straight line connecting (4,3) and the origin.

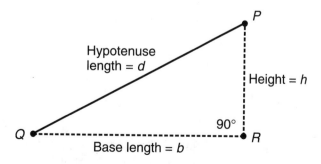

Figure 1-5   The Pythagorean theorem for right triangles.

The general formula

We can generalize the previous example to get a formula for the distance of *any* point from the origin in the Cartesian plane. In fact, we can repeat the explanation of the previous example almost verbatim, only with a few substitutions.

Consider a point P with coordinates (x_p, y_p). We want to calculate the straight-line distance d of the point P from the origin $(0,0)$, as shown in Fig. 1-6. Once again, we use the Pythagorean theorem. Turn back to Fig. 1-5 and follow along by comparing with Fig. 1-6:

- The origin in Fig. 1-6 corresponds to the point Q in Fig. 1-5.
- The point $(x_p, 0)$ in Fig. 1-6 corresponds to the point R in Fig. 1-5.
- The point (x_p, y_p) in Fig. 1-6 corresponds to the point P in Fig. 1-5.

The following facts are also visually evident:

- The line segment connecting the origin and $(x_p, 0)$ has length $b = x_p$.
- The line segment connecting $(x_p, 0)$ and (x_p, y_p) has height $h = y_p$.
- The line segment connecting the origin and (x_p, y_p) has length d (unknown).

The Pythagorean formula tells us that

$$d = (b^2 + h^2)^{1/2} = (x_p^2 + y_p^2)^{1/2}$$

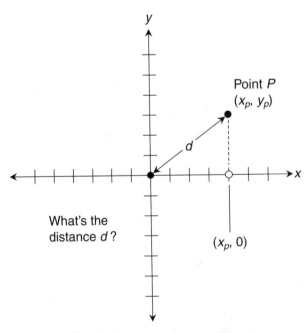

Figure 1-6 Using the Pythagorean theorem, we can derive a formula for the distance d of a generalized point $P = (x_p, y_p)$ from the origin.

That's it! The point (x_p, y_p) is $(x_p^2 + y_p^2)^{1/2}$ units away from the origin, as we would measure it along a straight line.

Are you confused?

You might ask, "Can the distance of a point from the origin ever be negative?" The answer is no. If you look at the formula and break down the process in your mind, you'll see why this is so. First, you square x_p, which is the x coordinate of P. Because x_p is a real number, its square must be a nonnegative real. Next, you square y_p, which is the y coordinate of P. This result must also be a nonnegative real. Next, you add these two nonnegative reals, which must produce another nonnegative real. Finally, you take the nonnegative square root, getting yet another nonnegative real. That's the distance of P from the origin. It can't be negative in a Cartesian plane whose axes represent real-number variables.

Here's a challenge!

Imagine a point $P = (x_p, y_p)$ in the Cartesian plane, where $x_p \neq 0$ and $y_p \neq 0$. Suppose that P is d units from the origin. What will happen to d if you multiply x_p by -1 and leave y unchanged? If you multiply y_p by -1 and leave x unchanged? If you multiply both x_p and y_p by -1?

Solution

This is a three-part challenge. Let's break each part down into steps and apply the distance formula in each case.

In the first situation, we change the x coordinate of P to its negative. Let's call the new point P_{x-}. Its coordinates are $(-x_p, y_p)$. Let d_{x-} be the distance of P_{x-} from the origin. Plugging the values into the formula, we obtain

$$d_{x-} = [(-x_p)^2 + y_p^2]^{1/2} = [(-1)^2 x_p^2 + y_p^2]^{1/2} = (x_p^2 + y_p^2)^{1/2} = d$$

In the second situation, we change the y coordinate of P to its negative. This time, let's call the new point P_{y-}. Its coordinates are $(x_p, -y_p)$. Let d_{y-} represent the distance of P_{y-} from the origin. Plugging the values into the formula, we obtain

$$d_{y-} = [(x_p)^2 + (-y_p)^2]^{1/2} = [x_p^2 + (-1)^2 y_p^2]^{1/2} = (x_p^2 + y_p^2)^{1/2} = d$$

In the third case, we change both the x and y coordinates of P to their negatives. We can call the new point P_{xy-} with coordinates $(-x_p, -y_p)$. If we let d_{xy-} represent the distance of P_{xy-} from the origin, we have

$$d_{xy-} = [(-x_p)^2 + (-y_p)^2]^{1/2} = [(-1)^2 x_p^2 + (-1)^2 y_p^2]^{1/2} = (x_p^2 + y_p^2)^{1/2} = d$$

We've shown that we can negate either or both of the coordinate values of a point in the Cartesian plane, and although the point's location will usually change, its distance from the origin will always stay the same.

Distance between Any Two Points

The distance between any two points on a number line is easy to calculate. We take the absolute value of the difference between the numbers corresponding to the points. In the Cartesian plane, each point needs two numbers to be defined, so the process is more complicated.

Setting up the problem

Figure 1-7 shows two generic points, *P* and *Q*, in the Cartesian plane. Their coordinates are

$$P = (x_p, y_p)$$

and

$$Q = (x_q, y_q)$$

Suppose we want to find the distance *d* between these points. We can construct a triangle by choosing a third point, *R* (which isn't on the line defined by *P* and *Q*) and then connecting *P*, *Q*, and *R* by line segments to get a triangle. The shape of triangle *PQR* depends on the location of *R*. If we choose certain coordinates for *R*, we can get a right triangle with the right angle at vertex *R*.

  With the help of Fig. 1-7, it's easy to see what the coordinates of *R* should be. If I travel "straight down" (parallel to the *y* axis) from *P*, and if you travel "straight to the right" (parallel to the

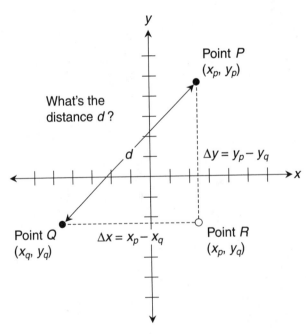

Figure 1-7   We can find the distance *d* between two points $P = (x_p, y_p)$ and $Q = (x_q, y_q)$ by choosing point *R* to get a right triangle, and then applying the Pythagorean theorem.

x axis) from *Q*, our paths will cross at a right angle when we reach the point whose coordinates are (x_p, y_q). Those are the coordinates that *R* must have if we want the two sides of the triangle to be perpendicular there.

Are you confused?

"Wait!" you say. "Isn't there another point besides *R* that we can choose to create a right triangle along with points *P* and *Q*?" Yes, there is. The situation is shown in Fig. 1-8. If I go "straight up" (parallel to the *y* axis) from *Q*, and if you go "straight to the left" (parallel to the *x* axis) from *P*, we will meet at a right angle when we reach the coordinates (x_q, y_p). In this case, we might call the right-angle vertex point *S*. We won't use this geometry in the derivation that follows. But we could, and the final distance formula would turn out the same.

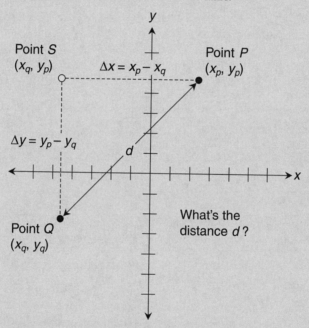

Figure 1-8   Alternative geometry for finding the distance between two points. In this case, the right angle appears at point *S*.

Dimensions and "deltas"

Mathematicians use the uppercase Greek letter delta (Δ) to stand for the phrase "the difference in" or "the difference between." Using this notation, we can say that

- The difference in the *x* values of points *R* and *Q* in Fig. 1-7 is $x_p - x_q$, or Δx. That's the length of the base of a right triangle.
- The difference in the *y* values of points *P* and *R* is $y_p - y_q$, or Δy. That's the height of a right triangle.

We can see from Fig. 1-7 that the distance d between points Q and P is the length of the hypotenuse of triangle PQR. We're ready to find a formula for d using the Pythagorean theorem.

The general formula

Look back once more at Fig. 1-5. The relative positions of points P, Q, and R here are similar to their positions in Fig. 1-7. (I've set things up that way on purpose, as you can probably guess.) We can define the lengths of the sides of the triangle in Fig. 1-7 as follows:

- The line segment connecting points Q and R has length $b = \Delta x = x_p - x_q$.
- The line segment connecting points R and P has height $h = \Delta y = y_p - y_q$.
- The line segment connecting points Q and P has length d (unknown).

The Pythagorean formula tells us that

$$d = (b^2 + h^2)^{1/2} = (\Delta x^2 + \Delta y^2)^{1/2} = [(x_p - x_q)^2 + (y_p - y_q)^2]^{1/2}$$

An example

Let's find the distance d between the following points in the Cartesian plane, using the formula we've derived:

$$P = (-5,-2)$$

and

$$Q = (7,3)$$

Plugging the values $x_p = -5$, $y_p = -2$, $x_q = 7$, and $y_q = 3$ into our formula, we get

$$d = [(x_p - x_q)^2 + (y_p - y_q)^2]^{1/2} = [(-5 - 7)^2 + (-2 - 3)^2]^{1/2}$$
$$= [(-12)^2 + (-5)^2]^{1/2} = (144 + 25)^{1/2} = 169^{1/2} = 13$$

- -

Here's a challenge!

It's reasonable to suppose that the distance between two points shouldn't depend on the direction in which we travel. But if you're a "show-me" person (as a mathematician should be), you might demand proof. Let's do it!

Solution

When we derived the distance formula previously, we traveled upward and to the right in Fig. 1-7 (from Q to P). When we work with directional displacement, it's customary to subtract the starting-point coordinates from the finishing-point coordinates. That's how we got

$$\Delta x = x_p - x_q$$

and

$$\Delta y = y_p - y_q$$

If we travel downward and to the left (from P to Q), we get

$$\Delta^* x = x_q - x_p$$

and

$$\Delta^* y = y_q - y_p$$

when we subtract the starting-point coordinates from the finishing-point coordinates. These new "star deltas" are the negatives of the original "plain deltas" because the subtractions are done in reverse. If we plug the "star deltas" straightaway into the derivation for d we worked out a few minutes ago, we can maneuver to get

$$d = (\Delta^* x^2 + \Delta^* y^2)^{1/2} = [(-\Delta x)^2 + (-\Delta y)^2]^{1/2} = [(-1)^2 \Delta x^2 + (-1)^2 \Delta y^2]^{1/2}$$
$$= (\Delta x^2 + \Delta y^2)^{1/2} = [(x_p - x_q)^2 + (y_p - y_q)^2]^{1/2}$$

That's the same distance formula we got when we went from Q to P. This proves that the direction of travel isn't important when we talk about the simple distance between two points in Cartesian coordinates. (When we work with *vectors* later in this book, the direction will matter. Directional distance is known as *displacement*.)

Finding the Midpoint

We can find the midpoint between two points on a number line by calculating the *arithmetic mean* (or average value) of the numbers corresponding to the points. In Cartesian xy coordinates, we must make two calculations. First, we average the x values of the two points to get the x value of the point midway between. Then, we average the y values of the points to get the y value of the point midway between.

A "mini theorem"

Once again, imagine points P and Q in the Cartesian plane with the coordinates

$$P = (x_p, y_p)$$

and

$$Q = (x_q, y_q)$$

Suppose we want to find the coordinates of the midpoint. That's the point that bisects a straight line segment connecting P and Q. As before, we start out by choosing the point R "below and

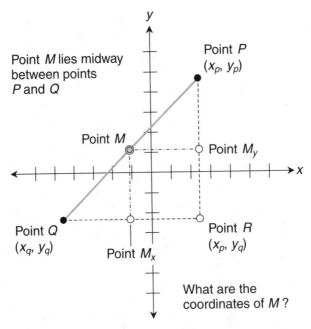

Figure 1-9 We can calculate the coordinates of the midpoint of a line segment whose endpoints are known.

to the right" that forms a right triangle *PQR*, as shown in Fig. 1-9. Imagine a movable point *M* that we can slide freely along line segment *PQ*. When we draw a perpendicular from *M* to side *QR*, we get a point M_x. When we draw a perpendicular from *M* to side *RP*, we get a point M_y.

Consider the three right triangles MQM_x, PMM_y, and *PQR*. The laws of basic geometry tell us that these triangles are *similar*, meaning that the lengths of their corresponding sides are in the same ratios. According to the definition of similarity for triangles, we know the following two facts:

- Point M_x is midway between *Q* and *R* if and only if *M* is midway between *P* and *Q*.
- Point M_y is midway between *R* and *P* if and only if *M* is midway between *P* and *Q*.

Now, instead of saying that *M* stands for "movable point," let's say that *M* stands for "midpoint." In this case, the *x* value of M_x (the midpoint of line segment *QR*) must be the *x* value of *M*, and the *y* value of M_y (the midpoint of line segment *RP*) must be the *y* value of *M*.

The general formula

We've reduced our Cartesian two-space midpoint problem to two separate number-line midpoint problems. Side *QR* of triangle *PQR* is parallel to the *x* axis, and side *RP* of triangle *PQR* is parallel to the *y* axis. We can find the *x* value of M_x by averaging the *x* values of *Q* and *R*. When we do this and call the result x_m, we get

$$x_m = (x_p + x_q)/2$$

In the same way, we can calculate the y value of M_y by averaging the y values of R and P. Calling the result y_m, we have

$$y_m = (y_p + y_q)/2$$

We can use the "mini theorem" we finished a few moments ago to conclude that the coordinates of point M, the midpoint of line segment PQ, are

$$(x_m, y_m) = [(x_p + x_q)/2, (y_p + y_q)/2]$$

An example

Let's find the coordinates (x_m, y_m) of the midpoint M between the same two points for which we found the separation distance earlier in this chapter:

$$P = (-5, -2)$$

and

$$Q = (7, 3)$$

When we plug $x_p = -5$, $y_p = -2$, $x_q = 7$, and $y_q = 3$ into the midpoint formula, we get

$$(x_m, y_m) = [(x_p + x_q)/2, (y_p + y_q)/2] = [(-5 + 7)/2, (-2 + 3)/2]$$
$$= (2/2, 1/2) = (1, 1/2)$$

- -

Are you a skeptic?

It seems reasonable to suppose the midpoint between points P and Q should not depend on whether we go from P to Q or from Q to P. We can prove this by showing that for all real numbers x_p, y_p, x_q, and y_q, we have

$$[(x_p + x_q)/2, (y_p + y_q)/2] = [(x_q + x_p)/2, (y_q + y_p)/2]$$

This demonstration is easy, but let's go through it step-by-step to completely follow the logic. For the x coordinates, the commutative law of addition tells us that

$$x_p + x_q = x_q + x_p$$

Dividing each side by 2 gives us

$$(x_p + x_q)/2 = (x_q + x_p)/2$$

For the y coordinates, the commutative law says that

$$y_p + y_q = y_q + y_p$$

Again dividing each side by 2, we get

$$(y_p + y_q)/2 = (y_q + y_p)/2$$

We've shown that the coordinates in the ordered pair on the left-hand side of the original equation are equal to the corresponding coordinates in the ordered pair on the right-hand side. The ordered pairs are identical, so the midpoint is the same in either direction.

Are you confused?

To find a midpoint of a line segment in Cartesian two-space, you simply average the coordinates of the endpoints. This method always works if the midpoint lies on a *straight line segment* between the two endpoints. But you might wonder, "How can we find the midpoint between two points along an *arc* connecting those points?" In a situation like that, we must determine the length of the arc. Depending on the nature of the arc, that can be fairly hard, very hard, or almost impossible! Arc-length problems are beyond the scope of this book, but you'll learn how to solve them in *Calculus Know-It-All*.

Here's a challenge!

Consider two points in the Cartesian plane, one of which is at the origin. Show that the coordinate values of the midpoint are exactly half the corresponding coordinate values of the point not on the origin.

Solution

We can plug in (0,0) as the coordinates of either point in the general midpoint formula, and work things out from there. First, let's suppose that point P is at the origin and the coordinates of point Q are (x_q, y_q). Then $x_p = 0$ and $y_p = 0$. If we call the coordinates of the midpoint (x_m, y_m), we have

$$(x_m, y_m) = [(x_p + x_q)/2, (y_p + y_q)/2] = [(0 + x_q)/2, (0 + y_q)/2]$$
$$= (x_q/2, y_q/2)$$

Now, let Q be at the origin and let the coordinates of P be (x_p, y_p). In that case, we have

$$(x_m, y_m) = [(x_p + x_q)/2, (y_p + y_q)/2] = [(x_p + 0)/2, (y_p + 0)/2]$$
$$= (x_p/2, y_p/2)$$

Practice Exercises

This is an open-book quiz. You may (and should) refer to the text as you solve these problems. Don't hurry! You'll find worked-out answers in App. A. The solutions in the appendix may not represent the only way a problem can be figured out. If you think you can solve a particular problem in a quicker or better way than you see there, by all means try it!

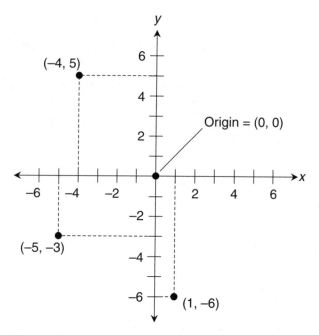

Figure 1-10 Illustration for Problems 1 through 7.

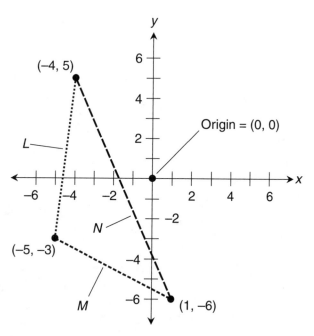

Figure 1-11 Illustration for Problems 8 through 10.

1. What are the x and y coordinates of the points shown in Fig. 1-10?

2. Determine the distance of the point (−4,5) from the origin in Fig. 1-10. Using a calculator, round off the answer to three decimal places.

3. Determine the distance of the point (−5,−3) from the origin in Fig. 1-10. Using a calculator, round it off to three decimal places.

4. Determine the distance of the point (1,−6) from the origin in Fig. 1-10. Using a calculator, round it off to three decimal places.

5. Determine the distance between the points (−4,5) and (−5,−3) in Fig. 1-10. Using a calculator, round it off to three decimal places.

6. Determine the distance between the points (−5,−3) and (1,−6) in Fig. 1-10. Using a calculator, round it off to three decimal places.

7. Determine the distance between the points (1,−6) and (−4,5) in Fig. 1-10. Using a calculator, round it off to three decimal places.

8. Determine the coordinates of the midpoint of line segment L in Fig. 1-11. Express the values in fractional and decimal form.

9. Determine the coordinates of the midpoint of line segment M in Fig. 1-11. Express the values in fractional and decimal form.

10. Determine the coordinates of the midpoint of line segment N in Fig. 1-11. Express the values in fractional and decimal form.

2

A Fresh Look at Trigonometry

Trigonometry (or "trig") involves the relationships between angles and distances. Traditional texts usually define the *trigonometric functions* of an angle as ratios between the lengths of the sides of a right triangle containing that angle. If you've done trigonometry with triangles, get ready for a new perspective!

Circles in the Cartesian Plane

In Cartesian *xy* coordinates, circles are represented by straightforward equations. The equation for a particular circle depends on its radius, and also on the location of its center point.

The unit circle

In trigonometry, we're interested in the circle whose center is at the origin and whose radius is 1. This is the simplest possible circle in the *xy* plane. It's called the *unit circle*, and is represented by the equation

$$x^2 + y^2 = 1$$

The unit circle gives us an elegant way to define the basic trigonometric functions. That's why these functions are sometimes called the *circular functions*. Before we get into the circular functions themselves, let's be sure we know how to define angles, which are the *arguments* (or inputs) of the trig functions.

Naming angles

Mathematicians often use Greek letters to represent angles. The italic, lowercase Greek letter *theta* is popular. It looks like an italic numeral *0* with a horizontal line through it (θ). When writing about two different angles, a second Greek letter is used along with θ. Most often, it's the italic, lowercase letter *phi*. This character looks like an italic lowercase English letter *o* with a forward slash through it (ϕ).

Sometimes the italic, lowercase Greek letters *alpha*, *beta*, and *gamma* are used to represent angles. These, respectively, look like the following symbols: α, β, and γ. When things get messy and there are a lot of angles to talk about, numeric subscripts may be used with Greek letters, so don't be surprised if you see text in which angles are denoted θ_1, θ_2, θ_3, and so on. If you read enough mathematical papers, you'll eventually come across angles that are represented by other lowercase Greek letters. Angle variables can also be represented by more familiar characters such as x, y, or z. As long as we know the context and stay consistent in a given situation, it really doesn't matter what we call an angle.

Radian measure

Imagine two rays pointing outward from the center of a circle. Each ray intersects the circle at a point. Suppose that the distance between these points, as measured along the arc of the circle, is equal to the radius of the circle. In that case, the measure of the angle between the rays is one *radian* (1 rad). There are always 2π rad in a full circle, where π (the lowercase, non-italic Greek letter pi) stands for the ratio of a circle's circumference to its diameter. The number π is irrational. Its value is approximately 3.14159.

Mathematicians prefer the radian as a standard unit of angular measure, and it's the unit we'll work with in this course. It's common practice to omit the "rad" after an angle when we know that we're working with radians. Based on that convention:

- An angle of $\pi/2$ represents 1/4 of a circle
- An angle of π represents 1/2 of a circle
- An angle of $3\pi/2$ represents 3/4 of a circle
- An angle of 2π represents a full circle

An *acute angle* has a measure of more than 0 but less than $\pi/2$, a *right angle* has a measure of exactly $\pi/2$, an *obtuse angle* has a measure of more than $\pi/2$ but less than π, a *straight angle* has a measure of exactly π, and a *reflex angle* has a measure of more than π but less than 2π.

Degree measure

The *angular degree* (°), also called the *degree of arc*, is the unit of angular measure familiar to lay people. One degree (1°) is 1/360 of a full circle. You probably know the following basic facts:

- An angle of 90° represents 1/4 of a circle
- An angle of 180° represents 1/2 of a circle
- An angle of 270° represents 3/4 of a circle
- An angle of 360° represents a full circle

An *acute angle* has a measure of more than 0 but less than 90°, a *right angle* has a measure of exactly 90°, an *obtuse angle* has a measure of more than 90° but less than 180°, a *straight angle* has a measure of exactly 180°, and a *reflex angle* has a measure of more than 180° but less than 360°.

Are you confused?

If you're used to measuring angles in degrees, the radian can seem unnatural at first. "Why," you might ask, "would we want to divide a circle into an *irrational* number of angular parts?" Mathematicians do this because it nearly always works out more simply than the degree-measure scheme in algebra, geometry, trigonometry, pre-calculus, and calculus. The radian is *more* natural than the degree, not less! We can define the radian in a circle without having to quote any numbers at all, just as we can define the diagonal of a square as the distance from one corner to the opposite corner. The radian is a purely geometric unit. The degree is contrived. (What's so special about the fraction 1/360, anyhow? To me, it would have made more sense if our distant ancestors had defined the degree as 1/100 of a circle.)

Here's a challenge!

The measure of a certain angle θ is $\pi/6$. What fraction of a complete circular rotation does this represent? What is the measure of θ in degrees?

Solution

A full circular rotation represents an angle of 2π. The value $\pi/6$ is equal to 1/12 of 2π. Therefore, the angle θ represents 1/12 of a full circle. In degree measure, that's 1/12 of 360°, which is 30°.

Primary Circular Functions

Let's look again at the equation of a unit circle in the Cartesian xy plane. We get it by adding the squares of the variables and setting the sum equal to 1:

$$x^2 + y^2 = 1$$

Imagine that θ is an angle whose vertex is at the origin, and we measure this angle in a counterclockwise sense from the x axis, as shown in Fig. 2-1. Suppose this angle corresponds to a ray that intersects the unit circle at a point P, where

$$P = (x_0, y_0)$$

We can define the three *basic circular functions*, also called the *primary circular functions*, of θ in a simple way. But before we get into that, let's extend our notion of angles to include negative values, and also to deal with angles larger than 2π.

Offbeat angles

In trigonometry, any *direction angle*, no matter how extreme, can always be reduced to something that's nonnegative but less than 2π. Even if the ray OP in Fig. 2-1 makes more than one complete revolution counterclockwise from the x axis, or if it turns clockwise instead, its

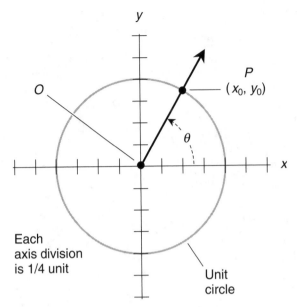

Figure 2-1   The unit circle, whose equation is $x^2 + y^2 = 1$, can serve as the basis for defining trigonometric functions. In this graph, each axis division represents 1/4 unit.

direction can always be defined by some counterclockwise angle of least 0 but less than 2π relative to the *x* axis.

Think of this situation another way. The point *P* must always be somewhere on the circle, no matter how many times or in what direction the ray *OP* rotates to end up in a particular position. Every point on the circle corresponds to exactly one nonnegative angle less than 2π counterclockwise from the *x* axis. Conversely, if we consider the continuous range of angles going counterclockwise over the half-open interval $[0,2\pi)$, we can account for every point on the circle.

Any offbeat direction angle such as $-9\pi/4$ can be reduced to a direction angle that measures at least 0 but less than 2π by adding or subtracting some whole-number multiple of 2π. But we must be careful about this. A direction angle specifies orientation only. The orientation of the ray *OP* is the same for an angle of 3π as for an angle of π, but the larger value carries with it the idea that the ray (also called a *vector*) *OP* has rotated one and a half times around, while the smaller angle implies that it has undergone only half of a rotation. For our purposes now, this doesn't matter. But in some disciplines and situations, it does!

Negative angles are encountered in trigonometry, especially in graphs of functions. Multiple revolutions of objects are important in physics and engineering. So if you ever hear or read about an angle such as $-\pi/2$ or 5π, you can be confident that it has meaning. The negative value indicates clockwise rotation. An angle larger than 2π indicates more than one complete rotation counterclockwise. An angle of less than -2π indicates more than one complete rotation clockwise.

The sine function

Look again at Fig. 2-1. Imagine that ray OP points along the x axis, and then starts to rotate counterclockwise at steady speed around its end point O, as if that point is a mechanical bearing. The point P, represented by coordinates (x_0, y_0), therefore revolves around O, following the unit circle.

Imagine what happens to the value of y_0 (the *ordinate* of point P) during one complete revolution of ray OP. The ordinate of P starts out at $y_0 = 0$, then increases until it reaches $y_0 = 1$ after P has gone 1/4 of the way around the circle (that is, the ray has turned through an angle of $\pi/2$). After that, y_0 begins to decrease, getting back to $y_0 = 0$ when P has gone 1/2 of the way around the circle (the ray has turned through an angle of π). As P continues in its orbit, y_0 keeps decreasing until the value of y_0 reaches its minimum of -1 when P has gone 3/4 of the way around the circle (the ray has turned through an angle of $3\pi/2$). After that, the value of y_0 rises again until, when P has gone completely around the circle, it returns to $y_0 = 0$ for $\theta = 2\pi$.

The value of y_0 is defined as the *sine* of the angle θ. The *sine function* is abbreviated as sin, so we can write

$$\sin \theta = y_0$$

Circular motion

Imagine that you attach a "glow-in-the-dark" ball to the end of a string, and then swing the ball around and around at a steady rate of one revolution per second. Suppose that you make the ball circle your head so the path of the ball lies in a horizontal plane. Imagine that you are in the middle of a flat, open field at night. The ball describes a circle as viewed from high above, as shown in Fig. 2-2A. If a friend stands far away with her eyes exactly in the plane of the ball's orbit, she sees a point of light that oscillates back and forth, from right-to-left and left-to-right, along what appears to be a straight-line path (Fig. 2-2B). Starting from its rightmost apparent position, the glowing point moves toward the left for 1/2 second, speeding up and then slowing down; then it reverses direction; then it moves toward the right for

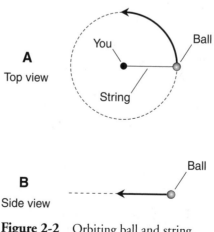

Figure 2-2   Orbiting ball and string. At A, as seen from above; at B, as seen edge-on.

1/2 second, speeding up and then slowing down; then turns around again. As seen by your friend, the ball reaches its extreme rightmost position at 1-second intervals, because its orbital speed is one revolution per second.

The sine wave

If you graph the apparent position of the ball as seen by your friend with respect to time, the result is a *sine wave*, which is a graphical plot of a sine function. Some sine waves "rise higher and lower" (corresponding to a longer string), some are "flatter" (the equivalent of a shorter string), some are "stretched out" (a slower rate of revolution), and some are "squashed" (a faster rate of revolution). But the characteristic shape of the wave, known as a *sinusoid*, is the same in every case.

You can whirl the ball around faster or slower than one revolution per second, thereby altering the *frequency* of the sine wave: the number of times a complete wave cycle repeats within a specified interval on the independent-variable axis. You can make the string longer or shorter, thereby adjusting the *amplitude* of the wave: the difference between the extreme values of its dependent variable. No matter what changes you might make of this sort, the sinusoid can always be defined in terms of a moving point that orbits a central point at a constant speed in a perfect circle.

If we want to graph a sinusoid in the Cartesian plane, the circular-motion analogy can be stated as

$$y = a \sin b\theta$$

where a is a constant that depends on the radius of the circle, and b is a constant that depends on the revolution rate. The angle θ is expressed counterclockwise from the positive x axis. Figure 2-3 illustrates a graph of the basic sine function; it's a sinusoid for which $a = 1$ and $b = 1$, and for which the angle is expressed in radians.

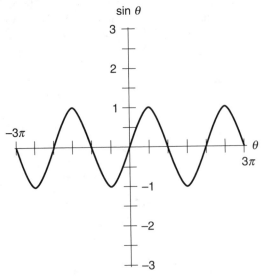

Figure 2-3   Graph of the sine function for values of θ between -3π and 3π. Each division on the horizontal axis represents $\pi/2$ units. Each division on the vertical axis represents 1/2 unit.

The cosine function

Look again at Fig. 2-1. Imagine, once again, a ray *OP* running outward from the origin through point *P* on the circle. Imagine that at first, the ray points along the *x* axis, and then it rotates steadily in a counterclockwise direction.

Now let's think about what happens to the value of x_0 (the *abscissa* of point *P*) during one complete revolution of ray *OP*. It starts out at $x_0 = 1$, then decreases until it reaches $x_0 = 0$ when $\theta = \pi/2$. Then x_0 continues to decrease, getting down to $x_0 = -1$ when $\theta = \pi$. As *P* continues counterclockwise around the circle, x_0 increases. When $\theta = 3\pi/2$, we get back up to $x_0 = 0$. After that, x_0 increases further until, when *P* has gone completely around the circle, it returns to $x_0 = 1$ for $\theta = 2\pi$.

The value of x_0 is defined as the *cosine* of the angle θ. The *cosine function* is abbreviated as cos, so we can write

$$\cos \theta = x_0$$

The cosine wave

Circular motion in the Cartesian plane can be defined in terms of the cosine function by means of the equation

$$y = a \cos b\theta$$

where *a* is a constant that depends on the radius of the circle, and *b* is a constant that depends on the revolution rate, just as is the case with the sine function. The angle θ is measured or defined counterclockwise from the positive *x* axis, as always.

The shape of a cosine wave is exactly the same as the shape of a sine wave. Both waves are sinusoids. But the entire cosine wave is shifted to the left by 1/4 of a *cycle* with respect to the sine wave. That works out to an angle of $\pi/2$. Figure 2-4 shows a graph of the basic cosine

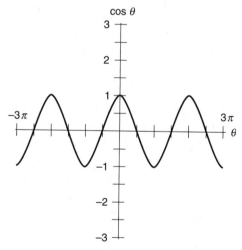

Figure 2-4   Graph of the cosine function for values of θ between -3π and 3π. Each division on the horizontal axis represents $\pi/2$ units. Each division on the vertical axis represents 1/2 unit.

function; it's a *cosine wave* for which $a = 1$ and $b = 1$. Because the cosine wave in Fig. 2-4 has the same frequency but a difference in horizontal position compared with the sine wave in Fig. 2-3, the two waves are said to differ in *phase*. For those of you who like fancy technical terms, a phase difference of 1/4 cycle (or $\pi/2$) is known in electrical engineering as *phase quadrature*.

The tangent function

Once again, refer to Fig. 2-1. The *tangent* (abbreviated as tan) of an angle θ can be defined using the same ray OP and the same point $P = (x_0, y_0)$ as we use when we define the sine and cosine functions. The definition is

$$\tan \theta = y_0/x_0$$

We've seen that $\sin \theta = y_0$ and $\cos \theta = x_0$, so we can express the tangent function as

$$\tan \theta = \sin \theta / \cos \theta$$

The tangent function is interesting because, unlike the sine and cosine functions, it "blows up" at certain values of θ. This is shown by a graph of the function (Fig. 2-5). Whenever $x_0 = 0$, the denominator of either quotient above becomes 0, so the tangent function is not defined for any angle θ such that $\cos \theta = 0$. This happens whenever θ is a positive or negative odd-integer multiple of $\pi/2$.

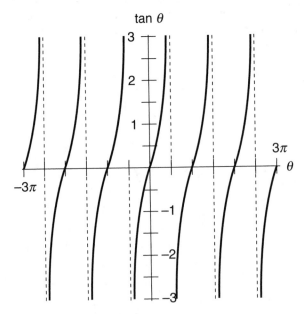

Figure 2-5 Graph of the tangent function for values of θ between -3π and 3π. Each division on the horizontal axis represents $\pi/2$ units. Each division on the vertical axis represents 1/2 unit.

Singularities

When a function "blows up" as the tangent function does at all the odd-integer multiples of $\pi/2$, we say that the function is *singular* for the affected values of the input variable. Such a "blow-up point" is called a *singularity*.

If you've read books or watched movies about space travel and black holes, maybe you've seen or heard the term *space-time singularity*. That's a place where all the familiar rules of the universe break down. In a mathematical singularity, things aren't quite so dramatic, but the output value of a function becomes meaningless. In Fig. 2-5, the singularities are denoted by vertical dashed lines. The dashed lines themselves are known as *asymptotes*.

Inflection points

Midway between the singularities, the graph of the tangent function crosses the θ axis, and the sense of the curvature changes. Below the θ axis, the curves are always concave to the right and convex to the left. Above the θ axis, the curves are always concave to the left and convex to the right. Whenever we have a point on a curve where the sense of the curvature reverses, we call that point an *inflection point* or a *point of inflection*. (Some texts spell the word "inflexion.")

Lots of graphs have inflection points. If you're astute, you'll look back in this chapter and notice that the sine and cosine waves also have them. From your algebra courses, you might also remember that the graphs of many higher-degree polynomial functions have inflection points.

Are you confused?

Some students wonder if there's a way to define a function at a singularity. If you scrutinize Fig. 2-5 closely, you might be tempted to say that

$$\tan (\pi/2) = \pm\infty$$

where the symbol $\pm\infty$ means *positive or negative infinity*. The graph suggests that the output of the tangent function might attain values of infinity at the singular input points, doesn't it? It's an interesting notion; the problem is that we don't have a formal definition for *infinity* as a number. Mathematicians have found it difficult, over the generations, to make up a rigorous, workable definition for infinity as a number.

Some mathematicians have grappled with the notion of infinity and come up with a way of doing arithmetic with it. Most notable among these people was Georg Cantor, a German mathematician who lived from 1845 to 1918. He discovered the apparent existence of "multiple infinities," which he called *transfinite numbers*. If you're interested in studying transfinite numbers, try searching the Internet using that term as a phrase.

Here's a challenge!

Figure out the value of $\tan (\pi/4)$. Don't do any calculations. You should be able to infer this on the basis of geometry alone.

Solution

Draw a diagram of a unit circle, such as the one in Fig. 2-1, and place ray *OP* so that it subtends an angle of $\pi/4$ with respect to the *x* axis. (That's exactly "northeast" if the positive *x* axis goes "east" and the positive *y* axis goes "north.") Note that the ray *OP* also subtends an angle of $\pi/4$ with respect to the *y* axis, because the *x* and *y* axes are mutually perpendicular (oriented at an angle of exactly $\pi/2$ with respect to each other), and $\pi/4$ is half of $\pi/2$. Every point on the ray *OP* is equally distant from the *x* and *y* axes, including the point (x_0, y_0) where the ray intersects the circle. It follows that $x_0 = y_0$. Neither of them is equal to 0, so you know that $y_0/x_0 = 1$. According to the definition of the tangent function, you can conclude that

$$\tan(\pi/4) = y_0/x_0 = 1$$

Secondary Circular Functions

The three primary circular functions, as already defined, form the cornerstone of trigonometry. Three more circular functions exist. Their values represent the reciprocals of the values of the primary circular functions.

The cosecant function

Imagine the ray *OP* in Fig. 2-1, oriented at a certain angle θ with respect to the *x* axis, pointing outward from the origin, and intersecting the unit circle at $P = (x_0, y_0)$. The reciprocal of the ordinate, $1/y_0$, is defined as the *cosecant* of the angle θ. The cosecant function is abbreviated as csc, so we can write

$$\csc\theta = 1/y_0$$

Because y_0 is the value of the sine function, the cosecant is the reciprocal of the sine. For any angle θ, the following equation is always true as long as $\sin\theta \neq 0$:

$$\csc\theta = 1/\sin\theta$$

The cosecant of an angle θ is undefined when θ is any integer multiple of π. That's because the sine of any such angle is 0, which would make the cosecant equal to 1/0. Figure 2-6 is a graph of the cosecant function for values of θ between -3π and 3π. The vertical dashed lines denote the singularities. There's also a singularity along the *y* axis.

The secant function

Consider the reciprocal of the abscissa, that is, $1/x_0$, in Fig. 2-1. This value is the *secant* of the angle θ. The secant function is abbreviated as sec, so we can write

$$\sec\theta = 1/x_0$$

The secant of an angle is the reciprocal of the cosine. When $\cos\theta \neq 0$, the following equation is true:

$$\sec\theta = 1/\cos\theta$$

The secant is undefined for any positive or negative odd-integer multiple of $\pi/2$. Figure 2-7 is a graph of the secant function for values of θ between -3π and 3π. Note the input values for which the function is singular (vertical dashed lines).

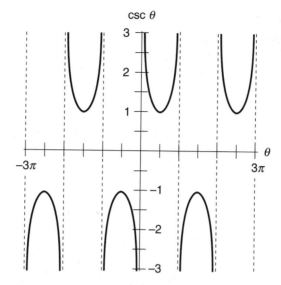

Figure 2-6 Graph of the cosecant function for
values of θ between -3π and 3π.
Each division on the horizontal axis
represents $\pi/2$ units. Each division on
the vertical axis represents $1/2$ unit.

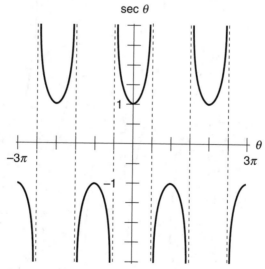

Figure 2-7 Graph of the secant function for
values of θ between -3π and 3π.
Each division on the horizontal axis
represents $\pi/2$ units. Each division on
the vertical axis represents $1/2$ unit.

The cotangent function

Now let's think about the value of x_0/y_0 at the point P where the ray OP crosses the unit circle. This ratio is called the *cotangent* of the angle θ. The cotangent function is abbreviated as cot, so we can write

$$\cot \theta = x_0/y_0$$

Because we already know that $\cos \theta = x_0$ and $\sin \theta = y_0$, we can express the cotangent function in terms of the cosine and the sine:

$$\cot \theta = \cos \theta/\sin \theta$$

The cotangent function is also the reciprocal of the tangent function:

$$\cot \theta = 1/\tan \theta$$

Whenever $y_0 = 0$, the denominators of all three quotients above become 0, so the cotangent function is not defined. Singularities occur at all integer multiples of π. Figure 2-8 is a graph of the cotangent function for values of θ between -3π and 3π. Singularities are, as in the other examples here, shown as vertical dashed lines.

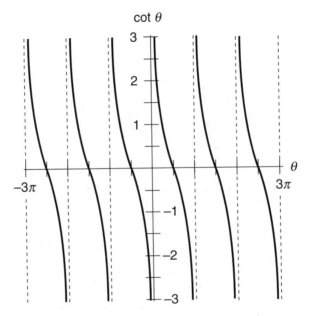

Figure 2-8   Graph of the cotangent function for values of θ between -3π and 3π. Each division on the horizontal axis represents $\pi/2$ units. Each division on the vertical axis represents 1/2 unit.

Are you confused?

Now that you know how the six circular functions are defined, you might wonder how you can determine the output values for specific inputs. The easiest way is to use a calculator. This approach will usually give you an approximation, not an exact value, because the output values of trigonometric functions are almost always irrational numbers. Remember to set the calculator to work for inputs in radians, not in degrees!

The values of the sine and cosine functions never get smaller than −1 or larger than 1. The values of the other four functions can vary wildly. Put a few numbers into your calculator and see what happens when you apply the circular functions to them. When you input a value for which a function is singular, you'll get an error message on the calculator.

Here's a challenge!

Figure out the value of cot $(5\pi/4)$. As in the previous challenge, you should be able to solve this problem entirely with geometry.

Solution

As you did before, draw a unit circle on a Cartesian coordinate grid. This time, orient the ray OP so that it subtends an angle of $5\pi/4$ with respect to the x axis. (That's exactly "southwest" if the positive x axis goes "east" and the positive y axis goes "north.") Every point on OP is equally distant from the x and y axes, including (x_0, y_0) where the ray intersects the circle. You can see that $x_0 = y_0$ and both of them are negative, so the ratio x_0/y_0 must be equal to 1. According to the definition of the cotangent function, you can therefore conclude that

$$\cot (5\pi/4) = 1$$

Pythagorean Extras

The Pythagorean theorem for right triangles, which we reviewed in Chap. 1, can be extended to cover three important *identities* (equations that always hold true) involving the circular functions.

Pythagorean identity for sine and cosine

The square of the sine of an angle plus the square of the cosine of the same angle is always equal to 1. We can write this fact as

$$(\sin \theta)^2 + (\cos \theta)^2 = 1$$

When the value of a trigonometric function is squared, the exponent 2 is customarily placed after the abbreviation of the function and before the input variable, so the parentheses can be eliminated from the expression. In that format, the above equation is written as

$$\sin^2 \theta + \cos^2 \theta = 1$$

Pythagorean identity for secant and tangent

The square of the secant of an angle minus the square of the tangent of the same angle is always equal to 1, as long as the angle is not an odd-integer multiple of $\pi/2$. We write this as

$$\sec^2 \theta - \tan^2 \theta = 1$$

Pythagorean identity for cosecant and cotangent

The square of the cosecant of an angle minus the square of the cotangent of the same angle is always equal to 1, as long as the angle is not an integer multiple of π. We write this as

$$\csc^2 \theta - \cot^2 \theta = 1$$

Are you confused?

You've probably seen the above formula for the sine and cosine in your algebra or trigonometry courses. If you haven't seen the other two formulas, you might wonder where they come from. They can both be derived from the first formula using simple algebra along with the facts we've reviewed in this chapter. You'll get a chance to work them out in Problems 9 and 10, later.

Here's a challenge!

Use a drawing of the unit circle to show that $\sin^2 \theta + \cos^2 \theta = 1$ for angles θ greater than 0 and less than $\pi/2$. (Here's a hint: A right triangle is involved.)

Solution

Figure 2-9 shows the unit circle with θ defined counterclockwise between the x axis and a ray emanating from the origin. When the angle is greater than 0 but less than $\pi/2$, a right triangle

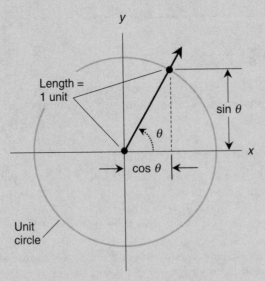

Figure 2-9 This drawing can help show that $\sin^2 \theta + \cos^2 \theta = 1$ when $0 < \theta < \pi/2$.

is formed, with a segment of the ray as the hypotenuse. The length of this segment is equal to the radius of the unit circle. This radius, by definition, is 1 unit. According to the Pythagorean theorem for right triangles, the square of the length of the hypotenuse is equal to the sum of the squares of the lengths of the other two sides. It is easy to see that the lengths of these other two sides are sin θ and cos θ. Therefore,

$$\sin^2 \theta + \cos^2 \theta = 1$$

Here's another challenge!

Use another drawing of the unit circle to show that $\sin^2 \theta + \cos^2 \theta = 1$ for angles θ greater than $3\pi/2$ and less than 2π. (Here's a hint: This range of angles is equivalent to the range of angles greater than $-\pi/2$ and less than 0.)

Solution

Figure 2-10 shows how this can be done. Draw a mirror image of Fig. 2-9, with the angle θ defined clockwise instead of counterclockwise. Again, you get a right triangle with a hypotenuse 1 unit long, while the other two sides have lengths of sin θ and cos θ. This triangle, like all right triangles, obeys the Pythagorean theorem. As in the previous challenge, you end up with

$$\sin^2 \theta + \cos^2 \theta = 1$$

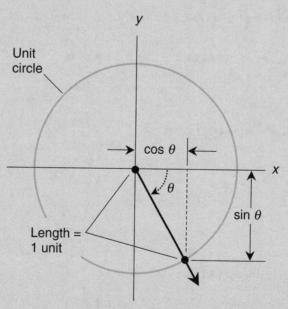

Figure 2-10   This drawing can help show that
$\sin^2 \theta + \cos^2 \theta = 1$ when $3\pi/2 < \theta < 2\pi$.

Practice Exercises

This is an open-book quiz. You may (and should) refer to the text as you solve these problems. Don't hurry! You'll find worked-out answers in App. A. The solutions in the appendix may not represent the only way a problem can be figured out. If you think you can solve a particular problem in a quicker or better way than you see there, by all means try it!

1. Approximately how many radians are there in 1°? Use a calculator and round the answer off to four decimal places, assuming that $\pi \approx 3.14159$.

2. What is the angle in radians representing 7/8 of a circular rotation counterclockwise? Express the answer in terms of π, not as a calculator-derived approximation.

3. What is the angle in radians corresponding to 120° counterclockwise? Express the answer in terms of π, not as a calculator-derived approximation.

4. Suppose that the earth is a perfectly smooth sphere with a circumference of 40,000 kilometers (km). Based on that notion, what is the angular separation (in radians) between two points $1000/\pi$ km apart as measured over the earth's surface along the shortest possible route?

5. Sketch a graph of the function $y = \sin x$ as a dashed curve in the Cartesian xy plane. Then sketch a graph of $y = 2 \sin x$ as a solid curve. How do the two functions compare?

6. Sketch a graph of the function $y = \sin x$ as a dashed curve in the Cartesian xy plane. Then sketch a graph of $y = \sin 2x$ as a solid curve. How do the two functions compare?

7. The secant of an angle can never be within a certain range of values. What is that range?

8. The cosecant of an angle can never be within a certain range of values. What is that range?

9. The Pythagorean formula for the sine and cosine is

$$\sin^2 \theta + \cos^2 \theta = 1$$

From this, derive the fact that

$$\sec^2 \theta - \tan^2 \theta = 1$$

10. Once again, consider the formula

$$\sin^2 \theta + \cos^2 \theta = 1$$

From this, derive the fact that

$$\csc^2 \theta - \cot^2 \theta = 1$$

3

Polar Two-Space

The Cartesian plane isn't the only tool for graphing on a flat surface. Instead of moving right-left and up-down from the origin, we can travel in a specified direction straight outward from the origin to reach a desired point. The *direction angle* is expressed in radians with respect to a *reference axis*. The outward distance is called the *radius*. This scheme gives us *polar two-space* or the *polar coordinate plane*.

The Variables

Figure 3-1 shows the basic polar coordinate plane. The independent variable is portrayed as an angle θ relative to a ray pointing to the right (or "east"). That ray is the reference axis. The dependent variable is portrayed as the radius r from the origin. In this way, we can define points in the plane as ordered pairs of the form (θ, r).

The radius

In the polar plane, the radial increments are concentric circles. The larger the circle, the greater the value of r. In Fig. 3-1, the circles aren't labeled in units. We can imagine each concentric circle, working outward, as increasing by any number of units we want. For example, each radial division might represent 1, 5, 10, or 100 units. Whatever size increments we choose, we must make sure that they stay the same size all the way out. That is, the relationship between the radius coordinate and the actual radius of the circle representing it must be linear.

The direction

As pure mathematicians, we express polar-coordinate direction angles in radians. We go counterclockwise from a reference axis pointing in the same direction as the positive x axis normally goes in the Cartesian xy plane. The angular scale must be linear. That is, the physical angle on the graph must be directly proportional to the value of θ.

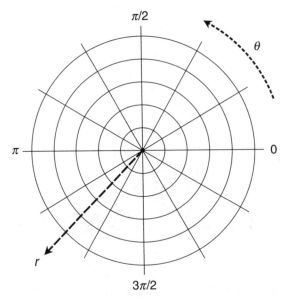

Figure 3-1 The polar coordinate plane. Angular divisions are straight lines passing through the origin. Each angular division represents π/6 units. Radial divisions are circles.

Strange values

In polar coordinates, it's okay to have nonstandard direction angles. If $\theta \geq 2\pi$, it represents at least one complete *counterclockwise* rotation from the reference axis. If the direction angle is $\theta < 0$, it represents *clockwise* rotation from the reference axis rather than counterclockwise rotation.

We can also have negative radius coordinates. If we encounter some point for which we're told that $r < 0$, we can multiply r by −1 so it becomes positive, and then add or subtract π to or from the direction. That's like saying "Proceed 10 km due east" instead of "Proceed −10 km due west."

Which variable is which?

If you read a lot of mathematics texts and papers, you'll sometimes see ordered pairs for polar coordinates with the radius listed first, and then the angle. Instead of the form (θ, r), the ordered pairs will take the form (r, θ). In this scheme, the radius is the independent variable, and the direction is the dependent variable. It works fine, but it's easier for most people to imagine that the radius depends on the direction.

Think of an old-fashioned radar display like the ones shown in war movies made in the middle of the last century. A bright radial ray rotates around a circular screen, revealing targets at various distances. The rotation continues at a steady rate; it's *independent*. Target distances are functions of the direction. Theoretically, a radar display could work in the opposite sense with an expanding bright circle instead of a rotating ray, and all of the targets would show up in the same places. But that geometry wasn't technologically practical when radar sets were first designed, and it was never used. Let's use the (θ, r) format for ordered pairs, where θ is the independent variable and r is the dependent variable.

Are you confused?

You ask, "How we can write down relations and functions intended for polar coordinates as opposed to those meant for Cartesian coordinates?" It's simple. When we want to denote a relation or function (call it f) in polar coordinates where the independent variable is θ and the dependent variable is r, we write

$$r = f(\theta)$$

We can read this out loud as "r equals f of θ." When we want to denote a relation or function (call it g) in Cartesian coordinates where the independent variable is x and the dependent variable is y, we can write

$$y = g(x)$$

We can read this out loud as "y equals g of x."

Here's a challenge!

Provide an example of a graphical object that represents a function in polar coordinates when θ is the independent variable, but not in Cartesian xy coordinates when x is the independent variable.

Solution

Consider a polar function that maps all inputs into the same output, such as

$$f(\theta) = 3$$

Because $f(\theta)$ is another way of denoting r, this function tells us that $r = 3$. The graph is a circle with a radius of 3 units. In Cartesian coordinates, the equation of the circle with radius of 3 units is

$$x^2 + y^2 = 9$$

(Note that $9 = 3^2$, the square of the radius.) If we let y be the dependent variable and x be the independent variable, we can rearrange this equation to get

$$y = \pm(9 - x^2)^{1/2}$$

We can't claim that $y = g(x)$ where g is a function of x in this case. There are values of x (the independent variable) that produce two values of y (the dependent variable). For example, if $x = 0$, then $y = \pm 3$. If we want to say that g is a relation, that's okay; but g is not a function.

Three Basic Graphs

Let's look at the graphs of three generalized equations in polar coordinates. In Cartesian coordinates, all equations of these forms produce straight-line graphs. Only one of them does it now!

Constant angle

When we set the direction angle to a numerical constant, we get a simple polar equation of the form

$$\theta = a$$

where a is the constant. As we allow the value of r to range over all the real numbers, the graph of any such equation is a straight line passing through the origin, subtending an angle of a with respect to the reference axis. Figure 3-2 shows two examples. In these cases, the equations are

$$\theta = \pi/3$$

and

$$\theta = 7\pi/8$$

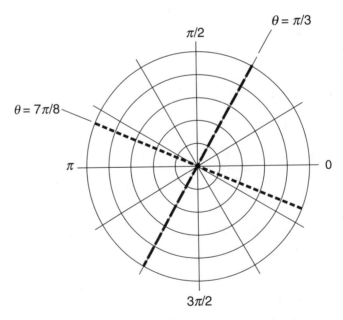

Figure 3-2   When we set the angle constant, the graph is a straight line through the origin. Here are two examples.

Constant radius

Imagine what happens if we set the radius to a numerical constant. This gives us a polar equation of the form

$$r = a$$

where a is the constant. The graph is a circle centered at the origin whose radius is a, as shown in Fig. 3-3, when we allow the direction angle θ to rotate through at least one full turn of 2π. If we allow the angle to span the entire set of real numbers, we trace around the circle infinitely many times, but that doesn't change the appearance of the graph.

Angle equals radius times positive constant

Now let's investigate a more interesting situation. Figure 3-4 shows an example of what happens in polar coordinates when we set the radius equal to a positive constant multiple of the angle. We get a pair of "mirror-image spirals."

To see how this graph arises, imagine a ray pointing from the origin straight out toward the right along the reference axis (labeled 0). The angle is 0, so the radius is 0. Now suppose the ray starts to rotate counterclockwise, like the sweep on an old-fashioned military radar screen. The angle *increases positively* at a constant rate. Therefore, the radius also increases at a constant rate, because the radius is a positive constant multiple of the angle. The resulting

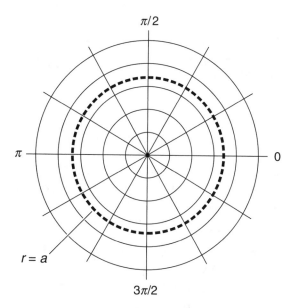

Figure 3-3   When we set the radius constant, the graph is a circle centered at the origin. In this case, the radius is an arbitrary value a.

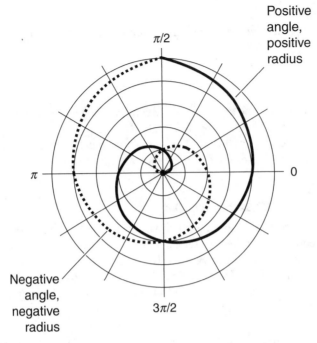

Figure 3-4 When we set the radius equal to a positive constant multiple of the angle, we get a pair of spirals.

graph is the solid spiral. The *pitch* (or "tightness") of the spiral depends on the value of the constant *a* in the equation

$$r = a\theta$$

Small positive values of *a* produce tightly curled-up spirals. Larger positive values of *a* produce more loosely pitched spirals.

Now suppose that the ray starts from the reference axis and rotates clockwise. At first, the angle is 0, so the radius is 0. As the ray turns, the angle *increases negatively* at a constant rate. That means the radius increases negatively at a constant rate, too, because we're multiplying the angle by a positive constant. We must plot the points in the exact opposite direction from the way the ray points. When we do that, we get the dashed spiral in Fig. 3-4. The pitch is the same as that of the heavy spiral, because we haven't changed the value of *a*. The entire graph of the equation consists of both spirals together.

Angle equals radius times negative constant

Figure 3-5 shows an example of what happens in polar coordinates when we set the radius equal to a negative constant multiple of the angle. As in the previous case, we get a pair of spirals, but they're "upside-down" with respect to the case when the constant is positive. To see

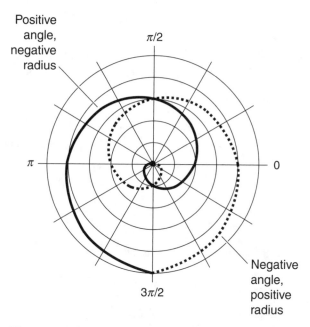

Figure 3-5 When we set the radius equal to a negative constant multiple of the angle, we get a pair of spirals "upside-down" relative to those for a positive constant multiple of the angle. Illustration for Problem 4.

how this works, you can trace around with rotating rays as we did in Fig. 3-4. Be careful with the signs and directions! Remember that negative angles go clockwise, and negative radii go in the opposite direction from the way the angle is defined.

Are you confused?

Look back at Fig. 3-2. If you ponder this graph for awhile, you might suspect that the indicated equations aren't the only ones that can represent these lines. You might ask, "If we allow r to range over all the real numbers, both positive and negative, can't the line for $\theta = \pi/3$ also be represented by other equations such as $\theta = 4\pi/3$ or $\theta = -2\pi/3$? Can't the line representing the $\theta = 7\pi/8$ also be represented by $\theta = 15\pi/8$ or $\theta = -\pi/8$?" The answers to these questions are "Yes." When we see an equation of the form $\theta = a$ representing a straight line through the origin in polar coordinates, we can add any integer multiple of π to the constant a, and we get another equation whose graph is the same line. In more formal terms, a particular line $\theta = a$ through the origin can be represented by

$$\theta = k\pi a$$

where k is any integer and a is a real-number constant.

Here's a challenge!

What's the value of the constant, *a*, in the function shown by the graph of Fig. 3-4? What's the equation of this pair of spirals? Assume that each radial division represents 1 unit.

Solution

Note that if $\theta = \pi$, then $r = 2$. You can solve for *a* by substituting this number pair in the general equation for the pair of spirals. Plugging in the numbers $(\theta, r) = (\pi, 2)$, proceed as follows:

$$r = a\theta$$

$$2 = a\pi$$

$$2/\pi = a$$

Therefore, $a = 2/\pi$, and the equation you seek is

$$r = (2/\pi)\theta$$

If you don't like parentheses, you can write it as

$$r = 2\theta/\pi$$

Here's another challenge!

What is the polar equation of a straight line running through the origin and ascending at an angle of $\pi/4$ as you move to the right, with the restriction that $0 \leq \theta < 2\pi$? If you drew this line on a standard Cartesian *xy* coordinate grid instead of the polar plane, what equation would it represent?

Solution

Two equations will work here. They are

$$\theta = \pi/4$$

and

$$\theta = 5\pi/4$$

Keep in mind that the value of *r* can be any real number: positive, negative, or zero.

First, look at the situation where $\theta = \pi/4$. When $r > 0$, you get a ray in the $\pi/4$ direction. When $r < 0$, you get a ray in the $5\pi/4$ direction. When $r = 0$, you get the origin point. The union of these two rays and the origin point forms the line running through the origin and ascending at an angle of $\pi/4$ as you move toward the right.

Now examine events with the equation $\theta = 5\pi/4$. When $r > 0$, you get a ray in the $5\pi/4$ direction. When $r < 0$, you get a ray in the $\pi/4$ direction. When $r = 0$, you get the origin point. The union of the two rays and the origin point forms the same line as in the first case. In the Cartesian *xy* plane, this line would be the graph of the equation $y = x$.

Coordinate Transformations

We can convert the coordinates of any point from polar to Cartesian systems and vice versa. Going from polar to Cartesian is easy, like floating down a river. Getting from Cartesian to polar is more difficult, like rowing up the same river. As you read along here, refer to Fig. 3-6, which shows a point in the polar grid superimposed on the Cartesian grid.

Polar to Cartesian

Suppose we have a point (θ,r) in polar coordinates. We can convert this point to Cartesian coordinates (x,y) using the formulas

$$x = r \cos \theta$$

and

$$y = r \sin \theta$$

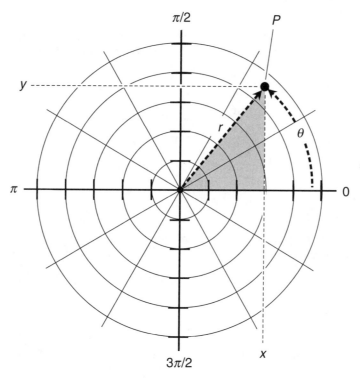

Figure 3-6 A point plotted in both polar and Cartesian coordinates. Each radial division in the polar grid represents 1 unit. Each division on the *x* and *y* axes of the Cartesian grid also represents 1 unit. The shaded region is a right triangle *x* units wide, *y* units tall, and having a hypotenuse *r* units long.

To understand how this works, imagine what happens when $r = 1$. The equation $r = 1$ in polar coordinates gives us a unit circle. We learned in Chap. 2 that when we have a unit circle in the Cartesian plane, then for any point (x,y) on that circle

$$x = \cos \theta$$

and

$$y = \sin \theta$$

Suppose that we double the radius of the circle. This makes the polar equation $r = 2$. The values of x and y in Cartesian coordinates both double, because when we double the length of the hypotenuse of a right triangle (such as the shaded region in Fig. 3-6), we also double the lengths of the other two sides. The new triangle is *similar* to the old one, meaning that its sides stay in the same ratio. Therefore

$$x = 2 \cos \theta$$

and

$$y = 2 \sin \theta$$

This scheme works no matter how large or small we make the circle, as long as it stays centered at the origin. If $r = a$, where a is some positive real number, the new right triangle is always similar to the old one, so we get

$$x = a \cos \theta$$

and

$$y = a \sin \theta$$

If our radius r happens to be negative, these formulas still work. (For "extra credit," can you figure out why?)

An example

Consider the point $(\theta,r) = (\pi,2)$ in polar coordinates. Let's find the (x,y) representation of this point in Cartesian coordinates using the polar-to-Cartesian conversion formulas

$$x = r \cos \theta$$

and

$$y = r \sin \theta$$

Plugging in the numbers gives us

$$x = 2 \cos \pi = 2 \times (-1) = -2$$

and

$$y = 2 \sin \pi = 2 \times 0 = 0$$

Therefore, $(x,y) = (-2,0)$.

Cartesian to polar: the radius

Figure 3-6 shows us that the radius r from the origin to our point $P = (x,y)$ is the length of the hypotenuse of a right triangle (the shaded region) that's x units wide and y units tall. Using the Pythagorean theorem, we can write the formula for determining r in terms of x and y as

$$r = (x^2 + y^2)^{1/2}$$

That's straightforward enough. Now it's time to work on the more difficult conversion: finding the polar angle for a point that's given to us in the Cartesian xy plane.

The Arctangent function

Before we can find the polar direction angle for a point that's given to us in Cartesian coordinates, we must be familiar with an *inverse trigonometric function* known as the *Arctangent*, which "undoes" the work of the tangent function. (The capital "A" is not a typo. We'll see why in a minute.) Consider, for example, the fact that

$$\tan (\pi/4) = 1$$

A true function that "undoes" the tangent must map an input value of 1 in the domain to an output value of $\pi/4$ in the range, but to no other values. In fact, no matter what we input to the function, we must never get more than one output.

To ensure that the inverse of the tangent behaves as a true function, we must restrict its range (output) to an open interval where we don't get any redundancy. By convention, mathematicians specify the open interval $(-\pi/2, \pi/2)$ for this purpose. When mathematicians make this sort of restriction in an inverse trigonometric function, they capitalize the "first letter" in the name of the function. That's a "code" to tell us that we're working with a true function, and not a mere relation. Some texts use the abbreviation $\tan^{-1}$ instead of Arctan to represent the inverse of the tangent function. We won't use this symbol here because some readers might confuse it with the reciprocal of the tangent, which is the cotangent, not the Arctangent!

If you're curious as to what the Arctangent function looks like when graphed, check out Fig. 3-7. This graph consists of the *principal branch* of the tangent function, tipped on its side and then flipped upside-down. Compare Fig. 3-7 with Fig. 2-5 on page 28. The principal branch of the tangent function is the one that passes through the origin.

Once we've made sure we won't run into any ambiguity, we can state the above fact using the Arctangent function, getting

$$\text{Arctan } 1 = \pi/4$$

For any real number u except odd-integer multiples of $\pi/2$ (for which the tangent function is undefined), we can always be sure that

$$\text{Arctan } (\tan u) = u$$

Going the other way, for any real number v, we can be confident that

$$\tan (\text{Arctan } v) = v$$

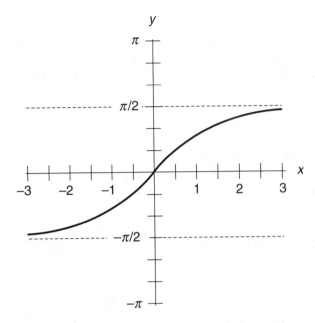

Figure 3-7   A graph of the Arctangent function. The domain extends over all the real numbers. The range is restricted to values larger than $-\pi/2$ and smaller than $\pi/2$. Each division on the *y* axis represents $\pi/6$ units.

Cartesian to polar: the angle

We now have the tools that we need to determine the polar angle θ for a point on the basis of its Cartesian coordinates *x* and *y*. We already know that

$$x = r \cos \theta$$

and

$$y = r \sin \theta$$

As long as $x \neq 0$, it follows that

$$y/x = (r \sin \theta)/(r \cos \theta) = (r/r)(\sin \theta)/(\cos \theta)$$

$$= (\sin \theta)/(\cos \theta) = \tan \theta$$

Simplifying, we get

$$\tan \theta = y/x$$

If we take the Arctangent of both sides, we obtain

$$\text{Arctan} (\tan \theta) = \text{Arctan} (y/x)$$

which can be rewritten as

$$\theta = \text{Arctan} (y/x)$$

Suppose the point $P = (x,y)$ happens to lie in the first or fourth quadrant of the Cartesian plane. In this case, we have

$$-\pi/2 < \theta < \pi/2$$

so we can directly use the conversion formula

$$\theta = \text{Arctan} (y/x)$$

If $P = (x,y)$ is in the second or third quadrant, then we have

$$\pi/2 < \theta < 3\pi/2$$

That's outside the range of the Arctangent function, but we can remedy this situation if we subtract π from θ. When we do this, we bring θ into the allowed range but we don't change its tangent, because the tangent function repeats itself every π radians. (If you look back at Fig. 2-5 again, you will notice that all of the branches in the graph are identical, and any two adjacent branches are π radians apart.) In this situation, we have

$$\theta - \pi = \text{Arctan} (y/x)$$

which can be rewritten as

$$\theta = \pi + \text{Arctan} (y/x)$$

Now we're ready to derive specific formulas for θ in terms of x and y. Let's break the scenario down into all possible general locations for $P = (x,y)$, and see what we get for θ in each case:

P at the origin. If $x = 0$ and $y = 0$, then θ is theoretically undefined. However, let's assign θ a default value of 0 at the origin. By doing that, we can "fill the hole" that would otherwise exist in our conversion scheme.

P on the +x axis. If $x > 0$ and $y = 0$, then we're on the positive x axis. We can see from Fig. 3-6 that $\theta = 0$.

P in the first quadrant. If $x > 0$ and $y > 0$, then we're in the first quadrant of the Cartesian plane where θ is larger than 0 but less than $\pi/2$. We can therefore directly apply the conversion formula

$$\theta = \text{Arctan} (y/x)$$

***P* on the +*y* axis.** If $x = 0$ and $y > 0$, then we're on the positive y axis. We can see from Fig. 3-6 that $\theta = \pi/2$.

***P* in the second quadrant.** If $x < 0$ and $y > 0$, then we're in the second quadrant of the Cartesian plane where θ is larger than $\pi/2$ but less than π. In this case, we must apply the modified conversion formula

$$\theta = \pi + \text{Arctan}\,(y/x)$$

***P* on the −*x* axis.** If $x < 0$ and $y = 0$, then we're on the negative x axis. We can see from Fig. 3-6 that $\theta = \pi$.

***P* in the third quadrant.** If $x < 0$ and $y < 0$, then we're in the third quadrant of the Cartesian plane where θ is larger than π but less than $3\pi/2$, so we apply the modified conversion formula

$$\theta = \pi + \text{Arctan}\,(y/x)$$

***P* on the −*y* axis.** If $x = 0$ and $y < 0$, then we're on the negative y axis. We can see from Fig. 3-6 that $\theta = 3\pi/2$.

***P* in the fourth quadrant.** If $x > 0$ and $y < 0$, then we're in the fourth quadrant of the Cartesian plane where θ is larger than $3\pi/2$ but smaller than 2π. That's the same thing as saying that $-\pi/2 < \theta < 0$. We'll get an angle in that range if we apply the original conversion formula

$$\theta = \text{Arctan}\,(y/x)$$

In the interest of elegance, we'd like the angle in the polar representation of a point to always be nonnegative but less than 2π. We can make this happen by adding in a complete rotation of 2π to the basic conversion formula, getting

$$\theta = 2\pi + \text{Arctan}\,(y/x)$$

We have taken care of all the possible locations for P. A summary of the nine-part conversion formula that we've developed is given in the following table.

$\theta = 0$	At the origin
$\theta = 0$	On the +x axis
$\theta = \text{Arctan}\,(y/x)$	In the first quadrant
$\theta = \pi/2$	On the +y axis
$\theta = \pi + \text{Arctan}\,(y/x)$	In the second quadrant
$\theta = \pi$	On the −x axis
$\theta = \pi + \text{Arctan}\,(y/x)$	In the third quadrant
$\theta = 3\pi/2$	On the −y axis
$\theta = 2\pi + \text{Arctan}\,(y/x)$	In the fourth quadrant

An example

Let's convert the Cartesian point (−5,−12) to polar form. Here, $x = -5$ and $y = -12$. When we plug these numbers into the formula for r, we get

$$r = [(-5)^2 + (-12)^2]^{1/2} = (25 + 144)^{1/2} = 169^{1/2} = 13$$

Our point is in the third quadrant of the Cartesian plane. To find the angle, we should use the formula

$$\theta = \pi + \text{Arctan } (y/x)$$

When we plug in $x = -5$ and $y = -12$, we get

$$\theta = \pi + \text{Arctan } [(-12)/(-5)] = \pi + \text{Arctan } (12/5)$$

That is a theoretically exact answer, but it's an irrational number. A calculator set to work in radians (not degrees) tells us that

$$\text{Arctan } (12/5) \approx 1.1760$$

rounded off to four decimal places. (Remember that the "wavy" equals sign means "is approximately equal to.") If we let $\pi \approx 3.1416$, also rounded off to four decimal places, we get

$$\theta \approx 3.1416 + 1.1760 \approx 4.3176$$

The polar equivalent of $(x,y) = (-5,-12)$ is therefore $(\theta,r) \approx (4.3176,13)$, where θ is approximated to four decimal places and r is exact.

Are you confused?

If the foregoing angle-conversion formula derivation baffles you, don't feel bad. It's complicated! If you don't grasp it to your satisfaction right now, set it aside for awhile. Read it again tomorrow, or the day after that. You might want to make up some problems with points in all four quadrants of the Cartesian plane, and then use these formulas to convert them to polar form. As you work out the arithmetic, you'll gain a better understanding of how (and why) the formulas work.

Here's a challenge!

Find the distance d in *radial units* between the points $P = (\pi,3)$ and $Q = (\pi/2,4)$ in polar coordinates, where a radial unit is equal to the radius of a unit circle centered at the origin.

Solution

Let's convert the polar coordinates of P and Q to Cartesian coordinates, and then employ the Cartesian distance formula to determine how far apart the two points are. Let's call the Cartesian versions of the points

$$P = (x_p, y_p)$$

and

$$Q = (x_q, y_q)$$

For *P*, we have

$$x_p = 3 \cos \pi = 3 \times (-1) = -3$$

and

$$y_p = 3 \sin \pi = 3 \times 0 = 0$$

The Cartesian coordinates of *P* are therefore $(x_p, y_p) = (-3, 0)$. For *Q*, we have

$$x_q = 4 \cos \pi/2 = 4 \times 0 = 0$$

and

$$y_q = 4 \sin \pi/2 = 4 \times 1 = 4$$

The Cartesian coordinates of *Q* are therefore $(x_q, y_q) = (0, 4)$. Using the Cartesian distance formula, we obtain

$$d = [(x_p - x_q)^2 + (y_p - y_q)^2]^{1/2} = [(-3 - 0)^2 + (0 - 4)^2]^{1/2}$$

$$= [(-3)^2 + (-4)^2]^{1/2} = (9 + 16)^{1/2} = 25^{1/2} = 5$$

We've found that the points $P = (\pi, 3)$ and $Q = (\pi/2, 4)$ are precisely 5 radial units apart in the polar coordinate plane.

- -

Practice Exercises

This is an open-book quiz. You may (and should) refer to the text as you solve these problems. Don't hurry! You'll find worked-out answers in App. A. The solutions in the appendix may not represent the only way a problem can be figured out. If you think you can solve a particular problem in a quicker or better way than you see there, by all means try it!

1. Is the relation $\theta = \pi/4$ a function in polar coordinates, where θ is the independent variable and *r* is the dependent variable? Why or why not? Is $\theta = \pi/2$ a function in the same polar system? Why or why not?

2. Suppose that we draw the lines representing the polar relations $\theta = \pi/4$ and $\theta = \pi/2$ directly onto the Cartesian *xy* plane, where *x* is the independent variable and *y* is the dependent variable. Do either of the resulting graphs represent functions in the Cartesian coordinate system? Why or why not?

3. Imagine a circle centered at the origin in polar coordinates. The equation for the circle is $r = a$, where a is a real-number constant. What other equation, if any, represents the same circle?

4. In Fig. 3-5 on page 43, suppose that each radial increment is π units. What's the value of the constant a in this case? What's the equation of the pair of spirals? (Here are a couple of reminders: The radial increments are the concentric circles. The value of a in this situation turns out negative.)

5. Figure 3-8 shows a line L and a circle C in polar coordinates. Line L passes through the origin, and every point on L is equidistant from the horizontal and vertical axes. Circle C is centered at the origin. Each radial division represents 1 unit. What's the polar equation representing L when we restrict the angles to positive values smaller than 2π? What's the polar equation representing C? (Here's a hint: Both equations can be represented in two ways.)

6. When we examine Fig. 3-8, we can see that L and C intersect at two points P and Q. What are the polar coordinates of P and Q, based on the information given in Problem 5? (Here's a hint: Both points can be represented in two ways.)

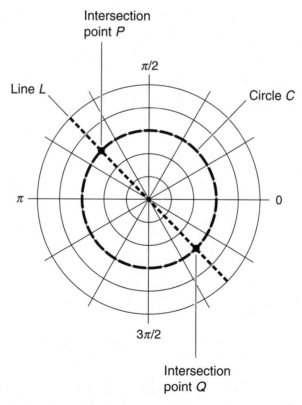

Figure 3-8   Illustration for Problems 5 through 10.
Each radial division is 1 unit.

7. Solve the system of equations from the solution to Problem 5, verifying the polar coordinates of points P and Q in Fig. 3-8.

8. Based on the information given in Problem 5, what are the Cartesian xy-coordinate equations of line L and circle C in Fig. 3-8?

9. Solve the system of equations from the solution to Problem 8 to determine the Cartesian coordinates of the intersection points P and Q in Fig. 3-8.

10. Based on the polar coordinates of points P and Q in Fig. 3-8 (the solutions to Problems 6 and 7), use the conversion formulas to derive the Cartesian coordinates of those two points.

4

Vector Basics

We can define the length of a line segment that connects two points, but the direction is ambiguous. If we want to take the direction into account, we must make a line segment into a *vector*. Mathematicians write vector names as bold letters of the alphabet. Alternatively, a vector name can be denoted as a letter with a line or arrow over it.

The "Cartesian Way"

In diagrams and graphs, a vector is drawn as a *directed line segment* whose direction is portrayed by putting an arrow at one end. When working in two-space, we can describe vectors in Cartesian coordinates or in polar coordinates. Let's look at the "Cartesian way" first.

Endpoints, locations, and notations

Figure 4-1 shows four vectors drawn on a Cartesian coordinate grid. Each vector has a beginning (the *originating point*) and an end space (the *terminating point*). In this situation, any of the four vectors can be defined according to two independent quantities:

- The length (*magnitude*)
- The way it points (*direction*)

It doesn't matter where the originating or terminating points actually are. The important thing is how the two points are located with respect to each other. Once a vector has been defined as having a specific magnitude and direction, we can "slide it around" all over the coordinate plane without changing its essential nature.

We can always think of the originating point for a vector as being located at the coordinate origin (0,0). When we place a vector so that its originating point is at (0,0), we say that the vector is in *standard form*. The standard form is convenient in Cartesian

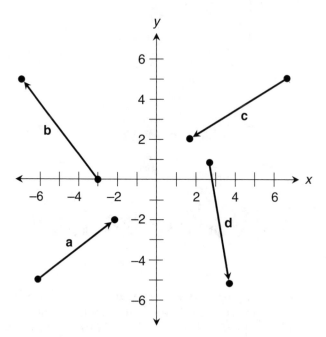

Figure 4-1 Four vectors in the Cartesian plane. In each case, the magnitude corresponds to the length of the line segment, and the direction is indicated by the arrow.

coordinates, because it allows us to uniquely define any vector as an ordered pair corresponding to

- The x coordinate of its terminating point (*x component*)
- The y coordinate of its terminating point (*y component*)

Figure 4-2 shows the same four vectors as Fig. 4-1 does, but all of the originating points have been moved to the coordinate origin. The magnitudes and directions of the corresponding vectors in Figs. 4-1 and 4-2 are identical. That's how we can tell that the vectors **a**, **b**, **c**, and **d** in Fig. 4-2 represent the same mathematical objects as the vectors **a**, **b**, **c**, and **d** in Fig. 4-1.

Cartesian magnitude

Imagine an arbitrary vector **a** in the Cartesian *xy plane*, extending from the origin (0,0) to the point (x_a, y_a) as shown in Fig. 4-3. The magnitude of **a** (which can be denoted as r_a, as $|\mathbf{a}|$, or as a) can be found by applying the formula for the distance of a point from the origin. We learned that formula in Chap. 1. Here it is, modified for the vector situation:

$$r_a = (x_a^2 + y_a^2)^{1/2}$$

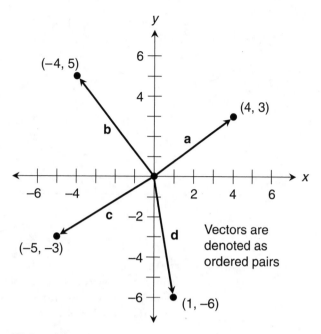

Figure 4-2   These are the same four vectors as shown in Fig. 4-1, positioned so that their originating points correspond to the coordinate origin (0,0).

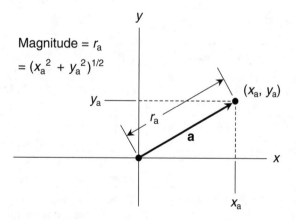

Figure 4-3   The magnitude of a vector can be defined as its length in the Cartesian plane.

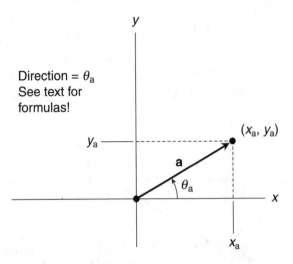

Direction = θ_a
See text for
formulas!

Figure 4-4   The direction of a vector can be
defined as its angle, in radians, going
counterclockwise from the positive x
axis in the Cartesian plane.

Cartesian direction

Now let's think about the direction of **a**, as shown in Fig. 4-4. We can denote it as an angle θ_a
or by writing dir **a**. To define θ_a in terms of its terminating-point coordinates (x_a, y_a), we must
go back to the polar-coordinate direction-finding system in Chap. 3. The following table has
those formulas, modified for our vector situation.

$\theta_a = 0$	When $x_a = 0$ and $y_a = 0$ so **a** terminates at the origin
$\theta_a = 0$	When $x_a > 0$ and $y_a = 0$ so **a** terminates on the $+x$ axis
$\theta_a = \text{Arctan } (y_a/x_a)$	When $x_a > 0$ and $y_a > 0$ so **a** terminates in the first quadrant
$\theta_a = \pi/2$	When $x_a = 0$ and $y_a > 0$ so **a** terminates on the $+y$ axis
$\theta_a = \pi + \text{Arctan } (y_a/x_a)$	When $x_a < 0$ and $y_a > 0$ so **a** terminates in the second quadrant
$\theta_a = \pi$	When $x_a < 0$ and $y_a = 0$ so **a** terminates on the $-x$ axis
$\theta_a = \pi + \text{Arctan } (y_a/x_a)$	When $x_a < 0$ and $y_a < 0$ so **a** terminates in the third quadrant
$\theta_a = 3\pi/2$	When $x_a = 0$ and $y_a < 0$ so **a** terminates on the $-y$ axis
$\theta_a = 2\pi + \text{Arctan } (y_a/x_a)$	When $x_a > 0$ and $y_a < 0$ so **a** terminates in the fourth quadrant

Cartesian vector sum

Let's consider two arbitrary vectors **a** and **b** in the Cartesian plane, in standard form with terminating-point coordinates

$$\mathbf{a} = (x_a, y_a)$$

and

$$\mathbf{b} = (x_b, y_b)$$

We calculate the sum vector **a** + **b** by adding the x and y terminating-point coordinates separately and then combining the sums to get a new ordered pair. When we do that, we get

$$\mathbf{a} + \mathbf{b} = [(x_a + x_b),(y_a + y_b)]$$

This sum can be illustrated geometrically by constructing a parallelogram with the two vectors **a** and **b** as adjacent sides, as shown in Fig. 4-5. The sum vector, **a** + **b**, corresponds to the *directional diagonal* of the parallelogram going away from the coordinate origin.

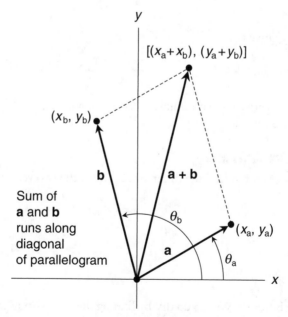

Figure 4-5   We can determine the sum of two vectors **a** and **b** by finding the directional diagonal of a parallelogram with **a** and **b** as adjacent sides.

An example

Consider two vectors in the Cartesian plane. Suppose they're both in standard form. (From now on, let's agree that all vectors are in standard form so they "begin" at the coordinate origin, unless we specifically state otherwise.) The vectors are defined according to the ordered pairs

$$\mathbf{a} = (4,0)$$

and

$$\mathbf{b} = (3,4)$$

In this case, we have $x_a = 4$, $x_b = 3$, $y_a = 0$, and $y_b = 4$. We find the sum vector by adding the corresponding coordinates to get

$$\mathbf{a} + \mathbf{b} = [(x_a + x_b),(y_a + y_b)] = [(4 + 3),(0 + 4)] = (7,4)$$

Cartesian negative of a vector

To find the *Cartesian negative* of a vector, we take the *additive inverses* (that is, the negatives) of both coordinate values. Given the vector

$$\mathbf{a} = (x_a,y_a)$$

its Cartesian negative is

$$-\mathbf{a} = (-x_a,-y_a)$$

The Cartesian negative of a vector always has the same magnitude as the original, but points in the opposite direction.

Cartesian vector difference

Suppose that we want to find the difference between the two vectors

$$\mathbf{a} = (x_a,y_a)$$

and

$$\mathbf{b} = (x_b,y_b)$$

by subtracting $\mathbf{b}$ from $\mathbf{a}$. We can do this by finding the Cartesian negative of $\mathbf{b}$ and then adding $-\mathbf{b}$ to $\mathbf{a}$ to get

$$\mathbf{a} - \mathbf{b} = \mathbf{a} + (-\mathbf{b}) = \{[(x_a + (-x_b)],[(y_a + (-y_b)]\}$$
$$= [(x_a - x_b),(y_a - y_b)]$$

We can skip the step where we find the negative of the second vector and directly subtract the coordinate values, but we must be sure we keep the vectors and coordinate values in the correct order if we take that shortcut.

An example

Let's look again at the same two vectors for which we found the Cartesian sum a few moments ago:

$$\mathbf{a} = (4,0)$$

and

$$\mathbf{b} = (3,4)$$

As before, we have $x_a = 4$, $x_b = 3$, $y_a = 0$, and $y_b = 4$. We can find $\mathbf{a} - \mathbf{b}$ by taking the differences of the corresponding coordinates, as long as we keep the vectors in the correct order. Then we get

$$\mathbf{a} - \mathbf{b} = [(x_a - x_b),(y_a - y_b)] = [(4 - 3),(0 - 4)] = (1,-4)$$

Are you confused?

If you have trouble with the notion of a vector, here are three real-world examples of vector quantities in two dimensions. If you like, draw diagrams to help your mind's eye envision what's happening in each case:

- When the wind blows at 5 meters per second from east to west, you can say that the magnitude of its *velocity vector* is 5 and the direction is toward the west. In Cartesian coordinates where the $+x$ axis goes east, the $+y$ axis goes north, the $-x$ axis goes west, and the $-y$ axis goes south, you would assign this vector the ordered pair $(-5,0)$.
- When you push on a rolling cart with a force of 10 newtons toward the north, you're applying a *force vector* to the cart with a magnitude of 10 and a direction toward the north. In Cartesian coordinates where the $+x$ axis goes east, the $+y$ axis goes north, the $-x$ axis goes west, and the $-y$ axis goes south, you would assign this vector the ordered pair $(0,10)$.
- When you accelerate a car at 5 feet per second per second in a direction somewhat to the east of north, the magnitude of the car's *acceleration vector* is 5 and the direction is somewhat to the east of north. A "neat" situation of this sort occurs when the x (or eastward) component is 3 and the y (or northward) component is 4, so you get the ordered pair $(3,4)$. These components form the two shorter sides of a 3:4:5 right triangle whose hypotenuse measures 5 units (the magnitude).

Here's a challenge!

Show that Cartesian vector addition is *commutative*. That is, show that for any two vectors $\mathbf{a}$ and $\mathbf{b}$ expressed as ordered pairs in the Cartesian plane,

$$\mathbf{a} + \mathbf{b} = \mathbf{b} + \mathbf{a}$$

Solution

This fact is easy, although rather tedious, to demonstrate *rigorously*. (In pure mathematics, the term *rigor* refers to the process of proving something in a series of absolutely logical steps. It has nothing to do with the physical condition called *rigor mortis*.) We must define the two vectors by coordinates, and then work through the arithmetic with those coordinates. Let's call the two vectors

$$\mathbf{a} = (x_a, y_a)$$

and

$$\mathbf{b} = (x_b, y_b)$$

As defined earlier in this chapter, the Cartesian sum $\mathbf{a} + \mathbf{b}$ is

$$\mathbf{a} + \mathbf{b} = [(x_a + x_b), (y_a + y_b)]$$

Using the same definition, the Cartesian sum $\mathbf{b} + \mathbf{a}$ is

$$\mathbf{b} + \mathbf{a} = [(x_b + x_a), (y_b + y_a)]$$

All four of the coordinate values x_a, x_b, y_a, and y_b are real numbers. We know from basic algebra that addition of real numbers is commutative. Therefore, we can reverse both of the sums in the elements of the ordered pair above, getting

$$\mathbf{b} + \mathbf{a} = [(x_a + x_b), (y_a + y_b)]$$

That's the ordered pair that defines $\mathbf{a} + \mathbf{b}$. We have just shown that

$$\mathbf{a} + \mathbf{b} = \mathbf{b} + \mathbf{a}$$

for any two Cartesian vectors $\mathbf{a}$ and $\mathbf{b}$.

The "Polar Way"

In the polar coordinate plane, we draw a vector as a ray going straight outward from the origin to a point defined by a specific angle and a specific radius. Figure 4-6 shows two vectors $\mathbf{a}$ and $\mathbf{b}$ with originating points at $(0,0)$ and terminating points at (θ_a, r_a) and (θ_b, r_b), respectively.

Polar magnitude and direction

The magnitude and direction of a vector $\mathbf{a} = (\theta_a, r_a)$ in the polar coordinate plane are defined directly by the coordinates. The magnitude is r_a, the straight-line distance of the terminating point from the origin. The direction angle is θ_a, the angle that the ray subtends in a

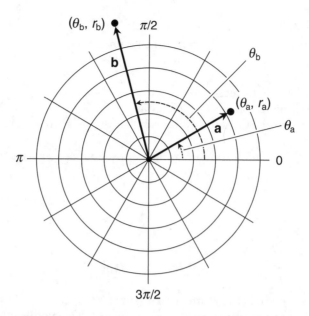

Figure 4-6   Vectors in the polar plane are defined by
ordered pairs for their terminating points,
denoting the direction angle (relative
to the reference axis marked 0) and the
radius (the distance from the origin).

counterclockwise sense from the reference axis (labeled 0 here). By convention, we restrict the
vector magnitude and direction to the ranges

$$r_a \geq 0$$

and

$$0 \leq \theta_a < 2\pi$$

If a vector's magnitude is 0, then the direction angle doesn't matter; the usual custom is to set
it equal to 0.

Special constraints

When defining polar vectors, we must be more particular about what's "legal" and what's
"illegal" than we were when defining polar points in Chap. 3. With polar vectors:

- We don't allow negative magnitudes
- We don't allow negative direction angles
- We don't allow direction angles of 2π or larger

These constraints ensure that the set of all polar-plane vectors can be paired off in a *one-to-one
correspondence* (also called a *bijection*) with the set of all Cartesian-plane vectors.

Polar vector sum

If we have two vectors in polar form, their sum can be found by following these steps, in order:

1. Convert both vectors to Cartesian coordinates
2. Add the vectors the Cartesian way
3. Convert the Cartesian vector sum back to polar coordinates

Let's look at the situation in more formal terms. Suppose we have two vectors expressed in polar form as

$$\mathbf{a} = (\theta_a, r_a)$$

and

$$\mathbf{b} = (\theta_b, r_b)$$

To convert these vectors to Cartesian coordinates, we can use formulas adapted from the polar-to-Cartesian conversion we learned in Chap. 3. The modified formulas are

$$(x_a, y_a) = [(r_a \cos \theta_a), (r_a \sin \theta_a)]$$

and

$$(x_b, y_b) = [(r_b \cos \theta_b), (r_b \sin \theta_b)]$$

Once we have obtained the Cartesian ordered pairs, we add their elements individually to get

$$\mathbf{a} + \mathbf{b} = [(x_a + x_b), (y_a + y_b)]$$

Let's call this Cartesian sum vector **c**, and say that

$$\mathbf{c} = \mathbf{a} + \mathbf{b} = [(x_a + x_b), (y_a + y_b)] = (x_c, y_c)$$

To convert **c** from Cartesian coordinates into polar coordinates, we can use the formulas given earlier in this chapter for the magnitude and direction angle of a vector in the xy plane. If we call the magnitude r_c and the direction angle θ_c, we can write down the polar coordinates of sum vector as

$$\mathbf{c} = (\theta_c, r_c)$$

An example

Let's find the polar sum of the vectors

$$\mathbf{a} = (\pi/4, 2)$$

and

$$\mathbf{b} = (7\pi/4, 2)$$

Using the formulas for conversion stated earlier in this chapter, we find that the Cartesian equivalents are

$$\mathbf{a} = \{[2 \cos (\pi/4)], [2 \sin (\pi/4)]\}$$

and

$$\mathbf{b} = \{[2 \cos (7\pi/4)], [2 \sin (7\pi/4)]\}$$

From trigonometry, we (hopefully) recall that the cosines and sines of these particular angles have values that are easy to denote, even though they're irrational:

$$\cos (\pi/4) = 2^{1/2}/2$$
$$\sin (\pi/4) = 2^{1/2}/2$$
$$\cos (7\pi/4) = 2^{1/2}/2$$
$$\sin (7\pi/4) = -2^{1/2}/2$$

Substituting these values in the ordered pairs for the Cartesian vectors, we get

$$\mathbf{a} = [(2 \times 2^{1/2}/2), (2 \times 2^{1/2}/2)] = (2^{1/2}, 2^{1/2})$$

and

$$\mathbf{b} = \{(2 \times 2^{1/2}/2), [(2 \times (-2^{1/2}/2)]\} = (2^{1/2}, -2^{1/2})$$

When we add these Cartesian vectors, we obtain

$$\mathbf{a} + \mathbf{b} = \{(2^{1/2} + 2^{1/2}), [2^{1/2} + (-2^{1/2})]\} = [(2 \times 2^{1/2}), 0]$$

Let's call this sum vector $\mathbf{c} = (x_c, y_c)$. Then we have

$$x_c = 2 \times 2^{1/2}$$

and

$$y_c = 0$$

Using the Cartesian-to-polar conversion formulas, we get

$$\theta_c = 0$$

and

$$r_c = (x_c^2 + y_c^2)^{1/2} = [(2 \times 2^{1/2})^2 + 0^2]^{1/2} = 2 \times 2^{1/2}$$

Putting these coordinates into an ordered pair, we derive our final answer as

$$\mathbf{a} + \mathbf{b} = [0,(2 \times 2^{1/2})]$$

That's the polar sum of our original two polar vectors. The first coordinate is the angle in radians. The second coordinate is the magnitude in linear units.

Polar vector difference

When we want to subtract a polar vector from another polar vector, we follow these steps in order:

1. Convert both vectors to Cartesian coordinates
2. Find the Cartesian negative of the second vector
3. Add the first vector to the negative of the second vector the Cartesian way
4. Convert the resultant back to polar coordinates

Once again, imagine that we have

$$\mathbf{a} = (\theta_a, r_a)$$

and

$$\mathbf{b} = (\theta_b, r_b)$$

The Cartesian equivalents are

$$(x_a, y_a) = [(r_a \cos \theta_a),(r_a \sin \theta_a)]$$

and

$$(x_b, y_b) = [(r_b \cos \theta_b),(r_b \sin \theta_b)]$$

We find the Cartesian negative of **b** as

$$-\mathbf{b} = (-x_b, -y_b)$$

The difference vector **a** − **b** is therefore

$$\mathbf{a} - \mathbf{b} = \mathbf{a} + (-\mathbf{b}) = \{[(x_a + (-x_b)],[(y_a + (-y_b)]\}$$
$$= [(x_a - x_b),(y_a - y_b)]$$

Let's call this difference vector **d**. We can say that

$$\mathbf{d} = \mathbf{a} - \mathbf{b} = [(x_a - x_b),(y_a - y_b)] = (x_d, y_d)$$

We can skip the step where we find the Cartesian negative of the second vector and directly subtract the coordinate values, but we must take special care to keep the vectors and coordinate values in the correct order if we do it that way. To convert **d** from Cartesian coordinates into polar coordinates, we can take advantage of the same formulas that we use to complete the process of polar vector addition.

Polar negative of a vector

Once in awhile, we'll want to find the negative of a vector the polar way. To do that, we reverse its direction and leave the magnitude the same. We can do this by adding π to the angle if it's at least 0 but less than π to begin with, or by subtracting π if it's at least π but less than 2π to begin with. In formal terms, suppose we have a polar vector

$$\mathbf{a} = (\theta_a, r_a)$$

If $0 \leq \theta_a < \pi$, then the polar negative is

$$-\mathbf{a} = [(\theta_a + \pi), r_a]$$

If $\pi \leq \theta_a < 2\pi$, then the polar negative is

$$-\mathbf{a} = [(\theta_a - \pi), r_a]$$

An example

Let's find the polar difference $\mathbf{a} - \mathbf{b}$ between the vectors

$$\mathbf{a} = (\pi/4, 2)$$

and

$$\mathbf{b} = (7\pi/4, 2)$$

In the addition example we finished a few minutes ago, we found that the Cartesian ordered pairs for these vectors are

$$\mathbf{a} = (2^{1/2}, 2^{1/2})$$

and

$$\mathbf{b} = (2^{1/2}, -2^{1/2})$$

The negative of **b** is

$$-\mathbf{b} = (-2^{1/2}, 2^{1/2})$$

When we add **a** to **−b**, we get

$$\mathbf{a} + (-\mathbf{b}) = \{[2^{1/2} + (-2^{1/2})], (2^{1/2} + 2^{1/2})\} = [0, (2 \times 2^{1/2})]$$

That's the same as $\mathbf{a} - \mathbf{b}$. Let's call this Cartesian difference vector $\mathbf{d} = (x_d, y_d)$. Then

$$x_d = 0$$

and

$$y_d = 2 \times 2^{1/2}$$

Using the Cartesian-to-polar conversion table, we can see that

$$\theta_d = \pi/2$$

and

$$r_d = (x_d^2 + y_d^2)^{1/2} = [0^2 + (2 \times 2^{1/2})^2]^{1/2} = 2 \times 2^{1/2}$$

The polar ordered pair is therefore

$$\mathbf{a} - \mathbf{b} = [(\pi/2), (2 \times 2^{1/2})]$$

The first coordinate is the angle in radians. The second coordinate is the magnitude in linear units.

Are you confused?

By now you might wonder, "What's the difference between a polar vector sum and a Cartesian vector sum? Or a polar vector negative and a Cartesian vector negative? Or a polar vector difference and a Cartesian vector difference? If we start with the same vector or vectors, shouldn't we get the same vector when we're finished calculating, whether we do it the polar way or the Cartesian way?" That's an excellent question. The answer is yes. The mathematical methods differ, but the resultant vectors are equivalent whether we work them out the polar way or the Cartesian way.

Here's a challenge!

Draw polar coordinate diagrams of the vector addition and subtraction facts we worked out in this section.

Solution

The original two polar vectors were

$$\mathbf{a} = (\theta_a, r_a) = (\pi/4, 2)$$

and

$$\mathbf{b} = (\theta_b, r_b) = (7\pi/4, 2)$$

We found their polar sum to be

$$\mathbf{a} + \mathbf{b} = [0, (2 \times 2^{1/2})]$$

and their polar difference to be

$$\mathbf{a} - \mathbf{b} = [(\pi/2), (2 \times 2^{1/2})]$$

When we converted the two vectors to Cartesian form, we got

$$\mathbf{a} = (2^{1/2}, 2^{1/2})$$

and

$$\mathbf{b} = (2^{1/2}, -2^{1/2})$$

We found their Cartesian sum to be

$$\mathbf{a} + \mathbf{b} = [(2 \times 2^{1/2}), 0]$$

and their Cartesian difference to be

$$\mathbf{a} - \mathbf{b} = [0, (2 \times 2^{1/2})]$$

We can illustrate the original vectors, the vector sum, the negative of the second vector, and the vector difference in four diagrams:

- Figure 4-7 shows the polar sum, including **a**, **b**, and **a** + **b**.

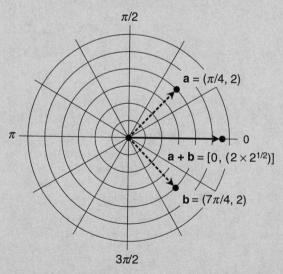

Figure 4-7   Polar sum of two vectors. Each radial division represents 1/2 unit.

- Figure 4-8 shows the polar difference, including **a**, **b**, **−b**, and **a − b**.

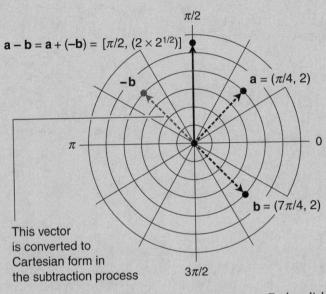

Figure 4-8   Polar difference between two vectors. Each radial division represents 1/2 unit.

- Figure 4-9 shows the Cartesian sum, including **a**, **b**, and **a + b**.

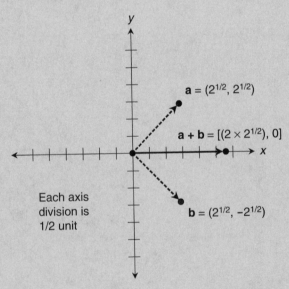

Figure 4-9   Cartesian sum of two vectors. Each axis division represents 1/2 unit.

- Figure 4-10 shows the Cartesian difference, including **a**, **b**, **−b**, and **a − b**.

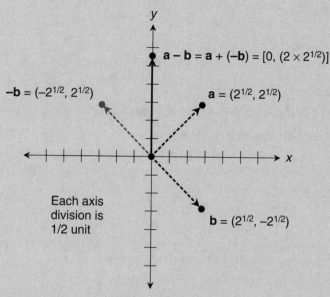

Figure 4-10. Cartesian difference between two vectors. Each axis division represents 1/2 unit.

Practice Exercises

This is an open-book quiz. You may (and should) refer to the text as you solve these problems. Don't hurry! You'll find worked-out answers in App. A. The solutions in the appendix may not represent the only way a problem can be figured out. If you think you can solve a particular problem in a quicker or better way than you see there, by all means try it!

1. Consider two vectors **a** and **b** in the Cartesian plane, with coordinates defined as follows:

$$\mathbf{a} = (-3, 6)$$

and

$$\mathbf{b} = (2, 5)$$

Work out, in strict detail, the Cartesian vector sums **a** + **b**, **b** + **a**, **a** − **b**, and **b** − **a**.

2. A vector is defined as the *zero vector* (denoted by a bold numeral **0**) if and only if its magnitude is equal to 0. In the Cartesian plane, the zero vector is expressed as the ordered pair (0,0). Show that when a vector is added to its Cartesian negative in either order, the result is the zero vector.

3. Imagine two arbitrary vectors **a** and **b** in the Cartesian plane, with coordinates defined as follows:

$$\mathbf{a} = (x_a, y_a)$$

and

$$\mathbf{b} = (x_b, y_b)$$

Show that the vector **b** − **a** is the Cartesian negative of the vector **a** − **b**.

4. Find the Cartesian sum of the vectors

$$\mathbf{a} = (4,5)$$

and

$$\mathbf{b} = (-2,-3)$$

Compare this with the sum of their negatives

$$-\mathbf{a} = (-4,-5)$$

and

$$-\mathbf{b} = (2,3)$$

5. Prove that Cartesian vector negation distributes through Cartesian vector addition. That is, show that for two Cartesian vectors **a** and **b**, it's always true that

$$-(\mathbf{a} + \mathbf{b}) = -\mathbf{a} + (-\mathbf{b})$$

6. Find the polar sum of the vectors

$$\mathbf{a} = (\pi/2,4)$$

and

$$\mathbf{b} = (\pi,3)$$

7. Find the polar negative of the vector **a** + **b** from the solution to Problem 6.

8. Find the polar negatives −**a** and −**b** of the vectors stated in Problem 6.

9. Find the polar sum of the vectors −**a** and −**b** from the solution to Problem 8. Compare this with the solution to Problem 7.

10. Find the polar differences **a** − **b** and **b** − **a** between the vectors stated in Problem 6.

5

Vector Multiplication

We've seen how vectors add and subtract in two dimensions. In this chapter, we'll learn how to multiply a vector by a real number. Then we'll explore two different ways in which vectors can be multiplied by each other.

Product of Scalar and Vector

The simplest form of vector multiplication involves changing the magnitude by a real-number factor called a *scalar*. A scalar is a one-dimensional quantity that can be positive, negative, or zero. If the scalar is positive, the vector direction stays the same. If the scalar is negative, the vector direction reverses. If the scalar is zero, the vector disappears.

Cartesian vector times positive scalar

Imagine a standard-form vector **a** in the Cartesian xy plane, defined by an ordered pair whose coordinates are x_a and y_a, so that

$$\mathbf{a} = (x_a, y_a)$$

Suppose that we multiply a positive scalar k_+ by each of the vector coordinates individually, getting two new coordinates. Mathematically, we write this as

$$k_+\mathbf{a} = (k_+x_a, k_+y_a)$$

This vector is called the *left-hand Cartesian product* of k_+ and **a.** If we multiply both original coordinates on the right by k_+ instead, we get

$$\mathbf{a}\,k_+ = (x_a k_+, y_a k_+)$$

That's the *right-hand Cartesian product* of **a** and k_+. The individual coordinates of $k_+\mathbf{a}$ and $\mathbf{a}k_+$ are products of real numbers. We learned in pre-algebra that real-number multiplication is commutative, so it follows that

$$k_+\mathbf{a} = (k_+x_a, k_+y_a) = (x_a k_+, y_a k_+) = \mathbf{a}\,k_+$$

We've just shown that multiplication of a Cartesian-plane vector by a positive scalar is commutative. We don't have to worry about whether we multiply on the left or the right; we can simply talk about the *Cartesian product* of the vector and the positive scalar.

An example

Figure 5-1 illustrates the Cartesian vector $(-1,-2)$ as a solid, arrowed line segment. If we multiply this vector by 3 on the left, we get

$$3 \times (-1,-2) = \{[3 \times (-1)],[3 \times (-2)]\} = (-3,-6)$$

If we multiply the original vector by 3 on the right, we get

$$(-1,-2) \times 3 = [(-1 \times 3)],(-2 \times 3)] = (-3,-6)$$

The new vector is shown as a dashed, gray, arrowed line segment pointing in the same direction as the original vector, but 3 times as long.

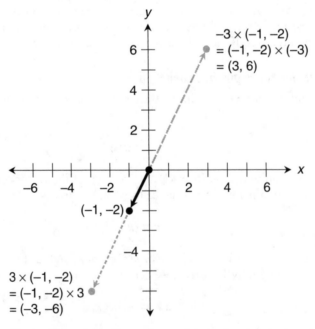

Figure 5-1   Cartesian products of the scalars 3 and −3 with the vector (−1,−2).

Cartesian vector times negative scalar

Now suppose we want to multiply $\mathbf{a}$ by a negative scalar instead of a positive scalar. Let's call the scalar k_-. The left-hand Cartesian product of k_- and $\mathbf{a}$ is

$$k_-\mathbf{a} = (k_-x_a, k_-y_a)$$

The right-hand Cartesian product is

$$\mathbf{a}k_- = (x_a k_-, y_a k_-)$$

As with the positive constant, the commutative property of real-number multiplication tells us that

$$k_-\mathbf{a} = (k_-x_a, k_-y_a) = (x_a k_-, y_a k_-) = \mathbf{a}k_-$$

We don't have to worry about whether we multiply on the left or the right. We get the same result either way.

An example

Once again, look at Fig. 5-1 with the vector $(-1,-2)$ shown as a solid, arrowed line segment. When we multiply it by the scalar -3 on the left, we obtain

$$-3 \times (-1,-2) = \{[-3 \times (-1)], [-3 \times (-2)]\} = (3,6)$$

Multiplying by the scalar on the right, we get

$$(-1,-2) \times (-3) = \{[-1 \times (-3)], [-2 \times (-3)]\} = (3,6)$$

This result is shown as a dashed, gray, arrowed line segment pointing in the opposite direction from the original vector, and 3 times as long.

Polar vector times positive scalar

Imagine some vector $\mathbf{a}$ in the polar-coordinate plane whose direction angle is θ_a and whose magnitude is r_a. If it's in standard form, we can express it as the ordered pair

$$\mathbf{a} = (\theta_a, r_a)$$

When we multiply $\mathbf{a}$ on the left by a positive scalar k_+, the angle remains the same, but the magnitude becomes k_+r_a. This gives us the *left-hand polar product* of k_+ and $\mathbf{a}$, which is

$$k_+\mathbf{a} = (\theta_a, k_+r_a)$$

If we multiply $\mathbf{a}$ on the right by k_+, we get the *right-hand polar product* of $\mathbf{a}$ and k_+, which is

$$\mathbf{a}k_+ = (\theta_a, r_a k_+)$$

Because real-number multiplication is commutative, we know that

$$k_+\mathbf{a} = (\theta_a, k_+r_a) = (\theta_a, r_a k_+) = \mathbf{a}\,k_+$$

As in the Cartesian case, we don't have to worry about whether we multiply on the left or the right. The *polar product* of the vector and the positive scalar is the same either way.

An example

In Fig. 5-2, the polar vector $(7\pi/4, 3/2)$ is shown as a solid, arrowed line segment. When we multiply this vector by 3 on the left, we get

$$3 \times (7\pi/4, 3/2) = [7\pi/4, (3 \times 3/2)] = (7\pi/4, 9/2)$$

Multiplying by 3 on the right yields

$$(7\pi/4, 3/2) \times 3 = \{7\pi/4, [(3/2) \times 3]\} = (7\pi/4, 9/2)$$

This polar product vector is represented by a dashed, gray, arrowed line segment pointing in the same direction as the original vector, but 3 times as long.

Polar vector times negative scalar

Again, consider our polar vector $\mathbf{a} = (\theta_a, r_a)$. Suppose that we want to multiply $\mathbf{a}$ on the left by a negative scalar k_-. It's tempting to suppose that we can leave the angle the same and make

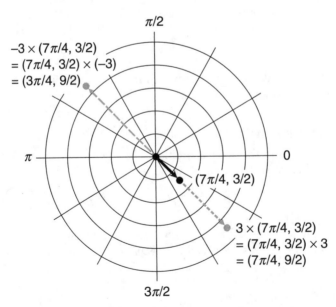

Figure 5-2 Polar products of the scalars 3 and −3 with the vector $(7\pi/4, 3/2)$. Each radial division represents 1 unit.

the magnitude equal to $k_- r_a$. But that gives us a negative magnitude, which is forbidden by the rules we've accepted for polar vectors. The proper approach is to multiply the original vector magnitude r_a by the absolute value of k_-. In this situation, that's $-k_-$. Then we reverse the direction of the vector by either adding or subtracting π to get a direction angle that's nonnegative but smaller than 2π. We define the result as the *left-hand polar product* of k_- and **a**, and write it as

$$k_- \mathbf{a} = [(\theta_a + \pi),(-k_- r_a)]$$

if $0 \leq \theta_a < \pi$, and

$$k_- \mathbf{a} = [(\theta_a - \pi),(-k_- r_a)]$$

if $\pi \leq \theta_a < 2\pi$. Because k_- is negative, $-k_-$ is positive; therefore $-k_- r_a$ is positive, which ensures that our scalar-vector product has positive magnitude. If we multiply **a** on the right by k_-, we get the *right-hand polar product* of **a** and k_-, which is

$$\mathbf{a}k_- = [(\theta_a + \pi),r_a(-k_-)]$$

if $0 \leq \theta_a < \pi$, and

$$\mathbf{a}k_- = [(\theta_a - \pi),r_a(-k_-)]$$

if $\pi \leq \theta_a < 2\pi$. As before, k_- is negative so $-k_-$ is positive; that means $r_a(-k_-)$ is positive, ensuring that our vector-scalar product has positive magnitude. The commutative law assures us that for any negative scalar $k-$ and any polar vector **a**, it's always true that

$$k_- \mathbf{a} = \mathbf{a}k_-$$

As before, we can leave out the left-hand and right-hand jargon, and simply talk about the polar product of the vector and the scalar.

An example

Look again at Fig. 5-2. When we multiply the original polar vector $(7\pi/4,3/2)$ by -3 on the left, we get

$$-3 \times (7\pi/4,3/2) = [(7\pi/4 - \pi),(3 \times 3/2)] = (3\pi/4,9/2)$$

Multiplying the original polar vector by -3 on the right yields

$$(7\pi/4,3/2) \times (-3) = \{(7\pi/4 - \pi),[(3/2) \times 3]\} = (3\pi/4,9/2)$$

This result is shown as a dashed, gray, arrowed line segment pointing in the opposite direction from the original vector, and 3 times as long.

Are you confused?

You ask, "What happens when our positive scalar k_+ is between 0 and 1? What happens when our negative scalar k_- is between −1 and 0? What do we get if the scalar constant is 0?" If $0 < k_+ < 1$, the product vector points in the same direction as the original, but it's shorter. If $−1 < k_- < 0$, the product vector points in the opposite direction from the original, and it's shorter. If we multiply a vector by 0, we get the zero vector. In all of these cases, it doesn't matter whether we work in the Cartesian plane or in the polar plane.

Here's a challenge!

Prove that the multiplication of a Cartesian-plane vector by a positive scalar is left-hand distributive over vector addition. That is, if k_+ is a positive constant, and if **a** and **b** are Cartesian-plane vectors, then

$$k_+(\mathbf{a} + \mathbf{b}) = k_+\mathbf{a} + k_+\mathbf{b}$$

Solution

At first glance, this might seem like one of those facts that's intuitively obvious and difficult to prove. But all we have to do is work out some arithmetic with fancy characters. Let's start with

$$k_+(\mathbf{a} + \mathbf{b})$$

where k_+ is a positive real number, $\mathbf{a} = (x_a, y_a)$, and $\mathbf{b} = (x_b, y_b)$. We can expand the vector sum into an ordered pair, writing the above expression as

$$k_+(\mathbf{a} + \mathbf{b}) = k_+[(x_a + x_b),(y_a + y_b)]$$

The definition of left-hand scalar multiplication of a Cartesian vector tells us that we can rewrite this as

$$k_+(\mathbf{a} + \mathbf{b}) = \{[k_+(x_a + x_b)],[k_+(y_a + y_b)]\}$$

In pre-algebra, we learned that real-number multiplication is left-hand distributive over real-number addition, so we can morph the above equation to get

$$k_+(\mathbf{a} + \mathbf{b}) = [(k_+x_a + k_+x_b),(k_+y_a + k_+y_b)]$$

Let's set this equation aside for a little while. We shouldn't forget about it, however, because we're going to come back to it shortly.

Now, instead of the product of the scalar and the sum of the vectors, let's start with the sum of the scalar products

$$k_+\mathbf{a} + k_+\mathbf{b}$$

We can expand the individual vectors into ordered pairs to get

$$k_+\mathbf{a} + k_+\mathbf{b} = k_+(x_a,y_a) + k_+(x_b,y_b)$$

The definition of left-hand scalar multiplication lets us rewrite this equation as

$$k_+\mathbf{a} + k_+\mathbf{b} = (k_+x_a, k_+y_a) + (k_+x_b, k_+y_b)$$

According to the definition of the Cartesian sum of vectors, we can add the elements of these ordered pairs individually to get a new ordered pair. That gives us

$$k_+\mathbf{a} + k_+\mathbf{b} = [(k_+x_a + k_+x_b), (k_+y_a + k_+y_b)]$$

Take a close look at the right-hand side of this equation. It's the same as the right-hand side of the equation we put into "brain memory" a minute ago. That equation was

$$k_+(\mathbf{a} + \mathbf{b}) = [(k_+x_a + k_+x_b), (k_+y_a + k_+y_b)]$$

Taken together, the above two equations show us that

$$k_+(\mathbf{a} + \mathbf{b}) = k_+\mathbf{a} + k_+\mathbf{b}$$

Dot Product of Two Vectors

Mathematicians define two ways in which a vector can be multiplied by another vector. The simpler operation is called the *dot product* and is symbolized by a large dot (•). Sometimes it's called the *scalar product* because the end result is a scalar. Some texts refer to it as the *inner product*.

Cartesian dot product

Suppose we're given two standard-form vectors **a** and **b** in Cartesian coordinates, defined by the ordered pairs

$$\mathbf{a} = (x_a, y_a)$$

and

$$\mathbf{b} = (x_b, y_b)$$

The *Cartesian dot product* **a** • **b** is the real number we get when we multiply the x values by each other, multiply the y values by each other, and then add the two results. The formula is

$$\mathbf{a} \bullet \mathbf{b} = x_a x_b + y_a y_b$$

An example

Consider two standard-form vectors in the Cartesian xy plane, given by the ordered pairs

$$\mathbf{a} = (4, 0)$$

and

$$\mathbf{b} = (3,4)$$

In this case, $x_a = 4$, $x_b = 3$, $y_a = 0$, and $y_b = 4$. We calculate the dot product by plugging the numbers into the formula, getting

$$\mathbf{a} \cdot \mathbf{b} = (4 \times 3) + (0 \times 4) = 12 + 0 = 12$$

Polar dot product

Now let's work in the polar-coordinate plane. Imagine two vectors defined by the ordered pairs

$$\mathbf{a} = (\theta_a, r_a)$$

and

$$\mathbf{b} = (\theta_b, r_b)$$

Let $\theta_b - \theta_a$ be the angle between vectors $\mathbf{a}$ and $\mathbf{b}$, expressed in a rotational sense starting at $\mathbf{a}$ and finishing at $\mathbf{b}$ as shown in Fig. 5-3. We calculate the *polar dot product* $\mathbf{a} \cdot \mathbf{b}$ by multiplying the magnitude of $\mathbf{a}$ by the magnitude of $\mathbf{b}$, and then multiplying that result by the cosine of $\theta_b - \theta_a$ to get

$$\mathbf{a} \cdot \mathbf{b} = r_a r_b \cos (\theta_b - \theta_a)$$

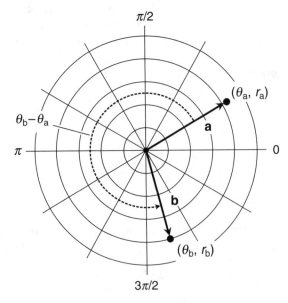

Figure 5-3   To find the polar dot product of two vectors, we must know the angle between them as we rotate from the first vector (in this case $\mathbf{a}$) to the second vector (in this case $\mathbf{b}$).

An example

Suppose that we're given two vectors **a** and **b** in the polar plane, and told that their coordinates are

$$\mathbf{a} = (\pi/6, 3)$$
$$\mathbf{b} = (5\pi/6, 2)$$

In this situation, $r_a = 3$, $r_b = 2$, $\theta_a = \pi/6$, and $\theta_b = 5\pi/6$. We have

$$\theta_b - \theta_a = 5\pi/6 - \pi/6 = 2\pi/3$$

Therefore, the dot product is

$$\mathbf{a} \bullet \mathbf{b} = r_a r_b \cos(\theta_b - \theta_a) = 3 \times 2 \times \cos(2\pi/3)$$
$$= 3 \times 2 \times (-1/2) = -3$$

Are you confused?

Do you wonder if the dot product of two polar-plane vectors is always equal to the dot product of the same vectors in the Cartesian plane when expressed in standard form? The answer is yes. Let's find out why.

Here's a challenge!

Prove that for any two vectors **a** and **b** in two-space, the polar dot product **a** • **b** is the same as the Cartesian dot product **a** • **b** when both vectors are in standard form.

Solution

We will start with the polar versions of the vectors, calling them

$$\mathbf{a} = (\theta_a, r_a)$$

and

$$\mathbf{b} = (\theta_b, r_b)$$

Let's convert these vectors to Cartesian form. We can use the formulas for conversion of points from polar to Cartesian coordinates (from Chap. 3). When we apply them to vector **a**, we get

$$x_a = r_a \cos \theta_a$$

and

$$y_a = r_a \sin \theta_a$$

so the standard Cartesian form of the vector is

$$\mathbf{a} = [(r_a \cos \theta_a), (r_a \sin \theta_a)]$$

When we apply the same conversion formulas to **b**, we obtain

$$x_b = r_b \cos \theta_b$$

and

$$y_b = r_b \sin \theta_b$$

so the standard Cartesian form is

$$\mathbf{b} = [(r_b \cos \theta_b),(r_b \sin \theta_b)]$$

The Cartesian dot product of the two vectors is

$$\mathbf{a} \bullet \mathbf{b} = x_a x_b + y_a y_b$$

Substituting the values we found for the individual vector coordinates, we get

$$\mathbf{a} \bullet \mathbf{b} = (r_a \cos \theta_a)(r_b \cos \theta_b) + (r_a \sin \theta_a)(r_b \sin \theta_b)$$
$$= r_a r_b (\cos \theta_a \cos \theta_b + \sin \theta_a \sin \theta_b)$$

As we think back to our trigonometry courses, we recall that there's a *trigonometric identity* telling us how to expand the cosine of the difference between two angles. When we name the angles so they apply to our situation here, that formula becomes

$$\cos (\theta_b - \theta_a) = \cos \theta_a \cos \theta_b + \sin \theta_a \sin \theta_b$$

We can substitute the left-hand side of this identity in the last part of the long equation we got a minute ago for the dot product, obtaining

$$\mathbf{a} \bullet \mathbf{b} = r_a r_b \cos (\theta_b - \theta_a)$$

This is the formula for the polar dot product! We've taken the polar versions of **a** and **b**, found their Cartesian dot product, and then found that it's identical to the polar dot product. We can now say, *Quod erat demonstradum*. That's Latin for "Which was to be proved." Some mathematicians write the abbreviation for this expression, "QED," when they've finished a proof.

- -

Cross Product of Two Vectors

The more complicated (and interesting) way to multiply two vectors by each other gives us a third vector that "jumps" out of the coordinate plane. This operation is known as the *cross product*. Some mathematicians call it the *vector product*. The cross product of two vectors **a** and **b** is written as $\mathbf{a} \times \mathbf{b}$.

Polar cross product

Imagine two arbitrary vectors in the polar-coordinate plane, expressed in standard form as ordered pairs

$$\mathbf{a} = (\theta_a, r_a)$$

and

$$\mathbf{b} = (\theta_b, r_b)$$

The magnitude of $\mathbf{a} \times \mathbf{b}$ is always nonnegative by default, and is easy to define. When $\mathbf{a}$ and $\mathbf{b}$ are in standard form, the originating point of $\mathbf{a} \times \mathbf{b}$ is at the coordinate origin, so all three vectors "start" at the same spot. The direction of $\mathbf{a} \times \mathbf{b}$ is always along the line passing through the origin at a right angle to the plane containing $\mathbf{a}$ and $\mathbf{b}$. But it's quite a trick to figure out in which direction the cross vector product points along this line!

Suppose that the difference $\theta_b - \theta_a$ between the direction angles is positive but less than π, as shown in the example of Fig. 5-4. If we start at vector $\mathbf{a}$ and rotate until we get to vector $\mathbf{b}$, we turn through an angle of $\theta_b - \theta_a$. To calculate the magnitude of $\mathbf{a} \times \mathbf{b}$ (which we will denote

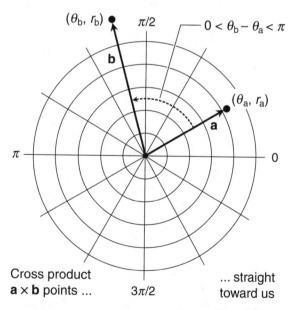

Figure 5-4 If $\theta_a < \theta_b$ and the two angles differ by less than π, then $\mathbf{a} \times \mathbf{b}$ points straight toward us as we look down on the plane containing $\mathbf{a}$ and $\mathbf{b}$.

as r_{axb}), we multiply the original vector magnitudes by each other, and then multiply by the sine of the difference angle. Mathematically,

$$r_{axb} = r_a r_b \sin(\theta_b - \theta_a)$$

In a situation of the sort shown in Fig. 5-4, the vector $\mathbf{a} \times \mathbf{b}$ points from the coordinate origin straight out of the page toward us.

If $\theta_b - \theta_a$ is larger than π, then things get a little bit complicated. To be sure that we assign the correct direction to the vector $\mathbf{a} \times \mathbf{b}$, we must always rotate counterclockwise, and we're never allowed to turn through more than a half circle. Figure 5-5 shows an example. We rotate through one full circular turn minus $\theta_b - \theta_a$, so the difference angle is

$$2\pi - (\theta_b - \theta_a)$$

which can be more simply written as

$$2\pi + \theta_a - \theta_b$$

In a situation like this, $\mathbf{a} \times \mathbf{b}$ points straight away from us.

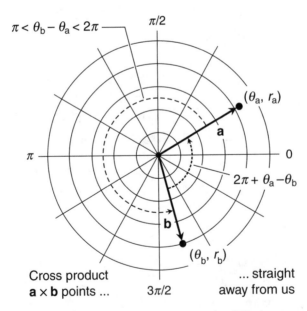

Figure 5-5   If $\theta_a < \theta_b$ and the two angles differ by more than π, then $\mathbf{a} \times \mathbf{b}$ points straight away from us as we look down on the plane containing $\mathbf{a}$ and $\mathbf{b}$.

If the vectors **a** and **b** point in exactly the same direction or in exactly opposite directions, then $\theta_b - \theta_a = 0$ or $\theta_b - \theta_a = \pi$. In these cases, the cross product is the zero vector. We'll see why in the next "challenge."

An example

Consider the following two polar vectors **a** and **b** in standard form:

$$\mathbf{a} = (\pi/4, 7)$$

and

$$\mathbf{b} = (\pi, 6)$$

Let's find the cross product, $\mathbf{a} \times \mathbf{b}$. We have

$$\theta_b - \theta_a = \pi - \pi/4 = 3\pi/4$$

Because $0 < \theta_b - \theta_a < \pi$, we know that $\mathbf{a} \times \mathbf{b}$ points toward us. Its magnitude is

$$r_a x_b = r_a r_b \sin(\theta_b - \theta_a) = 7 \times 6 \times \sin(3\pi/4)$$
$$= 7 \times 6 \times (2^{1/2}/2) = 21 \times 2^{1/2}$$

Another example

Now let's look at these two polar vectors **a** and **b** in standard form and find their cross product $\mathbf{a} \times \mathbf{b}$:

$$\mathbf{a} = (\pi/4, 7)$$

and

$$\mathbf{b} = (7\pi/4, 6)$$

This time, $\pi < \theta_b - \theta_a < 2\pi$, so $\mathbf{a} \times \mathbf{b}$ points away from us. To calculate the magnitude, we consider the difference angle to be

$$2\pi + \theta_a - \theta_b = 2\pi + \pi/4 - 7\pi/4 = \pi/2$$

Therefore

$$r_{axb} = r_a r_b \sin(2\pi + \theta_a - \theta_b) = 7 \times 6 \times \sin(\pi/2)$$
$$= 7 \times 6 \times 1 = 42$$

Are you confused?

We haven't discussed how to directly calculate the cross product of two Cartesian-plane vectors. There's a way to do it, but we must know how to work with vectors in *Cartesian three-space*. We'll learn those techniques in Chap. 8. Meanwhile, we can indirectly find the cross product of two Cartesian-plane vectors by converting them both to polar form and then finding their cross product the polar way.

Are you still confused?

Here's a game that can help you find the direction of the cross product $\mathbf{a} \times \mathbf{b}$ (in that order) between two vectors $\mathbf{a}$ and $\mathbf{b}$. It involves some maneuvers with your right hand. Some mathematicians, engineers, and physicists call this the *right-hand rule for cross products*.

If $0 < \theta_b - \theta_a < \pi$ (as in Fig. 5-4), point your right thumb out as if you're making a thumbs-up sign. Curl your fingers in the *counterclockwise* rotational sense from $\mathbf{a}$ to $\mathbf{b}$. Your thumb will point in the general direction of $\mathbf{a} \times \mathbf{b}$. If the page on which the vectors are printed is horizontal, your thumb should point straight up.

If $\pi < \theta_b - \theta_a < 2\pi$ (as in Fig. 5-5), curl your right-hand fingers in the *clockwise* rotational sense from $\mathbf{a}$ to $\mathbf{b}$. If the page on which the vectors are printed is horizontal, you'll have to twist your wrist in a clumsy fashion so that your thumb points straight down in the general direction of $\mathbf{a} \times \mathbf{b}$.

Remember that $\mathbf{a} \times \mathbf{b}$ always comes out of the origin *precisely perpendicular* to the plane containing $\mathbf{a}$ and $\mathbf{b}$.

Here's a challenge!

A few moments ago, it was mentioned that if two vectors point in the same direction or in opposite directions, then their cross product is the zero vector. Prove it!

Solution

First, consider two vectors $\mathbf{a}$ and $\mathbf{b}$ that have the same direction angle θ but different magnitudes r_a and r_b, so that

$$\mathbf{a} = (\theta, r_a)$$

and

$$\mathbf{b} = (\theta, r_b)$$

The magnitude of $\mathbf{a} \times \mathbf{b}$ is

$$r_{a \times b} = r_a r_b \sin(\theta - \theta) = r_a r_b \sin 0 = r_a r_b \times 0 = 0$$

Whenever a vector has a magnitude of zero, then it's the zero vector by definition, so

$$\mathbf{a} \times \mathbf{b} = \mathbf{0}$$

Now look at the case where **a** and **b** have angles that differ by π, so they point in opposite directions. As before, you can assign the coordinates

$$\mathbf{a} = (\theta, r_a)$$

Two possibilities exist for the direction angle of **a**. You can have

$$0 \leq \theta < \pi$$

or

$$\pi \leq \theta < 2\pi$$

If $0 \leq \theta < \pi$, then

$$\mathbf{b} = [(\theta + \pi), r_b]$$

and the magnitude of $\mathbf{a} \times \mathbf{b}$ is

$$r_{a \times b} = r_a r_b \sin [(\theta + \pi) - \theta] = r_a r_b \sin \pi = r_a r_b \times 0 = 0$$

Therefore

$$\mathbf{a} \times \mathbf{b} = 0$$

If $\pi \leq \theta < 2\pi$, then

$$\mathbf{b} = [(\theta - \pi), r_b]$$

In this case, the magnitude of $\mathbf{a} \times \mathbf{b}$ is

$$r_{a \times b} = r_a r_b \sin [(\theta - \pi) - \theta] = r_a r_b \sin (-\pi) = r_a r_b \times 0 = 0$$

so again,

$$\mathbf{a} \times \mathbf{b} = 0$$

Here's an "extra credit" challenge!

Prove that the cross product of two vectors is *anticommutative*. That is, show that for any two polar-plane vectors **a** and **b**, the magnitudes of $\mathbf{a} \times \mathbf{b}$ and $\mathbf{b} \times \mathbf{a}$ are the same, but they point in opposite directions.

Solution

You're on your own. That's what makes this is an "extra credit" problem!

Practice Exercises

This is an open-book quiz. You may (and should) refer to the text as you solve these problems. Don't hurry! You'll find worked-out answers in App. A. The solutions in the appendix may not represent the only way a problem can be figured out. If you think you can solve a particular problem in a quicker or better way than you see there, by all means try it!

1. Consider two standard-form vectors **a** and **b** in the Cartesian plane, represented by the ordered pairs

    $$\mathbf{a} = (5,-5)$$

    and

    $$\mathbf{b} = (-5,5)$$

    Calculate and compare the Cartesian products 4**a** and −4**b**.

2. Convert the original two vectors from Problem 1 into polar form. Then calculate and compare the polar products 4**a** and −4**b**.

3. Prove that the multiplication of a standard-form vector by a positive scalar is right-hand distributive over Cartesian-plane vector subtraction. That is, if k_+ is a positive constant, and if **a** and **b** are vectors in the *xy* plane, then

    $$(\mathbf{a} - \mathbf{b})k_+ = \mathbf{a}k_+ - \mathbf{b}k_+$$

4. Consider two standard-form vectors **a** and **b** in the Cartesian plane, represented by

    $$\mathbf{a} = (4,4)$$

    and

    $$\mathbf{b} = (-7,7)$$

    Calculate and compare the Cartesian dot products **a** • **b** and **b** • **a**.

5. Convert the original two vectors from Problem 4 into polar form. Then calculate and compare the polar dot products **a** • **b** and **b** • **a**.

6. Prove that the dot product is commutative for standard-form vectors in the Cartesian plane.

7. Prove that the dot product is commutative for vectors in the polar plane.

8. Prove that if k_+ is a positive constant, and if **a** and **b** are standard-form vectors in Cartesian or polar two-space, then

    $$k_+\mathbf{a} \bullet k_+\mathbf{b} = k_+^{2}(\mathbf{a} \bullet \mathbf{b})$$

    Demonstrate the Cartesian case first, and then the polar case.

9. Consider the two polar vectors

$$\mathbf{a} = (\pi/3, 4)$$

and

$$\mathbf{b} = (3\pi/2, 1)$$

Determine the polar cross product $\mathbf{a} \times \mathbf{b}$.

10. Consider the two polar vectors

$$\mathbf{a} = (\pi, 8)$$

and

$$\mathbf{b} = (7\pi/6, 5)$$

Determine the polar cross product $\mathbf{a} \times \mathbf{b}$.

6

Complex Numbers and Vectors

If you've had a comprehensive algebra course such as the predecessor to this book, *Algebra Know-It-All*, then you've been exposed to *imaginary numbers* and *complex numbers*. In this chapter, we'll take a closer look at how these quantities behave.

Numbers with Two Parts

A complex number consists of two components, the *real part* and the *imaginary part*. Complex numbers can be defined as ordered pairs and mapped one-to-one onto the points of a coordinate plane. They can also be represented as vectors.

The unit imaginary number

The set of imaginary numbers arises when we ask, "What is the square root of a negative real number?" This question poses a mystery to anyone who is familiar only with the real numbers. Unless we come up with some new sort of quantity, we have to say, "It's undefined."

In order to define the square root of a negative real number, mathematicians invented the *unit imaginary number*, called it i, and defined it on the basis of the equation

$$i^2 = -1$$

Once they had set down this rule, mathematicians explored how this strange new number behaved, and a new branch of number theory evolved.

Engineers and physicists use j instead of i to denote the unit imaginary number. That's what we'll use, because the lowercase italic i is found in other mathematical contexts, particularly in sequences and series. The unit imaginary number j is equal to the positive square root of -1. That is,

$$j = (-1)^{1/2}$$

When we use the symbol j to represent the unit imaginary number, we can also call it the *j operator*, a term commonly used by engineers.

The set of imaginary numbers

We can multiply j by any real number, known as a *real-number coefficient*, and the result is an *imaginary number*. The real coefficient is customarily written after j if it is positive or 0, and after $-j$ if it is negative. Examples are

$$j3 = j \times 3 = 3 \times j$$
$$-j5 = j \times (-5) = -5 \times j$$
$$-j2/3 = j \times (-2/3) = -2/3 \times j$$
$$j0 = j \times 0 = 0 \times j = 0$$

The set of all possible real-number multiples of j composes the *set of imaginary numbers*. For practical purposes, the elements of this set can be depicted along a number line corresponding one-to-one with the real-number line. By convention, the *imaginary-number line* is oriented vertically, as shown in Fig. 6-1.

When either j or $-j$ is multiplied by 0, the result is equal to the real number 0. Therefore, the intersection of the sets of imaginary and real numbers contains one element, namely, 0.

Figure 6-1   Imaginary numbers can be depicted as points on a vertical line. As we go upward, we get more positive-imaginary numbers; as we go downward, we get more negative-imaginary numbers.

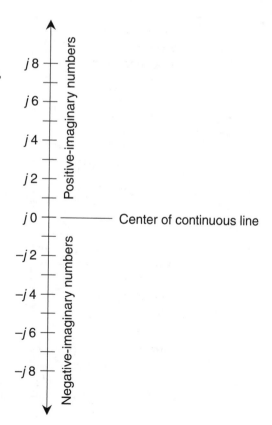

Complex numbers

When we add a real number to an imaginary number, we get a complex number. The general form for a complex number is

$$a + jb$$

where a and b are real numbers. If the real-number coefficient of j happens to be negative, then its absolute value is written following j, and a minus sign is used instead of a plus sign in the composite expression. So instead of

$$a + j(-b)$$

we should write

$$a - jb$$

Individual complex numbers can be depicted as points on a Cartesian coordinate plane as shown in Fig. 6-2. The intersection point between the real- and imaginary-number lines corresponds

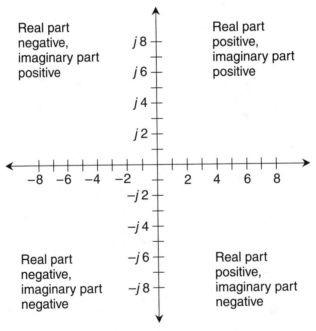

Figure 6-2   Complex numbers can be depicted as points on a plane, which is defined by the intersection of perpendicular real- and imaginary-number lines.

to 0 on the real-number line and $j0$ on the imaginary-number line. This plane is called the *Cartesian complex-number plane.*

An example

If the imaginary part of a complex quantity is 0, we have a *pure real* quantity. When the real part of a complex quantity is 0 and the imaginary part is something other than $j0$, we have a *pure imaginary* quantity. Figure 6-3 shows nine complex numbers plotted as points on the Cartesian complex-number plane, as follows.

- $0 + j0$, whose ordered pair is $(0,j0)$ and which is equal to the pure real 0 and the pure imaginary $j0$.
- $5 + j0$, whose ordered pair is $(5,j0)$ and which is equal to the pure real 5.
- $0 + j7$, whose ordered pair is $(0,j7)$ and which is equal to the pure imaginary $j7$.
- $-2 + j0$, whose ordered pair is $(-2,j0)$ and which is equal to the pure real -2.
- $0 - j8$, whose ordered pair is $(0,-j8)$ and which is equal to the pure imaginary $-j8$.
- $7 + j6$, whose ordered pair is $(7,j6)$.
- $-8 + j5$, whose ordered pair is $(-8,j5)$.
- $-5 - j5$, whose ordered pair is $(-5,-j5)$.
- $3 - j7$, whose ordered pair is $(3,-j7)$.

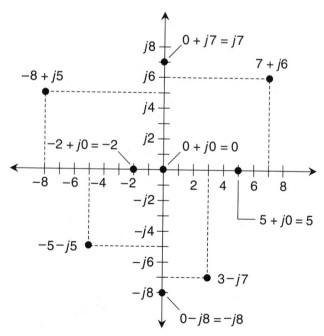

Figure 6-3   Some points in the Cartesian complex-number plane.

Are you confused?

We have learned that $(-1)^{1/2} = j$. You might now ask, "What about the square root of a negative real number other than -1, such as -4 or -100?" The positive square root of any negative real number is equal to j times the positive square root of the absolute value of that real number. For example,

$$(-4)^{1/2} = j \times 4^{1/2} = j2$$

and

$$(-100)^{1/2} = j \times 100^{1/2} = j10$$

We can also have negative square roots of negative reals. That's because $-j$ is not the same quantity as j. (You'll get a chance to prove this fact in Problem 1 at the end of this chapter.) Negating the above examples, we get

$$-(-4)^{1/2} = -j \times 4^{1/2} = -j2$$

and

$$-(-100)^{1/2} = -j \times 100^{1/2} = -j10$$

Here's a challenge!

Demonstrate what happens when $-j$ is raised to successively higher positive-integer powers.

Solution

Keep in mind that $-j$ is the negative square root of -1, which is $-(-1)^{1/2}$. By definition, we know that $j^2 = -1$, so we can calculate the square of $-j$ as

$$(-j)^2 = (-1 \times j)^2$$
$$= (-1)^2 \times j^2$$
$$= 1 \times j^2$$
$$= 1 \times (-1)$$
$$= -1$$

Now for the cube:

$$(-j)^3 = (-j)^2 \times (-j)$$
$$= -1 \times (-j)$$
$$= j$$

The fourth power:

$$(-j)^4 = (-j)^3 \times (-j)$$
$$= j \times (-j)$$
$$= -j^2$$
$$= -(-1)$$
$$= 1$$

The fifth power:

$$(-j)^5 = (-j)^4 \times (-j)$$
$$= 1 \times (-j)$$
$$= -j$$

The sixth power:

$$(-j)^6 = (-j)^5 \times (-j)$$
$$= -j \times (-j)$$
$$= (-j)^2$$
$$= -1$$

Can you see what will happen if we keep going like this, increasing the integer power by 1 over and over? We'll cycle endlessly through $-j$, -1, j, and 1. If you grind things out, you'll see that $j^7 = j$, $j^8 = 1$, $j^9 = -j$, $j^{10} = -1$, and so on. In general, if n is a positive integer, then

$$(-j)^n = (-j)^{n+4}$$

How Complex Numbers Behave

Complex numbers have properties that resemble those of the real numbers to some extent. But there are some major differences as well. Let's review the basic operations involving complex numbers, in case you've forgotten them. As we go along, we'll imagine two arbitrary complex numbers

$$a + jb$$

and

$$c + jd$$

where a, b, c, and d are real numbers, and $j = (-1)^{1/2}$.

Complex number sum

When we want to find the sum of two complex numbers, we add the real and imaginary parts independently to get the real and imaginary components of the result. The general formula is

$$(a + jb) + (c + jd) = (a + c) + j(b + d)$$

Complex number difference

We can find the difference between two complex numbers if we multiply the second complex number by −1, and then add it to the first complex number. The general formula is

$$(a + jb) - (c + jd) = (a + jb) + [-1(c + jd)]$$
$$= (a - c) + j(b - d)$$

Complex number product

When we want to multiply two complex numbers by each other, we can treat them individually as *binomials*. We multiply the binomials and then simplify their product, remembering that $j^2 = -1$. The general formula works out as

$$(a + jb)(c + jd) = ac + jad + jbc + j^2bd$$
$$= (ac - bd) + j(ad + bc)$$

Complex number ratio

Suppose that we want to find the ratio (quotient) of two complex numbers

$$(a + jb) / (c + jd)$$

Multiplying both the numerator and the denominator by $(c - jd)$, we obtain

$$[(a + jb)(c - jd)] / (c + jd)(c - jd)$$

which multiplies out to

$$(ac - jad + jbc - j^2bd) / (c^2 - jcd + jcd - j^2d^2)$$

This expression can be simplified to

$$[(ac + bd) + j(bc - ad)] / (c^2 + d^2)$$

When we separate out the real and imaginary parts, we get

$$[(ac + bd) / (c^2 + d^2)] + j[(bc - ad) / (c^2 + d^2)]$$

The square brackets, while technically superfluous, are included to visually set apart the real and imaginary parts of the result. We have just derived a general complex-number ratio formula that we can always use:

$$(a + jb)/(c + jd)$$
$$= [(ac + bd)/(c^2 + d^2)] + j\,[(bc - ad)/(c^2 + d^2)]$$

For this formula to work, the denominator must not be equal to $0 + j0$. That means we cannot have both $c = 0$ and $d = 0$. If both of these coefficients are 0, then we end up dividing by 0. That operation, unlike the square root of a negative real, remains undefined, at least as far as this book is concerned!

Complex number raised to positive-integer power

If $a + jb$ is a complex number and n is a positive integer, then $(a + jb)^n$ is the result of multiplying $(a + jb)$ by itself n times.

Complex conjugates

Suppose we encounter two complex numbers that have the same coefficients, but opposite signs between the real and imaginary parts, as in

$$a + jb$$

and

$$a - jb$$

We call any two such quantities *complex conjugates*. They have some interesting properties. When we add a complex number to its conjugate, we get twice the real coefficient. In general, we have

$$(a + jb) + (a - jb) = 2a$$

When we multiply a complex number by its conjugate, we get the sum of the squares of the coefficients. In general, we have

$$(a + jb)(a - jb) = a^2 + b^2$$

Complex conjugates are often encountered in engineering. They're especially useful in alternating-current (AC) circuit, radio-frequency (RF) antenna, and transmission-line theories.

Sum example

Let's find the sum of the two complex numbers $5 + j4$ and $2 - j3$. When we add the real parts, we get

$$5 + 2 = 7$$

When we add the imaginary parts, we get

$$j4 + (-j3) = j1 = j$$

The sum can be expressed directly as

$$(5 + j4) + (2 - j3) = 7 + j$$

The parentheses are not technically necessary, but they help to set the individual complex-number addends apart on the left-hand side of the equation.

Difference example

To find the difference between $5 + j4$ and $2 - j3$, we first multiply the second complex quantity by -1. That gives us

$$-1 \times (2 - j3) = -2 + j3$$

Now we can simply add $5 + j4$ and $-2 + j3$. Adding the real parts, we obtain

$$5 + (-2) = 3$$

Adding the imaginary parts gives us

$$j4 + j3 = j7$$

The difference can be expressed directly as

$$(5 + j4) - (2 - j3) = 3 + j7$$

Product example

Let's multiply the complex numbers $5 + j4$ and $2 - j3$ by each other. When we treat them as binomials, the problem works out in a straightforward fashion, but we have to be careful with the signs. We get

$$\begin{aligned}
(5 + j4)(2 - j3) &= 5 \times 2 + 5 \times (-j3) + j4 \times 2 + j4 \times (-j3) \\
&= 10 + (-j15) + j8 + j \times (-j) \times 4 \times 3 \\
&= 10 + (-j7) + 12 \\
&= 22 - j7
\end{aligned}$$

The product can be expressed directly as

$$(5 + j4)(2 - j3) = 22 - j7$$

Ratio example

When we find the ratio of a complex number to another complex number, we should expect some messy arithmetic. Let's calculate

$$(5 + j4) / (2 - j3)$$

Keeping track of the coefficients can be confusing when we use the formula for a ratio. Here's the general formula again:

$$(a + jb) / (c + jd) = [(ac + bd) / (c^2 + d^2)] + j [(bc - ad) / (c^2 + d^2)]$$

The denominator in both addends is $c^2 + d^2$. Here, $c = 2$ and $d = -3$, so we have

$$c^2 + d^2 = 2^2 + (-3)^2 = 4 + 9 = 13$$

We can substitute 13 for the quantity $c^2 + d^2$ in our formula, giving us the expression

$$[(ac + bd) / 13] + j [(bc - ad) / 13]$$

Knowing that $a = 5$, $b = 4$, $c = 2$, and $d = -3$, the above equation becomes

$$[5 \times 2 + 4 \times (-3)] / 13 + j [4 \times 2 - 5 \times (-3)] / 13$$

which works out to

$$-2/13 + j(23/13)$$

Our ratio can be expressed directly as

$$(5 + j4) / (2 - j3) = -2/13 + j(23/13)$$

Power example

Let's find the cube of the complex number $2 - j3$. We square it first, multiplying by itself to get

$$
\begin{aligned}
(2 - j3)(2 - j3) &= 2 \times 2 + 2 \times(-j3) + (-j3) \times 2 + (-j3) \times (-j3) \\
&= 4 + (-j6) + (-j6) + (-j) \times (-j) \times 3 \times 3 \\
&= 4 + (-j12) + (-9) \\
&= -5 - j12
\end{aligned}
$$

We multiply this result by the original quantity $2 - j3$, obtaining

$$
\begin{aligned}
(-5 - j12)(2 - j3) &= -5 \times 2 + (-5) \times (-j3) + (-j12) \times 2 + (-j12) \times (-j3) \\
&= -10 + j15 + (-j24) + (-j) \times (-j) \times 12 \times 3 \\
&= -10 + (-j9) + (-36) \\
&= -46 - j9
\end{aligned}
$$

The cube can be expressed directly as

$$(2 - j3)^3 = -46 - j9$$

Are you confused?

When working with complex numbers, you should pay close attention to whether or not a numeral after the j operator is a superscript. The two notations are perilously similar! For example, if you see

$$5 + j2$$

it means 5 plus twice j, which is a complex number that's neither pure real nor pure imaginary. But if you see

$$5 + j^2$$

it means 5 plus j squared, which can be simplified to $5 + (-1)$ or 4, which is pure real.

Here's a challenge!

Prove that the square of a complex number is equal to the square of the negative of that complex number. That is, show that

$$(a + jb)^2 = (-a - jb)^2$$

for all real-number coefficients a and b.

Solution

First, let's work out the square of $a + jb$. We get

$$(a + jb)^2 = (a + jb)(a + jb)$$
$$= a^2 + jab + jba + j^2b^2$$
$$= a^2 + j2ab - b^2$$
$$= a^2 - b^2 + j2ab$$

Note that in the final term $j2ab$, the numeral 2 is a multiplier, not an exponent! Now let's find the square of $-a - jb$. This is a "nightmare of negatives," so we must be careful with the signs. We have

$$(-a - jb)^2 = (-a - jb)(-a - jb)$$
$$= (-a)^2 + (-a)(-jb) + (-jb)(-a) + (-jb)^2$$

$$= a^2 + jab + jba + (-j)^2 b^2$$

$$= a^2 + j2ab - b^2$$

$$= (a^2 - b^2) + j2ab$$

That's exactly what we got when we squared $a + jb$. Therefore, we've shown that

$$(a + jb)^2 = (-a - jb)^2$$

Complex Vectors

We've seen how points can be represented as standard-form vectors in the Cartesian or polar coordinate planes. Because complex numbers can be plotted as points in a plane, it's tempting to think that we might portray them as vectors. We can; and when we do, things can get mighty interesting.

Cartesian model

When we want to represent a complex quantity as a vector in the Cartesian complex-number plane, we draw an arrowed line segment from the origin to the point representing the quantity. Figure 6-4 shows a few examples.

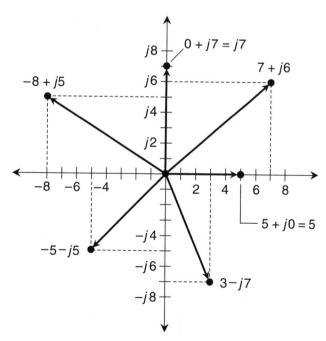

Figure 6-4 Some vectors in the Cartesian complex-number plane.

Polar model

Any vector in the Cartesian plane can also be represented as a vector in the polar coordinate plane. Figure 6-5 shows the vectors from Fig. 6-4 plotted on the polar plane. The radial increments (shown as concentric circles) are the same size as the horizontal- and vertical-axis increments in the Cartesian plane of Fig. 6-4 (that is, 1 unit). The polar scheme is not as common as the Cartesian scheme. But it's equally valid if we restrict the direction angles to positive values less than 2π, and if we forbid negative vector magnitudes.

The vectors in Fig. 6-5 theoretically represent the same complex numbers as those in Fig. 6-4. But the polar coordinates for a complex number differ from the Cartesian coordinates. The polar coordinates reflect the direction angle and magnitude of a vector, not the real and imaginary components. We can calculate the direction angle and magnitude of the polar vector if we know the real and imaginary parts of the equivalent complex number. We can also go the other way, and figure out the real and imaginary parts of the complex number if we know the polar vector direction angle and magnitude.

Cartesian-to-polar complex vector conversion

Imagine a complex number $t = a + jb$, represented as a vector $\mathbf{t_c}$ in the Cartesian complex-number plane, extending from the origin to the point (a,jb). We can derive the magnitude r of the equivalent polar vector $\mathbf{t_p}$ by applying the Pythagorean distance formula to get

$$r = (a^2 + b^2)^{1/2}$$

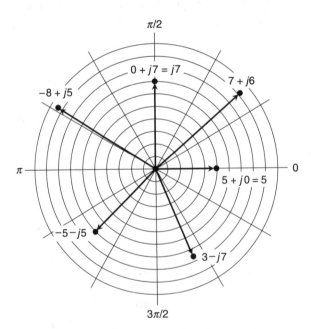

Figure 6-5 Complex numbers can be portrayed as vectors in the polar plane. Each radial division represents 1 unit. Cartesian coordinates are shown here. The polar coordinates are entirely different!

To determine the direction angle θ of the polar vector $\mathbf{t}_p$, we modify the polar-coordinate direction-finding system. Here's what we get. As we did in the Cartesian-to-polar coordinate-conversion scheme, we define $\theta = 0$ by default when we're at the origin. That way, we get a one-to-one correspondence between the set of Cartesian vectors and the set of polar vectors. (Keep that in mind, because we'll keep doing this whenever the situation comes up!)

$\theta = 0$ by default	When $a = 0$ and $jb = j0$ that is, at the origin
$\theta = 0$	When $a > 0$ and $jb = j0$
$\theta = \text{Arctan} \,(b/a)$	When $a > 0$ and $jb > j0$
$\theta = \pi/2$	When $a = 0$ and $jb > j0$
$\theta = \pi + \text{Arctan} \,(b/a)$	When $a < 0$ and $jb > j0$
$\theta = \pi$	When $a < 0$ and $jb = j0$
$\theta = \pi + \text{Arctan} \,(b/a)$	When $a < 0$ and $jb < j0$
$\theta = 3\pi/2$	When $a = 0$ and $jb < j0$
$\theta = 2\pi + \text{Arctan} \,(b/a)$	When $a > 0$ and $jb < j0$

Polar-to-Cartesian complex vector conversion

We can always convert a polar complex vector $\mathbf{t}_p$ into a Cartesian complex vector $\mathbf{t}_c$ that portrays a complex number $a + jb$ in the familiar form. If we have

$$\mathbf{t}_p = (\theta, r)$$

then the Cartesian vector equivalent is

$$\mathbf{t}_c = [(r \cos \theta), j(r \sin \theta)]$$

which represents the complex number

$$a + jb = r \cos \theta + j(r \sin \theta)$$

The parentheses are not strictly necessary here, but they keep the real and imaginary components clearly separated.

Absolute value

We can find the *absolute value* of a complex number $a + jb$, written $|\, a + jb\, |$, by calculating the magnitude of its vector. In the Cartesian complex plane, going from the origin $(0,0)$ to the point (a, jb), we have

$$|\, a + jb\, | = (a^2 + b^2)^{1/2}$$

as shown in Fig. 6-6. In the polar plane, the absolute value of a complex vector is the vector radius r.

Figure 6-6   The absolute value of a complex number is the magnitude of its vector.

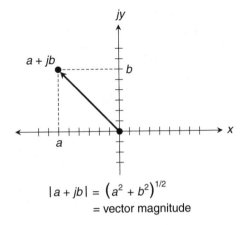

$$|a + jb| = \left(a^2 + b^2\right)^{1/2}$$
$$= \text{vector magnitude}$$

Complex vector sum and difference

When we want to add or subtract two complex vectors, we can work on the Cartesian real and imaginary parts separately. If the vectors are presented to us in polar form, we should convert them to Cartesian form and then add. We can always convert the resultant back to polar form after we're done with the addition process.

To find the difference between two complex vectors, we must be sure they're both in Cartesian form before we do any calculations. Once the vectors are in the Cartesian form, we take the negative of the second vector by negating both of its coordinates. Then we add the two resulting vectors. Again, if we want, we can convert the resultant back to polar form.

Are you confused?

You may ask, "Why isn't the addition and subtraction of polar coordinates directly done when we want to add or subtract complex vectors?" That's a good question. We can try to define vector sums and differences this way (adding or subtracting the polar angles and radii separately, for example), and we'll get output numbers when we grind out the arithmetic. But those numbers don't coincide with the geometric definitions of vector addition and subtraction. They don't give us the correct complex-number sums or differences. It's hard to say what those output numbers really mean, even though the idea is interesting! We should use Cartesian coordinates when we add or subtract complex vectors. We should use polar coordinates when we want to multiply or divide them.

Polar complex vector product

When we want to multiply two complex-number vectors, neither the dot product nor the cross product will give us the proper results. We must invent a new vector operation! Here's how it works.

1. We add the direction angles of the original two vectors to get the direction angle of the product vector.
2. If we end up with a direction angle larger than 2π, then we subtract 2π to get the correct angle for the product vector.

3. We multiply the original vector magnitudes by each other to get the magnitude of the product vector.

Polar complex vector ratio

When we want to find the ratio of two complex numbers, we can go through the complex vector product process "inside-out." Again, there are three steps.

1. We subtract the direction angle of the denominator vector from the direction angle of the numerator vector to get the direction angle of the ratio vector.
2. If we end up with a negative direction angle, then we add 2π to get the correct angle for the ratio vector.
3. We divide the magnitude of the numerator vector by the magnitude of the denominator vector to get the magnitude of the ratio vector.

Polar complex vector power

When we want to raise a complex number to a positive-integer power, we multiply the polar angle by that positive integer, and then take the power of the magnitude. If the angle of our resulting vector is 2π or larger, we subtract whatever multiple of 2π is necessary to bring the angle into the range where it's positive but less than 2π.

Absolute-value vector example

There are infinitely many vectors that represent complex numbers having an absolute value of 6. All the vectors have magnitudes of 6, and they all point outward from the origin. Figure 6-7 shows a few such vectors.

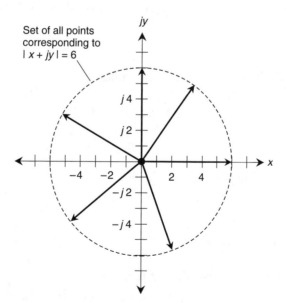

Figure 6-7   There are infinitely many complex numbers with an absolute value of 6. They all terminate on a circle of radius 6, centered at the origin.

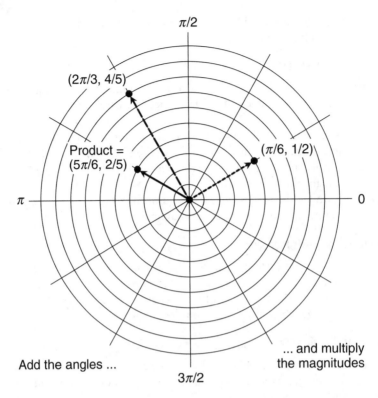

Figure 6-8   Product of the polar complex vectors $(\pi/6,1/2)$ and $(2\pi/3,4/5)$. Each radial division represents 0.1 unit.

Polar complex vector product example

Figure 6-8 shows the polar complex vectors $(\pi/6,1/2)$ and $(2\pi/3,4/5)$, along with their product. Each radial division is 0.1 unit. When we add the angles, we get

$$\pi/6 + 2\pi/3 = 5\pi/6$$

When we multiply the magnitudes, we get

$$1/2 \times 4/5 = 2/5$$

so the product vector is $(5\pi/6,2/5)$.

Polar complex vector ratio example

Figure 6-9 shows the ratio of the polar complex vectors $(7\pi/4,8)$ and $(\pi,2)$. Each radial division is 1 unit. When we subtract the angles, we get

$$7\pi/4 - \pi = 3\pi/4$$

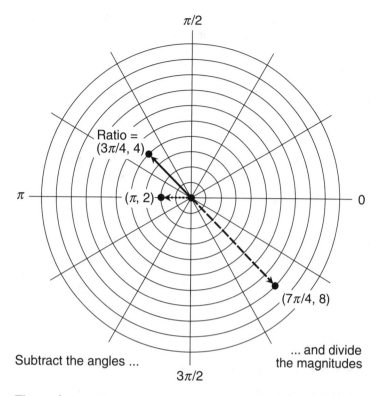

Figure 6-9   Ratio of the polar complex vector $(7\pi/4, 8)$ to the polar
complex vector $(\pi, 2)$. Each radial division represents
1 unit.

When we divide the magnitudes, we get

$$8/2 = 4$$

so the ratio vector is $(3\pi/4, 4)$.

De Moivre's theorem

The above schemes for finding products, ratios, and powers of polar complex numbers can
be summarized in a famous theorem attributed to the French mathematician *Abraham De
Moivre*, (pronounced "De Mwahvr"), who lived during the late 1600s and early 1700s. This
theorem can be found in two different versions, depending on which text you consult.

The first, and more general, version of *De Moivre's theorem* involves products and ratios.
Suppose we have two polar complex numbers c_1 and c_2, where

$$c_1 = r_1 \cos \theta_1 + j(r_1 \sin \theta_1)$$

and

$$c_2 = r_2 \cos \theta_2 + j(r_2 \sin \theta_2)$$

where r_1 and r_2 are real-number polar magnitudes, and θ_1 and θ_2 are real-number polar direction angles in radians. Then the product of c_1 and c_2 is

$$c_1 c_2 = r_1 r_2 \cos (\theta_1 + \theta_2) + j [r_1 r_2 \sin (\theta_1 + \theta_2)]$$

If r_2 is nonzero, the ratio of c_1 to c_2 is

$$c_1 / c_2 = (r_1/r_2) \cos (\theta_1 - \theta_2) + j [(r_1/r_2) \sin (\theta_1 - \theta_2)]$$

The second, and more commonly known, version of De Moivre's theorem can be derived from the first version. Suppose that we have a complex number c such that

$$c = r \cos \theta + j(r \sin \theta)$$

where r is the real-number polar magnitude and θ is the real-number polar direction angle. Also suppose that n is an integer. Then c to the nth power is

$$c^n = r^n \cos (n\theta) + j[r^n \sin (n\theta)]$$

I recommend that you enter this version of De Moivre's theorem into your "brain storage," and save it there forever!

- -

Are you confused?

Do you wonder why we haven't described how to find a root of a complex vector? You might think, "It ought to be simple, just like finding a power backward. Can't we divide the polar angle by the index of the root, and then take the root of the magnitude?" That's a good question. Doing that will indeed give us a root. But there are often two or more complex roots for any given complex number. We're about to see an example of this.

Here's a challenge!

Cube the polar complex vectors $(2\pi/3,1)$ and $(4\pi/3,1)$. Here's a warning: The solution might come as a surprise! What do you suppose these results imply?

Solution

To cube a polar complex vector, we multiply the direction angle by 3 (the value of the exponent) and cube the magnitude. Let's do this with the vectors we've been given here. In the case of $(2\pi/3,1)^3$, we get an angle of

$$(2\pi/3) \times 3 = 2\pi$$

That's outside the allowed range of angles, but if we subtract 2π, we get 0, which is okay. We get a magnitude of $1^3 = 1$. Now we know that

$$(2\pi/3,1)^3 = (0,1)$$

where the first coordinate represents the direction angle in radians, and the second coordinate represents the magnitude. If we draw this vector on a polar graph, we can see that this is the polar representation of the complex number $1 + j0$, which is equal to the pure real number 1. (If you like, you can use the conversion formulas to prove it.) In the case of $(4\pi/3,1)^3$, we get the direction angle

$$(4\pi/3) \times 3 = 4\pi$$

That's outside the allowed range of angles, but if we subtract 2π twice, then we get an angle of 0, and that's allowed. As before, we get a magnitude of $1^3 = 1$. Now we know that

$$(4\pi/3,1)^3 = (0,1)$$

where, again, the first coordinate represents the direction angle in radians, and the second coordinate represents the magnitude. This is the same as the previous result. It's the polar representation of $1 + j0$, which is the pure real number 1. We've found two cube roots of 1 in the realm of the complex numbers. Neither of these roots show up when we work with pure real numbers exclusively. There are three different complex cube roots of 1! They are

- The pure real number 1
- The complex number corresponding to the polar vector $(2\pi/3,1)$
- The complex number corresponding to the polar vector $(4\pi/3,1)$

Practice Exercises

This is an open-book quiz. You may (and should) refer to the text as you solve these problems. Don't hurry! You'll find worked-out answers in App. A. The solutions in the appendix may not represent the only way a problem can be figured out. If you think you can solve a particular problem in a quicker or better way than you see there, by all means try it!

1. Prove that $-j$ is not equal to j, even though, when squared, they both give us -1. Here's a hint: Use the tactic of *reductio ad absurdum*, where a statement is proved by assuming its opposite and then deriving a contradiction from that assumption.

2. Show that the reciprocal of j is equal to its negative; that is, $j^{-1} = -j$.

3. Find the sum and difference of the complex numbers $-3 + j4$ and $1 + j5$.

4. Find the ratio of the generalized complex conjugates $a + jb$ and $a - jb$. That is, work out a general formula for

$$(a + jb) \, / \, (a - jb)$$

where a and b are both nonzero real numbers.

5. Prove that if we take any two complex conjugates and square them individually, the results are complex conjugates. In other words, show that for all real-number coefficients a and b, $(a + jb)^2$ is the complex conjugate of $(a - jb)^2$.

6. Find the polar product of the polar complex vectors $(\pi/4, 2^{1/2})$ and $(3\pi/4, 2^{1/2})$. Then convert this product vector to Cartesian form and write down the "real-plus-imaginary" complex number that it represents.

7. Convert the polar complex vectors $(\pi/4, 2^{1/2})$ and $(3\pi/4, 2^{1/2})$ to the complex numbers they represent in "real-plus-imaginary" form. Multiply these numbers and compare with the solution to Problem 6.

8. Look at the results of the last "challenge," where we found these three cube roots of 1:

   - The pure real number 1
   - The complex number corresponding to the polar vector $(2\pi/3, 1)$
   - The complex number corresponding to the polar vector $(4\pi/3, 1)$

   Convert the polar vectors $(2\pi/3, 1)$ and $(4\pi/3, 1)$ to their "real-plus-imaginary" complex-number forms.

9. Graph the three cube roots of 1 as polar complex vectors. Label them as ordered pairs in the form (θ, r), where θ is the direction angle and r is the magnitude.

10. Graph the three cube roots of 1 as Cartesian complex vectors. Label them as complex numbers in the form $a + jb$, where a and b are real numbers. Also graph the unit circle, and note that the vectors all terminate on that circle.

7

Cartesian Three-Space

We can create three-dimensional graphs by adding a third axis perpendicular to the familiar x and y axes of the Cartesian plane. The new axis, usually called the z axis, passes through the xy plane at the origin, giving us *Cartesian three-space* or *Cartesian xyz space*.

How It's Assembled

Cartesian three-space has three real-number lines positioned so they all intersect at their zero points, and so each line is perpendicular to the other two. The point where the axes intersect constitutes the *origin*. Each axis portrays a real-number variable.

Axes and variables

Figure 7-1 is a perspective drawing of a Cartesian xyz space coordinate system. In a true-to-life three-dimensional portrayal, the positive x axis would run to the right, the negative x axis would run to the left, the positive y axis would run upward, the negative y axis would run downward, the positive z axis would project out from the page toward us, and the negative z axis would project behind the page away from us.

In Cartesian three-space, the axes are all linear, and they're all graduated in increments of the same size. For any axis, the change in value is always directly proportional to the physical displacement. If we move 3 millimeters along an axis and the value changes by 1 unit, then that's true all along the axis, and it's also true everywhere along both of the other axes. If the divisions differ in size between the axes, then we have *rectangular three-space*, but not true Cartesian three-space.

Cartesian three-space is often used to graph relations and functions having two independent variables. When this is done, x and y are usually the independent variables, and z is the dependent variable, whose value depends on both x and y.

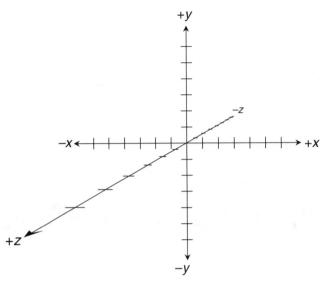

Figure 7-1 A pictorial rendition of Cartesian three-space. In this view, the *x* axis increases positively from left to right, the *y* axis increases positively from the bottom up, and the *z* axis increases positively from far to near.

Biaxial planes

Cartesian three-space contains three flat *biaxial (two-axis) planes* that intersect along the coordinate axes.

- The *xy* plane contains the axes for the variables *x* and *y*.
- The *xz* plane contains the axes for the variables *x* and *z*.
- The *yz* plane contains the axes for the variables *y* and *z*.

You'll see three rectangles in Fig. 7-2, one parallel to each of the three biaxial planes. Look closely at how these rectangles are oriented. They can help you envision the orientations of the three biaxial planes in space. Each of the three biaxial planes is perpendicular to both of the others.

Are you astute?

Figure 7-2 shows an alternative perspective on Cartesian three-space, in which we're looking "up" toward the *xz* plane from somewhere near the negative *y* axis. There's a difference between the *apparent* positions of the axes in Fig. 7-2 as compared with their positions in Fig. 7-1, but the orientations of the three axes are the same *with respect to each other*. You should get used to seeing Cartesian three-space from various points of view. I'll switch points of view often to keep you thinking!

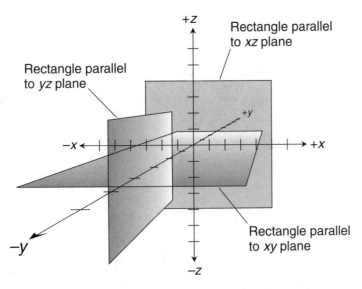

Figure 7-2   Cartesian three-space contains the *xy*, *xz*, and
yz planes. This drawing shows rectangles parallel to
each of these three biaxial planes. Note the difference
in the point of view between this illustration and
Figure 7-1.

Points and ordered triples

Figure 7-3 shows two specific points *P* and *Q*, plotted in Cartesian three-space. We've returned to the perspective of Fig. 7-1, with the positive *z* axis coming out of the page toward us. A point can always be denoted as an *ordered triple* in the form (*x,y,z*), according to the following scheme:

- The *x* coordinate represents the point's projection onto the *x* axis.
- The *y* coordinate represents the point's projection onto the *y* axis.
- The *z* coordinate represents the point's projection onto the *z* axis.

We get the projection of a point onto an axis by drawing a line from that point to the axis, and making sure that the line intersects that axis at a right angle. If this notion gives you trouble, you can think of the *x*, *y*, and *z* values for a particular point in the following way:

- The *x* coordinate is the point's perpendicular displacement (positive, negative, or zero) from the *yz* plane.
- The *y* coordinate is the point's perpendicular displacement (positive, negative, or zero) from the *xz* plane.
- The *z* coordinate is the point's perpendicular displacement (positive, negative, or zero) from the *xy* plane.

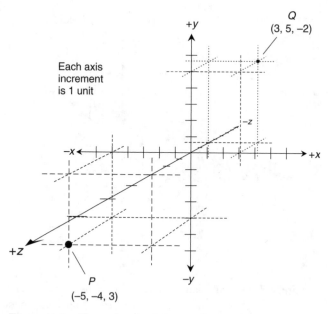

Figure 7-3   Two points in Cartesian three-space, along
with the corresponding ordered triples of the
form (x,y,z). On all three axes, each increment
represents 1 unit. Here, we've gone back to the
point of view shown in Figure 7-1.

In Fig. 7-3, the coordinates of point P are $(-5,-4,3)$, and the coordinates of point Q are
$(3,5,-2)$. As the system is portrayed here, we can get to point P from the origin by making the
following moves in any order:

- Go 5 units in the negative x direction (straight to the left).
- Go 4 units in the negative y direction (straight down).
- Go 3 units in the positive z direction (straight out of the page).

We can get from the origin to point Q by doing the following moves in any order:

- Go 3 units in the positive x direction (straight to the right).
- Go 5 units in the positive y direction (straight up).
- Go 2 units in the negative z direction (straight back behind the page).

If we were looking at the coordinate grid from a different viewpoint (that of Fig. 7-2, for
example), our movements would look different, but the points and their coordinates would
be the same.

A note for the picayune

An ordered triple represents the coordinates of a point in three-space, not the geometric point
itself. But we may talk or write as if an ordered triple actually is a point, just as we sometimes
think of a certain person when we read a name. That's okay, as long as we're aware of the
semantical difference between the name and the point.

Are you confused?

Some people have trouble envisioning three-dimensional situations "in the mind's eye." If you're having problems understanding exactly how the three axes should relate in Cartesian three-space, here's a "pool rule" for the orientation of the axes. Imagine the origin of the system resting on the surface of a swimming pool. Suppose that we align the positive x axis so that it runs along the water surface, pointing due east. Once we've done that, the other axes are oriented as follows:

- Negative values of x are west of the origin.
- Positive values of y are north of the origin.
- Negative values of y are south of the origin.
- Positive values of z are up in the air.
- Negative values of z are under the water.

You can look at the coordinate axes from any point you want, whether on the surface, in the sky, or under the water. No matter how your view of the system changes, the actual orientation of the axes with respect to each other always stays the same. This relative axis orientation is important. If it's not strictly followed, we'll get into trouble when we work with graphs and vectors in Cartesian three-space.

Here's a challenge!

Imagine an ordered triple (x,y,z) where all three variables are nonzero real numbers. Suppose that you've plotted a point P in xyz space. Because $x \neq 0$, $y \neq 0$, and $z \neq 0$, the point P doesn't lie on any of the axes. What will happen to the location of P if you

- Multiply x by -1 and leave y and z the same?
- Multiply y by -1 and leave x and z the same?
- Multiply z by -1 and leave x and y the same?

Solution

Here's what will take place in each of these three situations. You can use Fig. 7-2 as a visual aid. If you're a computer whiz, maybe you can program your machine to create an animated display for each of these three processes:

- If you multiply x by -1 and do not change the values of y or z, then point P will move parallel to the x axis to the opposite side of the yz plane, but P will end up at the same distance from the yz plane as it was before.
- If you multiply y by -1 and do not change the values of x or z, then point P will move parallel to the y axis to the opposite side of the xz plane, but P will end up at the same distance from the xz plane as it was before.
- If you multiply z by -1 and do not change the values of x or y, then point P will move parallel to the z axis to the opposite side of the xy plane, but P will end up at the same distance from the xy plane as it was before.

Distance of Point from Origin

In Cartesian three-space, the distance of a point from the origin depends on all three of the coordinates in the ordered triple representing the point. The formula for this distance resembles the formula for the distance of a point from the origin in Cartesian two-space.

The general formula

It's not difficult to derive a general formula for the distance of a point from the origin in Cartesian three-space, as long as we're willing to use our "spatial mind's eye." Suppose we name the point P, and assign it the coordinates

$$P = (x_p, y_p, z_p)$$

Figure 7-4A shows this situation, along with a point $P^* = (x_p, y_p, 0)$, which is the projection of P onto the xy plane. We've moved again back to the perspective of Fig. 7-2, looking in toward the origin from somewhere far out in space near the negative y axis. To find the distance of P^* from the origin, we can work entirely in the xy plane. This gives us a two-dimensional distance problem, which we learned how to handle in Chap. 1. Let's call the distance of P^* from the origin by the name a. Using the formula we learned in Chap. 1 for the distance of a point from the origin in Cartesian xy plane, we have

$$a = (x_p^2 + y_p^2)^{1/2}$$

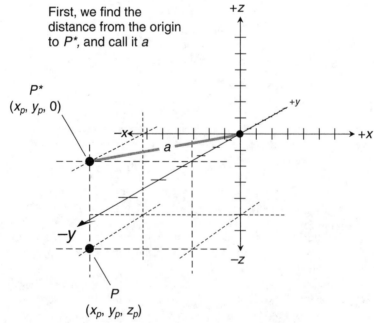

Figure 7-4A   Finding the distance of point P from the origin: step 1.

This completes the first step in a three-phase process. Figure 7-4B shows the second step. Here, we find the distance between P^* and P. Let's call that distance b. It's the perpendicular distance of P from the xy plane, which is simply the coordinate value z_p. Therefore, we have

$$b = z_p$$

That's the end of the second step. In Fig. 7-4C, the distance from the origin to P is labeled c. Note that we now have a right triangle with sides of lengths a, b, and c. The right angle is between the sides whose lengths are a and b. The Pythagorean theorem therefore allows us to make the claim that

$$a^2 + b^2 = c^2$$

Substituting the previously determined values for a and b into this formula gives us

$$[(x_p^2 + y_p^2)^{1/2}]^2 + z_p^2 = c^2$$

which simplifies to

$$x_p^2 + y_p^2 + z_p^2 = c^2$$

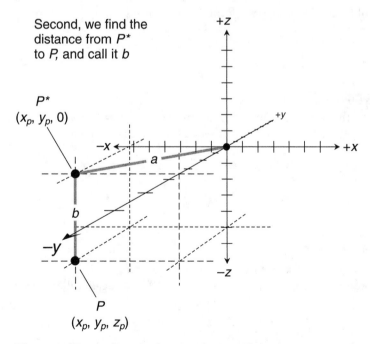

Figure 7-4B Finding the distance of point P from the origin: step 2.

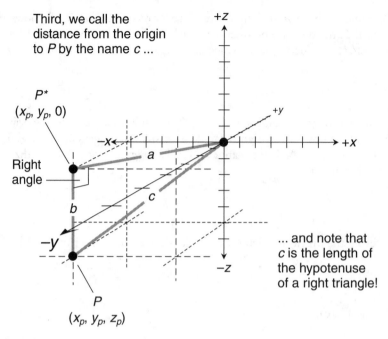

Third, we call the distance from the origin to *P* by the name *c* ...

P^*
$(x_p, y_p, 0)$

Right angle

a

b

c

−*y*

−*z*

... and note that *c* is the length of the hypotenuse of a right triangle!

P
(x_p, y_p, z_p)

Figure 7-4C   Finding the distance of point *P* from the origin: step 3.

When we switch the right-hand and left-hand sides of this equation and then take the 1/2 power of both sides, we get the formula we've been looking for, which is

$$c = (x_p^2 + y_p^2 + z_p^2)^{1/2}$$

An example

Let's find the distance from the origin to the point $P = (-5, -4, 3)$ as shown in Fig. 7-3. We have $x_p = -5$, $y_p = -4$, and $z_p = 3$. If we call the distance c, then

$$c = (x_p^2 + y_p^2 + z_p^2)^{1/2}$$

$$= [(-5)^2 + (-4)^2 + 3^2]^{1/2} = (25 + 16 + 9)^{1/2} = 50^{1/2}$$

Another example

Now let's find the distance from the origin to $Q = (3, 5, -2)$ as shown in Fig. 7-3. This time, the coordinates are $x_q = 3$, $y_q = 5$, and $z_q = -2$. We can again call the distance c, so

$$c = (x_q^2 + y_q^2 + z_q^2)^{1/2}$$

$$= [3^2 + 5^2 + (-2)^2]^{1/2} = (9 + 25 + 4)^{1/2} = 38^{1/2}$$

Are you confused?

You might ask, "Can the distance of a point from the origin in Cartesian three-space ever be undefined? Can it ever be negative?" The answers are no, and no! Imagine a point P in Cartesian three-space—anywhere you want—with the coordinates (x_p, y_p, z_p). To find the distance of P from the origin, you start by squaring x_p, which is the x coordinate of P. Because x_p is a real number, its square is a nonnegative real. Then you square y_p, which is the y coordinate of P. This result must also be a nonnegative real. Then you square z_p, which is the z coordinate of P. This square, too, is a nonnegative real. Next, you add the three nonnegative reals $x_p{}^2$, $y_p{}^2$, and $z_p{}^2$. That sum must be another nonnegative real. Finally, you take the nonnegative square root of the sum of the squares. The nonnegative square root of a nonnegative real number is always defined; and it's never negative itself, of course!

Are you still confused?

The formula we derived here is based on the idea that we start at the origin and go outward to point P. If we go inward from P to the origin, the distance is exactly the same. (If we were working with vectors, the vector displacements would be negatives of each other, but we're not there yet.)

Here's a challenge!

Suppose we're given a point $P = (x_p, y_p, z_p)$ in Cartesian three-space. Prove that if we negate any one, any two, or all three of the coordinates, the resulting point is the same distance from the origin as P.

Solution

For the point P, the distance c from the origin is

$$c = (x_p{}^2 + y_p{}^2 + z_p{}^2)^{1/2}$$

The square of any real number is always the same as the square of its negative. That tells us three things:

$$(-x_p)^2 = x_p{}^2$$
$$(-y_p)^2 = y_p{}^2$$
$$(-z_p)^2 = z_p{}^2$$

By substitution, all these quantities are identical:

$$(x_p{}^2 + y_p{}^2 + z_p{}^2)^{1/2}$$
$$[(-x_p)^2 + y_p{}^2 + z_p{}^2]^{1/2}$$
$$[x_p{}^2 + (-y_p)^2 + z_p{}^2]^{1/2}$$
$$[x_p{}^2 + y_p{}^2 + (-z_p)^2]^{1/2}$$
$$[(-x_p)^2 + (-y_p)^2 + z_p{}^2]^{1/2}$$
$$[(-x_p)^2 + y_p{}^2 + (-z_p)^2]^{1/2}$$
$$[x_p{}^2 + (-y_p)^2 + (-z_p)^2]^{1/2}$$
$$[(-x_p)^2 + (-y_p)^2 + (-z_p)^2]^{1/2}$$

These quantities represent the distances of the following points from the origin, respectively:

$$(x_p, y_p, z_p)$$

$$(-x_p, y_p, z_p)$$

$$(x_p, -y_p, z_p)$$

$$(x_p, y_p, -z_p)$$

$$(-x_p, -y_p, z_p)$$

$$(-x_p, y_p, -z_p)$$

$$(x_p, -y_p, -z_p)$$

$$(-x_p, -y_p, -z_p)$$

That's all the points we can get, in addition to P itself, by negating any one, any two, or all three of the coordinates of P. They're all the same distance c from the origin, where

$$c = (x_p^2 + y_p^2 + z_p^2)^{1/2}$$

Distance between Any Two Points

When we want to determine the distance between any two points in Cartesian three-space, we can expand the formula from Cartesian two-space that we learned in Chap. 1 into an extra dimension.

The general formula

Imagine two different points in Cartesian three-space, after the fashion of Fig. 7-5. Let's call the points and their coordinates

$$P = (x_p, y_p, z_p)$$

and

$$Q = (x_q, y_q, z_q)$$

where each coordinate can range over the entire set of real numbers. The distance d between these points, as we follow a straight-line path from P to Q, is

$$d = [(x_q - x_p)^2 + (y_q - y_p)^2 + (z_q - z_p)^2]^{1/2}$$

If we start at Q and finish at P, we reverse the orders of subtraction, so the formula becomes

$$d = [(x_p - x_q)^2 + (y_p - y_q)^2 + (z_p - z_q)^2]^{1/2}$$

We always subtract "starting coordinates" from "finishing coordinates."

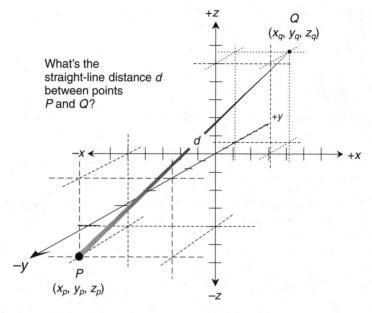

Figure 7-5   Distance between two points in Cartesian three-space.

An example

Let's calculate the distance between the points $P = (-5,-4,3)$ and $Q = (3,5,-2)$, starting at P and finishing at Q. We subtract the coordinates for P from those for Q in each term. Pairing off the coordinates for easy reference, we have

$$x_p = -5 \text{ and } x_q = 3$$
$$y_p = -4 \text{ and } y_q = 5$$
$$z_p = 3 \text{ and } z_q = -2$$

Plugging these values into the formula, we get

$$d = [(x_q - x_p)^2 + (y_q - y_p)^2 + (z_q - z_p)^2]^{1/2}$$
$$= \{[3 - (-5)]^2 + [5 - (-4)]^2 + (-2 - 3)^2\}^{1/2}$$
$$= [8^2 + 9^2 + (-5)^2]^{1/2} = (64 + 81 + 25)^{1/2} = 170^{1/2}$$

Another example

Now let's calculate the distance between these same two points, but starting at Q and finishing at P. We reverse the orders of the subtractions from the previous example. When we go through the arithmetic, we get

$$d = [(x_p - x_q)^2 + (y_p - y_q)^2 + (z_p - z_q)^2]^{1/2}$$
$$= \{(-5 - 3)^2 + (-4 - 5)^2 + [3 - (-2)]^2\}^{1/2}$$
$$= [(-8)^2 + (-9)^2 + 5^2]^{1/2} = (64 + 81 + 25)^{1/2} = 170^{1/2}$$

Are you confused?

You are probably not surprised that the distance between P and Q is the same in either direction. But you might ask, "Are there *any* situations where the distance between two points is different in one direction than in the other?" The answer is no, such a thing can never happen—as long as we always follow the same straight-line path through three-space to get from point to point. Let's prove that the direction doesn't matter when we want to express the distance between two points in space.

Here's a challenge!

Show that the distance between any two points in Cartesian three-space is the same, whichever direction we go.

Solution

It's sufficient to prove that for all real numbers x_p, y_p, z_p, x_q, y_q, and z_q, it's always the case that

$$[(x_q - x_p)^2 + (y_q - y_p)^2 + (z_q - z_p)^2]^{1/2} = [(x_p - x_q)^2 + (y_p - y_q)^2 + (z_p - z_q)^2]^{1/2}$$

Because $x_q - x_p$ and $x_p - x_q$ are negatives of each other, their squares are equal:

$$(x_q - x_p)^2 = (x_p - x_q)^2$$

Because $y_q - y_p$ and $y_p - y_q$ are negatives of each other, their squares are equal:

$$(y_q - y_p)^2 = (y_p - y_q)^2$$

Because $z_q - z_p$ and $z_p - z_q$ are negatives of each other, their squares are equal:

$$(z_q - z_p)^2 = (z_p - z_q)^2$$

Based on these three facts, we know that the squared differences on both sides of the original equation are equal, no matter what the values of the coordinates might be (as long as they're all real numbers). This tells us that the distance between any two points is the same in either direction.

Finding the Midpoint

We can find the midpoint along a straight-line path between two points in Cartesian three-space by averaging the corresponding coordinates.

The general formula

Suppose that we want to find the midpoint M along a straight-line segment connecting two points P and Q as shown in Fig. 7-6. We can assign the points the ordered triples

$$P = (x_p, y_p, z_p)$$

and

$$Q = (x_q, y_q, z_q)$$

Let's say that the coordinates of the midpoint M are

$$M = (x_m, y_m, z_m)$$

We find x_m by averaging x_p and x_q, getting

$$x_m = (x_p + x_q)/2$$

We find y_m by averaging y_p and y_q, getting

$$y_m = (y_p + y_q)/2$$

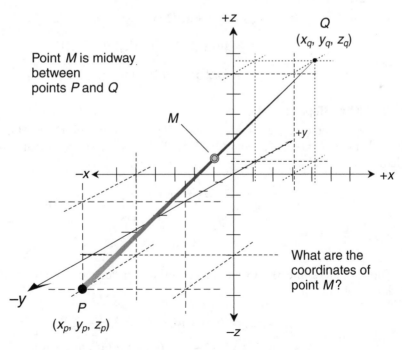

Figure 7-6 Midpoint of line segment connecting two points in Cartesian three-space.

We find z_m by averaging z_p and z_q, getting

$$z_m = (z_p + z_q)/2$$

The coordinates of M in terms of the coordinates of P and Q are therefore

$$(x_m,y_m,z_m) = [(x_p + x_q)/2,(y_p + y_q)/2,(z_p + z_q)/2]$$

An example

Let's find the midpoint between the origin and $P = (-5,-4,3)$ in Cartesian three-space. We can use the formula above with $Q = (0,0,0)$. The midpoint M has the coordinates

$$\begin{aligned}(x_m,y_m,z_m) &= [(x_p + x_q)/2,(y_p + y_q)/2,(z_p + z_q)/2]\\ &= [(-5 + 0)/2,(-4 + 0)/2,(3 + 0)/2]\\ &= (-5/2,-4/2,3/2) = (-5/2,-2,3/2)\end{aligned}$$

Another example

Now let's find the midpoint between the origin and $Q = (3,5,-2)$. This time, we let $P = (0,0,0)$, so the midpoint M has the coordinates

$$\begin{aligned}(x_m,y_m,z_m) &= [(x_p + x_q)/2,(y_p + y_q)/2,(z_p + z_q)/2]\\ &= \{(0 + 3)/2,(0 + 5)/2,[(0 + (-2)]/2\}\\ &= (3/2,5/2,-2/2) = (3/2,5/2,-1)\end{aligned}$$

Still another example

Now let's work out a tougher problem. Suppose we want to find the coordinates of the midpoint M between $P = (-5,-4,3)$ and $Q = (3,5,-2)$. Pairing off the coordinates for convenience, we have

$$x_p = -5 \text{ and } x_q = 3$$
$$y_p = -4 \text{ and } y_q = 5$$
$$z_p = 3 \text{ and } z_q = -2$$

Plugging the values into the formula and working through the arithmetic, we obtain the coordinates of M as

$$\begin{aligned}(x_m,y_m,z_m) &= [(x_p + x_q)/2,(y_p + y_q)/2,(z_p + z_q)/2]\\ &= \{(-5 + 3)/2,(-4 + 5)/2,[(3 + (-2)]/2\}\\ &= (-2/2,1/2,1/2) = (-1,1/2,1/2)\end{aligned}$$

Are you a skeptic?

Does it seem obvious that the midpoint between two points, say P and Q, doesn't depend on whether we go from P to Q or from Q to P? That's indeed the case; but if we demand proof, we must show that for real numbers x_p, y_p, z_p, x_q, y_q, and z_q, it's always true that

$$[(x_p + x_q)/2, (y_p + y_q)/2, (z_p + z_q)/2] = [(x_q + x_p)/2, (y_q + y_p)/2, (z_q + z_p)/2]$$

This proof is almost trivial, but it's good mental exercise to put it down in rigorous form. The commutative law for addition of real numbers tells us that

$$x_p + x_q = x_q + x_p$$

Dividing each side by 2 gives us

$$(x_p + x_q)/2 = (x_q + x_p)/2$$

Using the same logic with the y and z coordinates, we get

$$(y_p + y_q)/2 = (y_q + y_p)/2$$

and

$$(z_p + z_q)/2 = (z_q + z_p)/2$$

Based on these facts, we know that the coordinates on both sides of the original equation are identical. It follows that the midpoint along a straight-line segment connecting any two points in Cartesian three-space is the same, regardless of which way we go.

Here's a challenge!

Imagine two points in Cartesian three-space where corresponding coordinates are negatives of each other. Show that the midpoint is exactly at the origin.

Solution

We can choose any point P whose coordinates are all real numbers. Let's suppose that

$$P = (x_p, y_p, z_p)$$

Then the coordinates of Q are

$$Q = (-x_p, -y_p, -z_p)$$

The coordinates of the midpoint M are

$$(x_m, y_m, z_m) = \{[(x_p + (-x_p)]/2, [(y_p + (-y_p)]/2, [(z_p + (-z_p)/2]\}$$

$$= [(x_p - x_p)/2, (y_p - y_p)/2, (z_p - z_p)/2]$$

$$= (0/2, 0/2, 0/2) = (0,0,0)$$

The point $(0,0,0)$ is, of course, the origin of the coordinate system.

Practice Exercises

This is an open-book quiz. You may (and should) refer to the text as you solve these problems. Don't hurry! You'll find worked-out answers in App. A. The solutions in the appendix may not represent the only way a problem can be figured out. If you think you can solve a particular problem in a quicker or better way than you see there, by all means try it!

1. What are the individual x, y, and z coordinates of the three points P, Q, and R shown in Fig. 7-7?

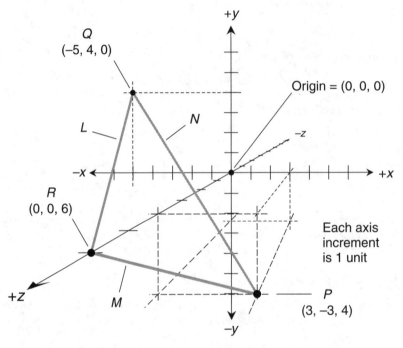

Figure 7-7   Illustration for Problems 1 through 10. Each axis division represents 1 unit.

2. Determine the distance of the point P from the origin in Fig. 7-7. Using a calculator, approximate the answer by rounding off to three decimal places.

3. Determine the distance of the point Q from the origin in Fig. 7-7. Using a calculator, approximate the answer by rounding off to three decimal places.

4. Determine the distance of the point R from the origin in Fig. 7-7. This should come out exact, so you won't need a calculator!

5. Determine the length of the line segment L in Fig. 7-7. Using a calculator, approximate the answer by rounding off to three decimal places.

6. Determine the length of the line segment M in Fig. 7-7. Using a calculator, approximate the answer by rounding off to three decimal places.

7. Determine the length of the line segment N in Fig. 7-7. Using a calculator, approximate the answer by rounding off to three decimal places.

8. Determine the coordinates of the midpoint of line segment L in Fig. 7-7.

9. Determine the coordinates of the midpoint of line segment M in Fig. 7-7.

10. Determine the coordinates of the midpoint of line segment N in Fig. 7-7.

Vectors in Cartesian Three-Space

We've learned how to work with Cartesian coordinates in two and three dimensions, and we've learned about vectors in two dimensions. Now it's time to explore how vectors behave in Cartesian *xyz* space.

How They're Defined

Imagine two vectors **a** and **b** in three-dimensional space. We have infinitely many more direction possibilities now than we did in two-space! We can denote our vectors as arrowed line segments, "starting" at the origin (0,0,0) and "ending" at points (x_a, y_a, z_a) and (x_b, y_b, z_b), as shown in Fig. 8-1.

Cartesian standard form

In Cartesian *xyz* space, vectors don't have to "start" at the coordinate origin, but there are advantages to putting them in that form. Any vector in this coordinate system, no matter where it "starts" and "ends," has an *equivalent vector* whose originating point is at (0,0,0). Such a vector is in Cartesian standard form.

Suppose that we have a vector **a'** that "starts" at a point P_1 and "ends" at another point P_2, with coordinates as

$$P_1 = (x_1, y_1, z_1)$$

and

$$P_2 = (x_2, y_2, z_2)$$

as shown in Fig. 8-2. The standard form of **a'**, denoted **a**, is defined by the terminating point P_a such that

$$P_a = (x_a, y_a, z_a) = [(x_2 - x_1), (y_2 - y_1), (z_2 - z_1)]$$

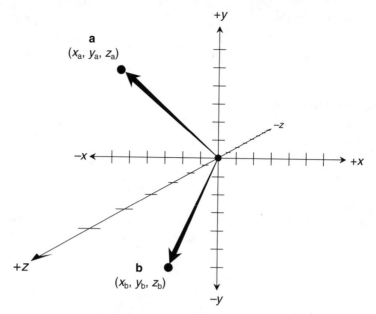

Figure 8-1 Two vectors in Cartesian *xyz* space. This is a perspective drawing (as are all three-space renditions in this book). In "real life," both vectors in this particular case would project generally toward us.

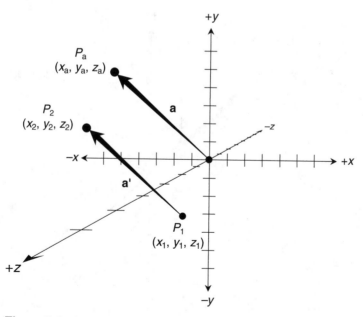

Figure 8-2 Two vectors in Cartesian *xyz* space. Vector **a** is in standard form because it "begins" at the origin (0,0,0). Vector **a′** is equivalent to **a**, because both vectors are equally long, and they both point in the same direction.

The two vectors **a** and **a′** are equivalent, because they're equally long and they point in the same direction.

Left-hand scalar multiplication

Imagine the vector **a** in standard form, defined by (x_a, y_a, z_a) as shown in Fig. 8-3. Suppose that we want to multiply a positive real scalar k_+ by the vector **a**. To do this, we multiply each coordinate by k_+, getting

$$k_+\mathbf{a} = k_+(x_a, y_a, z_a) = (k_+x_a, k_+y_a, k_+z_a)$$

The direction of our vector **a** does not change, but it becomes k_+ times as long. If we want to multiply a negative real scalar k_- by **a**, then we follow the same procedure with that constant, obtaining

$$k_-\mathbf{a} = k_-(x_a, y_a, z_a) = (k_-x_a, k_-y_a, k_-z_a)$$

We've just described how to get the *left-hand Cartesian product* of a vector and a scalar in Cartesian *xyz* space. Whenever we multiply a negative scalar by a vector, we reverse the direction in which the vector points. We also change its length by a factor equal to the absolute value of the scalar.

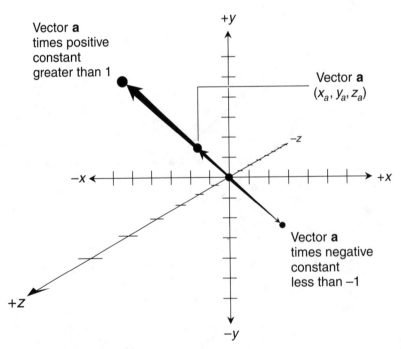

Figure 8-3   Multiplication of a standard-form vector by positive and negative real scalars in *xyz* space.

Right-hand scalar multiplication

Now suppose that we multiply all three of the original vector coordinates on the right by a scalar k_+. In this case, we get

$$\mathbf{a}k_+ = (x_a k_+, y_a k_+, z_a k_+)$$

That's the *right-hand Cartesian product* of $\mathbf{a}$ and k_+. If we multiply the original vector coordinates on the right by a negative constant k_-, we get

$$\mathbf{a}k_- = (x_a k_-, y_a k_-, z_a k_-)$$

That's the right-hand Cartesian product of $\mathbf{a}$ and k_-. As you might guess, it doesn't matter whether we multiply a vector by a constant on the left or the right; we get the same result either way. Scalar multiplication of a vector is *commutative*.

Magnitude

Let's keep thinking about our vector $\mathbf{a} = (x_a, y_a, z_a)$ in Cartesian *xyz* space. If we make sure that $\mathbf{a}$ is in standard form, we can calculate its magnitude, which we'll denote as r_a, by finding the distance of its terminating point from the origin. We learned how to do that in Chap. 7. We get

$$r_a = (x_a{}^2 + y_a{}^2 + z_a{}^2)^{1/2}$$

Here, the r stands for "radius." In some texts, vector magnitude is denoted by surrounding its name with absolute-value signs, or by changing the bold letter to a nonbold italic letter. Instead of r_a, you might see the magnitude of $\mathbf{a}$ written as $|\mathbf{a}|$ or a.

Direction

The x, y, and z coordinates contain all the information we need to fully and uniquely define the direction of a vector in Cartesian three-space, as long as the vector is in standard form. But there's a more explicit way to do it. We can define the direction of a Cartesian three-space vector if we know the measures of the angles θ_x, θ_y, and θ_z that the vector subtends relative to the $+x$, $+y$, and $+z$ axes, respectively, as shown in Fig. 8-4. These angles, expressed as an ordered triple $(\theta_x, \theta_y, \theta_z)$, are called *direction angles*. For any nonzero vector in *xyz* space, the direction angles are always nonnegative, and they're never larger than π. That means

$$0 \le \theta_x \le \pi$$
$$0 \le \theta_y \le \pi$$
$$0 \le \theta_z \le \pi$$

When we restrict the angles this way, we don't have to worry about whether we go clockwise or counterclockwise from the axes to the vectors.

An example

Imagine a nonstandard vector $\mathbf{c}'$ in Cartesian three-space. Suppose that the originating point is $(2,3,-7)$, and the terminating point is $(-1,4,-1)$. Let's convert it to standard form, and call

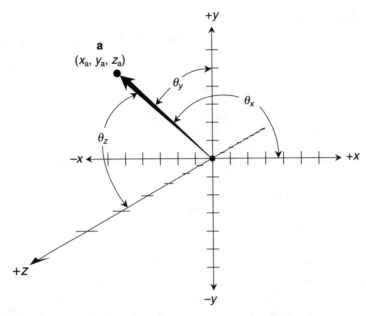

Figure 8-4   The direction of a vector in Cartesian *xyz* space is defined by the angles that the vector subtends with respect to each of the three positive axes.

the resulting vector **c**. To get the terminating points of **c**, we must individually subtract the originating coordinates of **c′** from the terminating coordinates of **c′**. The *x* coordinate of **c** is

$$x_c = -1 - 2 = -3$$

The *y* coordinate of **c** is

$$y_c = 4 - 3 = 1$$

The *z* coordinate of **c** is

$$z_c = -1 - (-7) = -1 + 7 = 6$$

Therefore, the standard form of **c′** is

$$\mathbf{c} = (x_c, y_c, z_c) = (-3, 1, 6)$$

Another example

Imagine the standard-form vector $\mathbf{a} = (2,3,4)$ in Cartesian *xyz* space. Suppose that we want to find the magnitude of this vector, accurate to three decimal places. We can assign it the coordinates $x_a = 2$, $y_a = 3$, and $z_a = 4$. Plugging these values into the magnitude formula, we get

$$\begin{aligned}|\mathbf{a}| &= (x_a^2 + y_a^2 + z_a^2)^{1/2} = (2^2 + 3^2 + 4^2)^{1/2} \\ &= (4 + 9 + 16)^{1/2} = 29^{1/2}\end{aligned}$$

Are you confused?

You might wonder, "When we want to do operations with vectors that aren't in standard form, must we *always* convert them to standard form first?" Not *always*. Sometimes we'll get a valid result from a vector operation if we leave the vector or vectors in nonstandard form. But sometimes our answer will turn out wrong, and sometimes we won't be able to figure out what to do at all. The safest course of action is to do operations on vectors in *xyz* space only after they've been converted to standard form.

Here's a challenge!

Imagine three standard-form vectors **a**, **b**, and **c** in *xyz* space, defined by ordered triples as

$$\mathbf{a} = (4,0,0)$$

$$\mathbf{b} = (0,-5,0)$$

$$\mathbf{c} = (0,0,3)$$

What are the direction angles of these vectors?

Solution

Figure 8-5 shows this situation. We can see that **a** lies along the positive *x* axis, **b** lies along the negative *y* axis, and **c** lies along the positive *z* axis. In Cartesian three-space, each of the coordinate

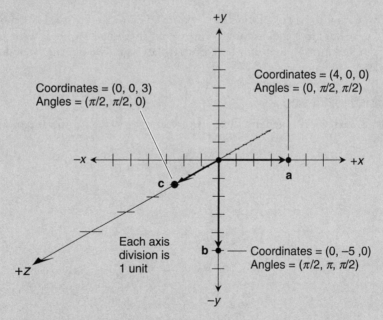

Figure 8-5   Three standard-form vectors and their direction angles. Each vector lies along one of the coordinate axes.

axes is perpendicular to the other two. This fact tells us that **a** subtends an angle of 0 with respect to the $+x$ axis, an angle of $\pi/2$ with respect to the $+y$ axis, and an angle of $\pi/2$ with respect to the $+z$ axis. The direction angles of **a** are therefore

$$(\theta_{xa}, \theta_{ya}, \theta_{za}) = (0, \pi/2, \pi/2)$$

We know that **b** subtends an angle of $\pi/2$ relative to the $+x$ axis, and an angle of $\pi/2$ with respect to the $+z$ axis. Because **b** points along the negative y axis, its angle is π relative to the $+y$ axis. The direction angles of **b** are therefore

$$(\theta_{xb}, \theta_{yb}, \theta_{zb}) = (\pi/2, \pi, \pi/2)$$

Finally, vector **c** subtends an angle of $\pi/2$ against the $+x$ axis, $\pi/2$ against the $+y$ axis, and 0 against the $+z$ axis, so the direction angles of **c** are

$$(\theta_{xc}, \theta_{yc}, \theta_{zc}) = (\pi/2, \pi/2, 0)$$

Remember that these ordered triples contain angle data in radians, not the x, y, and z coordinates for the terminating points!

Sum and Difference

When we want to add or subtract two vectors in Cartesian xyz space, we should make certain that they're both in standard form before we do anything else. Once we've gotten the vectors into standard form so that they both start at the origin, we can simply add or subtract the x, y, and z coordinates.

Cartesian vector sum

Suppose we have two generic three-space vectors in standard form, represented by ordered triples as

$$\mathbf{a} = (x_a, y_a, z_a)$$

and

$$\mathbf{b} = (x_b, y_b, z_b)$$

Their vector sum is

$$\mathbf{a} + \mathbf{b} = [(x_a + x_b), (y_a + y_b), (z_a + z_b)]$$

This sum can be found geometrically by constructing a parallelogram with vectors **a** and **b** as adjacent sides. The sum vector **a** + **b** is the diagonal of the parallelogram. An example is shown in Fig. 8-6. The figure doesn't look like a parallelogram because we're looking at it in

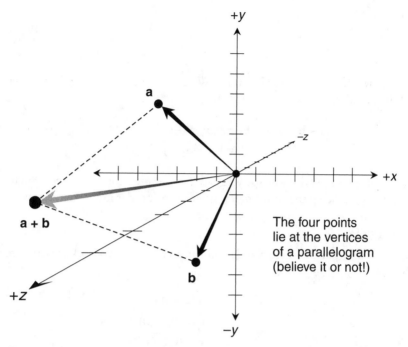

Figure 8-6 Vector addition in Cartesian *xyz* space. The terminating points of the three vectors **a**, **b**, and **a** + **b**, along with the origin, lie at the vertices of a parallelogram. Perspective distorts the view.

perspective, and from an oblique angle. All three vectors project generally in our direction; that is, they're all "coming out of the page."

Cartesian negative of a vector

To find the Cartesian negative of a standard-form vector in *xyz* space, we take the negatives of all three coordinate values. For example, if we have

$$\mathbf{a} = (x_a, y_a, z_a)$$

then the Cartesian negative vector is

$$-\mathbf{a} = (-x_a, -y_a, -z_a)$$

As in two-space, the Cartesian negative of a three-space vector always has the same magnitude as the original, but points in the opposite direction.

Cartesian vector difference

Let's look again at the two generic vectors

$$\mathbf{a} = (x_a, y_a, z_a)$$

and

$$\mathbf{b} = (x_b, y_b, z_b)$$

Suppose we want to subtract **b** from **a**. We can do this by finding the Cartesian negative of **b** and then adding that result to **a**, getting

$$\mathbf{a} - \mathbf{b} = \mathbf{a} + (-\mathbf{b}) = \{[(x_a + (-x_b)], [(y_a + (-y_b))], [(z_a + (-z_b))]\}$$
$$= [(x_a - x_b), (y_a - y_b), (z_a - z_b)]$$

We can skip the "find-the-negative" step and simply subtract the coordinate values, but we must be sure to keep the coordinates in the correct order if we do it that way.

An example

Let's look again at the three standard-form vectors that we worked with a few minutes ago. They are

$$\mathbf{a} = (4,0,0)$$
$$\mathbf{b} = (0,-5,0)$$
$$\mathbf{c} = (0,0,3)$$

Suppose we want to find the sum vector $\mathbf{a} + \mathbf{b}$. We add the x, y, and z coordinates individually to get

$$\mathbf{a} + \mathbf{b} = (4,0,0) + (0,-5,0) = \{(4 + 0), [0 + (-5)], (0 + 0)\}$$
$$= (4,-5,0)$$

If we add **c** to the right-hand side of this sum, we get

$$(\mathbf{a} + \mathbf{b}) + \mathbf{c} = (4,-5,0) + (0,0,3) = [(4 + 0), (-5 + 0), (0 + 3)]$$
$$= (4,-5,3)$$

Another example

Continuing with the same three vectors as previously, let's find the sum $\mathbf{b} + \mathbf{c}$. We add the x, y, and z coordinates individually to get

$$\mathbf{b} + \mathbf{c} = (0,-5,0) + (0,0,3) = [(0 + 0), (-5 + 0), (0 + 3)]$$
$$= (0,-5,3)$$

Adding **a** to the left-hand side of this sum, we obtain

$$\mathbf{a} + (\mathbf{b} + \mathbf{c}) = (4,0,0) + (0,-5,3) = \{(4 + 0), [0 + (-5)], (0 + 3)\}$$
$$= (4,-5,3)$$

Are you confused?

The previous example might lead you to ask, "Is vector addition associative in *xyz* space, just as real-number addition is associative in ordinary algebra?" The answer is yes. The following proof will show you why.

Here's a challenge!

Show that if **a**, **b**, and **c** are standard-form vectors in Cartesian *xyz* space, then addition among them is associative. That is

$$(\mathbf{a} + \mathbf{b}) + \mathbf{c} = \mathbf{a} + (\mathbf{b} + \mathbf{c})$$

Solution

Let's begin by assigning generic names to the coordinates of each vector. Using the same style as we've been working with all along, we can say that

$$\mathbf{a} = (x_a, y_a, z_a)$$

$$\mathbf{b} = (x_b, y_b, z_b)$$

$$\mathbf{c} = (x_c, y_c, z_c)$$

When we add **a** and **b** using the formula we've learned, we get

$$\mathbf{a} + \mathbf{b} = [(x_a + x_b), (y_a + y_b), (z_a + z_b)]$$

Adding **c** to this sum on the right, again using the formula we've learned, we obtain

$$(\mathbf{a} + \mathbf{b}) + \mathbf{c} = \{[(x_a + x_b) + x_c], [(y_a + y_b) + y_c], [(z_a + z_b) + z_c]\}$$

The associative law for addition of real numbers allows us to regroup each of the three coordinates in the ordered triple to get

$$(\mathbf{a} + \mathbf{b}) + \mathbf{c} = \{[x_a + (x_b + x_c)], [y_a + (y_b + y_c)], [z_a + (z_b + z_c)]\}$$

By definition, we know that

$$\{[x_a + (x_b + x_c)], [y_a + (y_b + y_c)], [z_a + (z_b + z_c)]\} = \mathbf{a} + (\mathbf{b} + \mathbf{c})$$

By substitution, we have

$$(\mathbf{a} + \mathbf{b}) + \mathbf{c} = \mathbf{a} + (\mathbf{b} + \mathbf{c})$$

Some Basic Properties

Here are some fundamental laws that apply to vectors and real-number scalars in *xyz* space. We won't delve into the proofs. Most of these facts are intuitive, and resemble similar laws in algebra. We've already seen a couple of them, but they're repeated here so you can use this section for reference in the future. Keep in mind that all of these rules assume that the vectors are in standard form.

Commutative law for vector addition

When we add any two vectors in *xyz* space, it doesn't matter in which order the addition is done. The resultant vector is the same either way. If **a** and **b** are vectors, then

$$\mathbf{a} + \mathbf{b} = \mathbf{b} + \mathbf{a}$$

Commutative law for vector-scalar multiplication

When we find the product of a vector and a scalar in *xyz* space, it doesn't matter which way we do it. If **a** is a vector and k is a scalar, then

$$k\mathbf{a} = \mathbf{a}k$$

Associative law for vector addition

When we add up three vectors in *xyz* space, it makes no difference how we group them. If **a**, **b**, and **c** are vectors, then

$$(\mathbf{a} + \mathbf{b}) + \mathbf{c} = \mathbf{a} + (\mathbf{b} + \mathbf{c})$$

Associative law for vector-scalar multiplication

Suppose that we have two scalars k_1 and k_2, along with some vector **a** in Cartesian *xyz* space. If we want to find the product $k_1 k_2 \mathbf{a}$, it makes no difference how we group the quantities. We can write this rule mathematically as

$$k_1 k_2 \mathbf{a} = (k_1 k_2)\mathbf{a} = k_1(k_2 \mathbf{a})$$

Distributive laws for scalar addition

Imagine that we have some vector **a** in *xyz* space, along with two real-number scalars k_1 and k_2. We can always be sure that

$$\mathbf{a}(k_1 + k_2) = \mathbf{a}k_1 + \mathbf{a}k_2$$

and

$$(k_1 + k_2)\mathbf{a} = k_1\mathbf{a} + k_2\mathbf{a}$$

The first rule is called the *left-hand distributive law* for multiplication of a vector by the sum of two scalars. The second law is called the *right-hand distributive law* for multiplication of the sum of two scalars by a vector.

Distributive laws for vector addition

Suppose we have two vectors **a** and **b** in *xyz* space, along with a real-number scalar *k*. We can always be certain that

$$k(\mathbf{a} + \mathbf{b}) = k\mathbf{a} + k\mathbf{b}$$

and

$$(\mathbf{a} + \mathbf{b})k = \mathbf{a}k + \mathbf{b}k$$

The first rule is called the left-hand distributive law for multiplication of a scalar by the sum of two vectors. The second law is called the right-hand distributive law for multiplication of the sum of two vectors by a scalar.

Unit vectors

Let's take a close look at the "structures" of two different vectors **a** and **b** in *xyz* space, both of which are expressed in the standard form. Suppose that their coordinates can be written as the familiar generic ordered triples

$$\mathbf{a} = (x_a, y_a, z_a)$$

and

$$\mathbf{b} = (x_b, y_b, z_b)$$

Either of these vectors can be split up into a sum of three *component vectors*, each of which lies along one of the coordinate axes. The component vectors are scalar multiples of mutually perpendicular vectors with magnitude 1. We have

$$\begin{aligned}
\mathbf{a} &= (x_a, y_a, z_a) \\
&= (x_a, 0, 0) + (0, y_a, 0) + (0, 0, z_a) \\
&= x_a(1, 0, 0) + y_a(0, 1, 0) + z_a(0, 0, 1)
\end{aligned}$$

and

$$\begin{aligned}
\mathbf{b} &= (x_b, y_b, z_b) \\
&= (x_b, 0, 0) + (0, y_b, 0) + (0, 0, z_b) \\
&= x_b(1, 0, 0) + y_b(0, 1, 0) + z_b(0, 0, 1)
\end{aligned}$$

The three vectors $(1,0,0)$, $(0,1,0)$, and $(0,0,1)$ are called *standard unit vectors*. (We can call them SUVs for short.) It's customary to name them **i**, **j**, and **k**, such that

$$\mathbf{i} = (1,0,0)$$
$$\mathbf{j} = (0,1,0)$$
$$\mathbf{k} = (0,0,1)$$

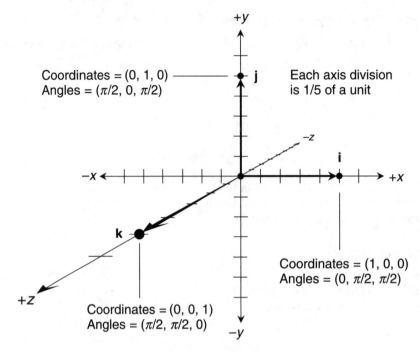

Figure 8-7   The three standard unit vectors **i**, **j**, and **k** in Cartesian *xyz* space.

Figure 8-7 illustrates the coordinates and direction angles of the three SUVs in Cartesian three-space, where each axis division represents 1/5 of a unit. Note that each SUV is perpendicular to the other two.

A generic example

Let's see what happens when we add two generic vectors component-by-component. Again, suppose we have

$$\mathbf{a} = (x_a, y_a, z_a)$$

and

$$\mathbf{b} = (x_b, y_b, z_b)$$

Expressed as sums of multiples of the SUVs, these two vectors are

$$\mathbf{a} = (x_a, y_a, z_a) = x_a\mathbf{i} + y_a\mathbf{j} + z_a\mathbf{k}$$

and

$$\mathbf{b} = (x_b, y_b, z_b) = x_b \mathbf{i} + y_b \mathbf{j} + z_b \mathbf{k}$$

When we add these components straightaway, we get

$$\mathbf{a} + \mathbf{b} = x_a \mathbf{i} + y_a \mathbf{j} + z_a \mathbf{k} + x_b \mathbf{i} + y_b \mathbf{j} + z_b \mathbf{k}$$

The commutative law for vector addition allows us to rearrange the addends on the right-hand side of this equation to get

$$\mathbf{a} + \mathbf{b} = x_a \mathbf{i} + x_b \mathbf{i} + y_a \mathbf{j} + y_b \mathbf{j} + z_a \mathbf{k} + z_b \mathbf{k}$$

Now let's use the right-hand distributive law for multiplication of the sum of two scalars by a vector to morph the previous equation into

$$\mathbf{a} + \mathbf{b} = (x_a + x_b)\mathbf{i} + (y_a + y_b)\mathbf{j} + (z_a + z_b)\mathbf{k}$$

That's the sum of the original vectors, expressed as a sum of multiples of SUVs.

A specific example

Suppose we're given a vector $\mathbf{b} = (-2,3,-7)$, and we're told to break it into a sum of multiples of $\mathbf{i}$, $\mathbf{j}$, and $\mathbf{k}$. We can imagine $\mathbf{i}$ as going 1 unit "to the right," $\mathbf{j}$ as going 1 unit "upward," and $\mathbf{k}$ as going 1 unit "toward us." The breakdown proceeds as follows:

$$\mathbf{b} = (-2,3,-7) = -2 \times (1,0,0) + 3 \times (0,1,0) + (-7) \times (0,0,1)$$

$$= -2\mathbf{i} + 3\mathbf{j} + (-7)\mathbf{k} = -2\mathbf{i} + 3\mathbf{j} - 7\mathbf{k}$$

Are you confused?

By now you might wonder, "Must I memorize all of the rules mentioned in this section?" Not necessarily. You can always come back to these pages for reference. But honestly, I recommend that you do memorize them. If you take a lot of physics or engineering courses later on, you'll be glad that you did.

Dot Product

As we've been doing throughout this chapter, let's revisit our generic standard-form vectors in *xyz* space, defined as

$$\mathbf{a} = (x_a, y_a, z_a)$$

and

$$\mathbf{b} = (x_b, y_b, z_b)$$

We can calculate the dot product $\mathbf{a} \cdot \mathbf{b}$ as a real number using the formula

$$\mathbf{a} \cdot \mathbf{b} = x_a x_b + y_a y_b + z_a z_b$$

Alternatively, it is

$$\mathbf{a} \cdot \mathbf{b} = r_a r_b \cos \theta_{ab}$$

where r_a is the magnitude of $\mathbf{a}$, r_b is the magnitude of $\mathbf{b}$, and θ_{ab} is the angle between $\mathbf{a}$ and $\mathbf{b}$ as determined in the plane containing them both, rotating from $\mathbf{a}$ to $\mathbf{b}$.

An example

Let's find the dot product of the two Cartesian vectors

$$\mathbf{a} = (2,3,4)$$

and

$$\mathbf{b} = (-1,5,0)$$

We can call the coordinates $x_a = 2$, $y_a = 3$, $z_a = 4$, $x_b = -1$, $y_b = 5$, and $z_b = 0$. Plugging these values into the formula, we get

$$\mathbf{a} \cdot \mathbf{b} = x_a x_b + y_a y_b + z_a z_b = 2 \times (-1) + 3 \times 5 + 4 \times 0$$
$$= -2 + 15 + 0 = 13$$

Another example

Suppose we want to find the dot product of the two Cartesian vectors

$$\mathbf{a} = (-4,1,-3)$$

and

$$\mathbf{b} = (-3,6,6)$$

This time, we have $x_a = -4$, $y_a = 1$, $z_a = -3$, $x_b = -3$, $y_b = 6$, and $z_b = 6$. When we substitute these coordinates into the formula, we have

$$\mathbf{a} \cdot \mathbf{b} = x_a x_b + y_a y_b + z_a z_b = -4 \times (-3) + 1 \times 6 + (-3) \times 6$$
$$= 12 + 6 + (-18) = 0$$

Are you confused?

Do you wonder how two nonzero vectors can have a dot product of 0? If we look closely at the alternative formula for the dot product, we can figure it out. That formula, once again, is

$$\mathbf{a} \bullet \mathbf{b} = r_a r_b \cos \theta_{ab}$$

The right-hand side of this equation will attain a value of 0 if at least one of the following is true:

- The magnitude of **a** is equal to 0
- The magnitude of **b** is equal to 0
- The cosine of the angle between **a** and **b** is equal to 0

Neither of the vectors in the preceding example has a magnitude of 0, so we must conclude that $\cos \theta_{ab} = 0$. That can happen only when **a** and **b** are perpendicular to each other, so θ_{ab} is either $\pi/2$ or $3\pi/2$. In the preceding example, the two vectors

$$\mathbf{a} = (-4,1,-3)$$

and

$$\mathbf{b} = (-3,6,6)$$

are mutually perpendicular. That's not obvious from the ordered triples, is it?

Here's a challenge!

Show that for any two vectors pointing in the same direction, their dot product is equal to the product of their magnitudes. Then show that for any two vectors pointing in opposite directions, their dot product is equal to the negative of the product of their magnitudes.

Solution

Imagine two vectors **a** and **b** that point in the same direction. In this situation, the angle θ_{ab} between the vectors is equal to 0. If the magnitude of **a** is r_a and the magnitude of **b** is r_b, then the dot product is

$$\mathbf{a} \bullet \mathbf{b} = r_a r_b \cos \theta_{ab} = r_a r_b \cos 0 = r_a r_b \times 1 = r_a r_b$$

Now think of two vectors **c** and **d** that point in opposite directions. The angle θ_{cd} between the vectors is equal to π. If the magnitude of **c** is r_c and the magnitude of **d** is r_d, then the dot product is

$$\mathbf{c} \bullet \mathbf{d} = r_c r_d \cos \theta_{cd} = r_c r_d \cos \pi = r_c r_d \times (-1) = -r_c r_d$$

Cross Product

The cross product $\mathbf{a} \times \mathbf{b}$ of two vectors $\mathbf{a}$ and $\mathbf{b}$ in three-dimensional space can be found according to the same rules we learned for finding a cross product in polar two-space. We get a vector perpendicular to the plane containing $\mathbf{a}$ and $\mathbf{b}$, and whose magnitude $r_{a\times b}$ is given by

$$r_{a\times b} = r_a r_b \sin \theta_{ab}$$

where r_a is the magnitude of $\mathbf{a}$, r_b is the magnitude of $\mathbf{b}$, and θ_{ab} is the angle between $\mathbf{a}$ and $\mathbf{b}$, expressed in the rotational sense going from $\mathbf{a}$ to $\mathbf{b}$.

When we want to figure out a cross product, it's always best to keep the angle between the vectors nonnegative, but not larger than π. That is, we should restrict the angle to the following range:

$$0 \leq \theta_{ab} \leq \pi$$

If we look at vectors $\mathbf{a}$ and $\mathbf{b}$ from some vantage point far away from the plane containing them, and if θ_{ab} turns through a half circle or less counterclockwise as we go from $\mathbf{a}$ to $\mathbf{b}$, then $\mathbf{a} \times \mathbf{b}$ points toward us. If θ_{ab} turns through a half circle or less clockwise as we go from $\mathbf{a}$ to $\mathbf{b}$, then $\mathbf{a} \times \mathbf{b}$ points away from us. In any case, the cross product vector is precisely perpendicular to both the original vectors.

An example

Consider two vectors $\mathbf{a}$ and $\mathbf{b}$ in three-space. Imagine that they both have magnitude 2, but their directions differ by $\pi/6$. We can plug the numbers into the formula for the magnitude of the cross product of two vectors, and calculate as follows:

$$r_{a\times b} = r_a r_b \sin \theta_{ab} = 2 \times 2 \times \sin (\pi/6) = 4 \times 1/2 = 2$$

If the $\pi/6$ angular rotation from $\mathbf{a}$ to $\mathbf{b}$ goes counterclockwise as we observe it, then $\mathbf{a} \times \mathbf{b}$ points toward us. If the $\pi/6$ angular rotation from $\mathbf{a}$ to $\mathbf{b}$ goes clockwise as we see it, then $\mathbf{a} \times \mathbf{b}$ points away from us.

Another example

Now think about two vectors $\mathbf{c}$ and $\mathbf{d}$, represented by ordered triples as

$$\mathbf{c} = (1,1,1)$$

and

$$\mathbf{d} = (-2,-2,-2)$$

Let's find the cross product $\mathbf{c} \times \mathbf{d}$. From the information we've been given, we can see immediately that $\mathbf{d} = -2\mathbf{c}$. That means the magnitude of $\mathbf{d}$ is twice the magnitude of $\mathbf{c}$, and the two vectors point in opposite directions. We can calculate the magnitude r_c of vector $\mathbf{c}$ as

$$r_c = (1^2 + 1^2 + 1^2)^{1/2} = (1 + 1 + 1)^{1/2} = 3^{1/2}$$

and the magnitude r_d of vector **d** as

$$r_d = [(-2)^2 + (-2)^2 + (-2)^2]^{1/2} = (4 + 4 + 4)^{1/2} = 12^{1/2}$$

When two vectors point in opposite directions, the angle between them is π, whether we go clockwise or counterclockwise. We now have all the information we need to figure out the magnitude $r_{c \times d}$ of the cross product **c** $\times$ **d** using the formula

$$r_{c \times d} = r_c r_d \sin \theta_{cd} = 3^{1/2} \times 12^{1/2} \times \sin \pi$$
$$= 3^{1/2} \times 12^{1/2} \times 0 = 0$$

The cross product **c** $\times$ **d** is the zero vector, because its magnitude is 0. Although we don't yet have a formula for figuring out cross products directly from ordered triples in *xyz* space, we can infer from this result that

$$(1,1,1) \times (-2,-2,-2) = (0,0,0)$$

where the bold times sign ($\times$) denotes the cross product, not ordinary multiplication.

Are you confused?

The preceding result might make you wonder, "If two vectors point in exactly the same direction or in exactly opposite directions is their cross product always the zero vector?" The answer is yes, and it doesn't depend on the magnitudes of the original two vectors. Let's prove this fact now.

Here's a challenge!

Show that the cross product of any two vectors that point in the same direction or in opposite directions, regardless of their magnitudes, is the zero vector.

Solution

When two vectors **a** and **b** point in the same direction, the angle θ_{ab} between them is 0. In such a situation, the magnitude $r_{a \times b}$ of the cross product is

$$r_{a \times b} = r_a r_b \sin \theta_{ab} = r_a r_b \sin 0 = r_a r_b \times 0 = 0$$

Therefore, **a** $\times$ **b** = 0, because if a vector has a magnitude of 0, then it's the zero vector by definition. When two vectors **c** and **d** point in opposite directions, the angle θ_{cd} between them is π, so the magnitude $r_{c \times d}$ of the cross product is

$$r_{c \times d} = r_c r_d \sin \theta_{cd} = r_c r_d \sin \pi = r_c r_d \times 0 = 0$$

Again, we have **c** $\times$ **d** = 0 by definition.

Some More Vector Laws

Here are some more rules involving vectors. You'll find these useful for future reference if you get serious about higher mathematics, physical science, or engineering.

Commutative law for dot product

When we figure out the dot product of two vectors, it doesn't matter in which order we work it. The result is the same either way. If **a** and **b** are vectors in three-space, then

$$\mathbf{a} \cdot \mathbf{b} = \mathbf{b} \cdot \mathbf{a}$$

Reverse-directional commutative law for cross product

Suppose θ_{ab} is the angle between two vectors **a** and **b** as defined in the plane containing **a** and **b**, such that $0 \leq \theta_{ab} \leq \pi$, and such that we're allowed to rotate in either direction. The magnitude of the cross-product vector is a nonnegative real number, and is independent of the order in which the operation is performed. This can be proven on the basis of the commutative property for multiplication of real numbers. We have

$$r_{a \times b} = r_a r_b \sin \theta_{ab}$$

and

$$r_{b \times a} = r_b r_a \sin \theta_{ab} = r_a r_b \sin \theta_{ab}$$

The direction of **b** × **a** in space is exactly opposite that of **a** × **b**. Figure 8-8 can help us see why this is true when we apply the right-hand rule for cross products (from Chap. 5) both ways.

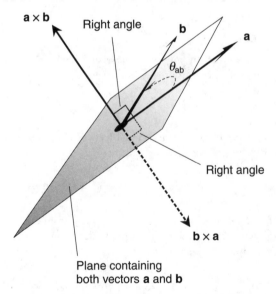

Figure 8-8 The vector **b** × **a** has the same magnitude as vector **a** × **b**, but points in the opposite direction.

Distributive laws for dot product over vector addition

Imagine that we have three vectors **a**, **b**, and **c** in three-space. We can always be sure that

$$\mathbf{a} \bullet (\mathbf{b} + \mathbf{c}) = (\mathbf{a} \bullet \mathbf{b}) + (\mathbf{a} \bullet \mathbf{c})$$

This fact is called the left-hand distributive law for a dot product over the sum of two vectors. It's also true that

$$(\mathbf{a} + \mathbf{b}) \bullet \mathbf{c} = (\mathbf{a} \bullet \mathbf{c}) + (\mathbf{b} \bullet \mathbf{c})$$

which, as you can probably guess, is the right-hand distributive law for the sum of two vectors over a dot product.

Distributive laws for cross product over vector addition

Suppose that **a**, **b**, and **c** are vectors in three-space. Then we can always be sure that

$$\mathbf{a} \times (\mathbf{b} + \mathbf{c}) = (\mathbf{a} \times \mathbf{b}) + (\mathbf{a} \times \mathbf{c})$$

This property is known as the left-hand distributive law for a cross product over the sum of two vectors. A similar rule exists when we cross multiply a sum of vectors on the right. The right-hand distributive law for the sum of two vectors over a cross product tells us that

$$(\mathbf{a} + \mathbf{b}) \times \mathbf{c} = (\mathbf{a} \times \mathbf{c}) + (\mathbf{b} \times \mathbf{c})$$

We can expand these rules to pairs of *polynomial vector sums*, each having *n* addends (where $n = 2$, $n = 3$, $n = 4$, etc.), in the same way as multiplication is distributive with respect to addition for polynomials in algebra. For example, for $n = 2$, we have the cross product of two *binomial vector sums*, getting

$$(\mathbf{a} + \mathbf{b}) \times (\mathbf{c} + \mathbf{d}) = (\mathbf{a} \times \mathbf{c}) + (\mathbf{a} \times \mathbf{d}) + (\mathbf{b} \times \mathbf{c}) + (\mathbf{b} \times \mathbf{d})$$

In the case of $n = 3$, the cross product of two *trinomial vector sums* expands as

$$(\mathbf{a} + \mathbf{b} + \mathbf{c}) \times (\mathbf{d} + \mathbf{e} + \mathbf{f}) = (\mathbf{a} \times \mathbf{d}) + (\mathbf{a} \times \mathbf{e}) + (\mathbf{a} \times \mathbf{f}) + (\mathbf{b} \times \mathbf{d}) + (\mathbf{b} \times \mathbf{e}) + (\mathbf{b} \times \mathbf{f})$$
$$+ (\mathbf{c} \times \mathbf{d}) + (\mathbf{c} \times \mathbf{e}) + (\mathbf{c} \times \mathbf{f})$$

Dot product of cross products

Imagine that we have four vectors **a**, **b**, **c**, and **d** in three-space. We can rearrange a dot product of cross products as

$$(\mathbf{a} \times \mathbf{b}) \bullet (\mathbf{c} \times \mathbf{d}) = (\mathbf{a} \bullet \mathbf{c})(\mathbf{b} \bullet \mathbf{d}) - (\mathbf{a} \bullet \mathbf{d})(\mathbf{b} \bullet \mathbf{c})$$

We always end up with a scalar quantity (that is, a real number).

Dot product of mixed vectors and scalars

Suppose that t and u are real numbers, and we have two three-space vectors **a** and **b**. We can rearrange a dot product of scalar multiples as

$$t\mathbf{a} \cdot u\mathbf{b} = tu(\mathbf{a} \cdot \mathbf{b})$$

The result is always a scalar.

Cross product of mixed vectors and scalars

Once again, imagine that t and u are real numbers, and we have two three-space vectors **a** and **b**. We can rearrange a cross product of scalar multiples as

$$t\mathbf{a} \times u\mathbf{b} = tu(\mathbf{a} \times \mathbf{b})$$

The result is always a vector quantity.

Here's a challenge!

Imagine two vectors in Cartesian *xyz* space whose coordinates are expressed as

$$\mathbf{a} = (x_a, y_a, z_a)$$

and

$$\mathbf{b} = (x_b, y_b, z_b)$$

Derive a general expression for **a** × **b** in the form of an ordered triple.

Solution

Let's go back to the concept of SUVs that we learned earlier in this chapter. These vectors are

$$\mathbf{i} = (1,0,0)$$
$$\mathbf{j} = (0,1,0)$$
$$\mathbf{k} = (0,0,1)$$

Now let's evaluate and list all the cross products we can get from these vectors. Using the right-hand rule for cross products (from Chap. 5) along with the formula for the magnitude of the cross product of vectors, we can deduce, along with the help of Fig. 8-7 on page 140, that

$$\mathbf{i} \times \mathbf{j} = \mathbf{k}$$
$$\mathbf{j} \times \mathbf{i} = -\mathbf{k}$$
$$\mathbf{i} \times \mathbf{k} = -\mathbf{j}$$
$$\mathbf{k} \times \mathbf{i} = \mathbf{j}$$
$$\mathbf{j} \times \mathbf{k} = \mathbf{i}$$
$$\mathbf{k} \times \mathbf{j} = -\mathbf{i}$$

We can write the cross product $\mathbf{a} \times \mathbf{b}$ as

$$\mathbf{a} \times \mathbf{b} = (x_a, y_a, z_a) \times (x_b, y_b, z_b)$$
$$= (x_a\mathbf{i} + y_a\mathbf{j} + z_a\mathbf{k}) \times (x_b\mathbf{i} + y_b\mathbf{j} + z_b\mathbf{k})$$

Using the left-hand distributive law for the cross product over vector addition as it applies to trinomials, we can expand this to

$$\mathbf{a} \times \mathbf{b} = (x_a\mathbf{i} \times x_b\mathbf{i}) + (x_a\mathbf{i} \times y_b\mathbf{j}) + (x_a\mathbf{i} \times z_b\mathbf{k}) + (y_a\mathbf{j} \times x_b\mathbf{i}) + (y_a\mathbf{j} \times y_b\mathbf{j}) + (y_a\mathbf{j} \times z_b\mathbf{k})$$
$$+ (z_a\mathbf{k} \times x_b\mathbf{i}) + (z_a\mathbf{k} \times y_b\mathbf{j}) + (z_a\mathbf{k} \times z_b\mathbf{k})$$

With our newfound knowledge of how scalar multiplication and cross products can be mixed (see "Cross product of mixed vectors and scalars"), we can morph each of the terms after the equals sign to get

$$\mathbf{a} \times \mathbf{b} = x_a x_b(\mathbf{i} \times \mathbf{i}) + x_a y_b(\mathbf{i} \times \mathbf{j}) + x_a z_b(\mathbf{i} \times \mathbf{k}) + y_a x_b(\mathbf{j} \times \mathbf{i}) + y_a y_b(\mathbf{j} \times \mathbf{j}) + y_a z_b(\mathbf{j} \times \mathbf{k})$$
$$+ z_a x_b(\mathbf{k} \times \mathbf{i}) + z_a y_b(\mathbf{k} \times \mathbf{j}) + z_a z_b(\mathbf{k} \times \mathbf{k})$$

A few moments ago, we proved that we always get the zero vector if we take the cross product of any vector with another vector pointing in the same direction. That means the cross product of any vector with itself is the zero vector. Because the zero vector has zero magnitude, we get the zero vector if we multiply it by any scalar. With all this information in mind, we can rewrite the previous equation as

$$\mathbf{a} \times \mathbf{b} = \mathbf{0} + x_a y_b(\mathbf{i} \times \mathbf{j}) + x_a z_b(\mathbf{i} \times \mathbf{k}) + y_a x_b(\mathbf{j} \times \mathbf{i}) + \mathbf{0} + y_a z_b(\mathbf{j} \times \mathbf{k}) + z_a x_b(\mathbf{k} \times \mathbf{i}) + z_a y_b(\mathbf{k} \times \mathbf{j}) + \mathbf{0}$$

Looking back at the six "factoids" involving pairwise cross products of $\mathbf{i}$, $\mathbf{j}$, and $\mathbf{k}$, and getting rid of the zero vectors in the previous equation, we can simplify it to

$$\mathbf{a} \times \mathbf{b} = x_a y_b\mathbf{k} + x_a z_b(-\mathbf{j}) + y_a x_b(-\mathbf{k}) + y_a z_b\mathbf{i} + z_a x_b\mathbf{j} + z_a y_b(-\mathbf{i})$$

Rearranging the signs, we obtain

$$\mathbf{a} \times \mathbf{b} = x_a y_b\mathbf{k} - x_a z_b\mathbf{j} - y_a x_b\mathbf{k} + y_a z_b\mathbf{i} + z_a x_b\mathbf{j} - z_a y_b\mathbf{i}$$

This can be morphed a little more, based on rules we've learned in this chapter, getting

$$\mathbf{a} \times \mathbf{b} = (y_a z_b - z_a y_b)\mathbf{i} + (z_a x_b - x_a z_b)\mathbf{j} + (x_a y_b - y_a x_b)\mathbf{k}$$

This SUV-based equation tells us three things:

- The x coordinate of $\mathbf{a} \times \mathbf{b}$ is $y_a z_b - z_a y_b$
- The y coordinate of $\mathbf{a} \times \mathbf{b}$ is $z_a x_b - x_a z_b$
- The z coordinate of $\mathbf{a} \times \mathbf{b}$ is $x_a y_b - y_a x_b$

Knowing these three facts, we can write the x, y, and z coordinates of $\mathbf{a} \times \mathbf{b}$ as an ordered triple to get

$$\mathbf{a} \times \mathbf{b} = [(y_a z_b - z_a y_b), (z_a x_b - x_a z_b), (x_a y_b - y_a x_b)]$$

We've found a formula that allows us to directly calculate the cross product of two vectors in *xyz* space when we're given both vectors as ordered triples.

Here's an extra-credit challenge!

The formulas for the seven laws in this section were stated straightaway. We didn't show how they are derived. If you're ambitious (and you have a good pen along with plenty of blank sheets of paper), derive these seven laws by working out the general arithmetic step by step. Following are the names of those laws again, for reference:

- Commutative law for dot product
- Reverse-directional commutative law for cross product
- Distributive laws for dot product over vector addition
- Distributive laws for cross product over vector addition
- Dot product of cross products
- Dot product of mixed vectors and scalars
- Cross product of mixed vectors and scalars

Solution

You're on your own. That's why you get extra credit! Here's a hint: The work is rather tedious, but it's straightforward.

Practice Exercises

This is an open-book quiz. You may (and should) refer to the text as you solve these problems. Don't hurry! You'll find worked-out answers in App. A. The solutions in the appendix may not represent the only way a problem can be figured out. If you think you can solve a particular problem in a quicker or better way than you see there, by all means try it!

1. Find the magnitude r_a of the standard-form vector

$$\mathbf{a} = (8, -1, -6)$$

   in Cartesian *xyz* space. Assume the values given are exact. Using a calculator, round off the answer to three decimal places.

2. Imagine a nonstandard vector $\mathbf{a}'$ that originates at $(-2, 0, 4)$ and terminates at the origin. Convert $\mathbf{a}'$ to standard form.

3. What's the standard form of the product $4\mathbf{b}'$, where $\mathbf{b}'$ originates at $(2, 3, 4)$ and terminates at $(6, 7, 8)$? Here's a hint: Convert $\mathbf{b}'$ to its standard form $\mathbf{b}$ first, and then multiply that vector by 4.

4. Consider the two standard-form vectors

$$\mathbf{a} = (-7, -10, 0)$$

and

$$\mathbf{b} = (8, -1, -6)$$

in *xyz* space. What is their dot product?

5. Consider the two standard-form vectors

$$\mathbf{a} = (2, 6, 0)$$

and

$$\mathbf{b} = (7, 4, 3)$$

in *xyz* space. What is their cross product?

6. Imagine two standard-form vectors **f** and **g** that point in the same direction in three-space. Suppose that the magnitude r_f of **f** is equal to 4, and the magnitude r_g of **g** is equal to 7. What is **f** • **g**?

7. Imagine two standard-form vectors **f** and **g** that point in opposite directions in three-space. Suppose that the magnitude r_f of vector **f** is equal to 4, and the magnitude r_g of vector **g** is equal to 7. What is **f** • **g**?

8. Imagine two standard-form vectors **f** and **g** that are perpendicular to each other in three-space, and we're looking at them from a point of view such that we see the angle going counterclockwise from **f** to **g** as $\pi/2$. Suppose that the magnitude r_f of vector **f** is equal to 4, and the magnitude r_g of vector **g** is equal to 7. What is **f** • **g**? What is **g** • **f**?

9. Imagine two standard-form vectors **f** and **g** that are perpendicular to each other in three-space, and we're looking at them from a point of view such that we see the angle going counterclockwise from **f** to **g** as $\pi/2$. Suppose that the magnitude r_f of vector **f** is equal to 4, and the magnitude r_g of vector **g** is equal to 7. What is **f** ✕ **g**? What is **g** ✕ **f**?

10. Consider two standard-form vectors **a** and **b** that both lie in the *xy* plane within Cartesian *xyz* space. Suppose that **a** = (2,0,0), so it points along the +*x* axis. Suppose that **b** has magnitude 2 and rotates counterclockwise in the *xy* plane, starting at (2,0,0), then going around through (0,2,0), (−2,0,0), and (0,−2,0), finally ending up back at (2,0,0). Now imagine that we watch all this activity from somewhere high above the *xy* plane, near the +*z* axis. Describe what happens to the cross product vector **a** ✕ **b** as vector **b** goes through a complete counterclockwise rotation. What will we see if **b** keeps rotating counterclockwise indefinitely?

9

Alternative Three-Space

We can define the locations of points in three dimensions by methods other than the Cartesian system. In this chapter, we'll learn about the two most common alternative coordinate schemes for three-space.

Cylindrical Coordinates

Figure 9-1 is a functional diagram of a system of *cylindrical coordinates*. It's basically a polar coordinate plane of the sort we learned about in Chap. 3, with the addition of a height axis to define the third dimension.

How it works

To set up a cylindrical coordinate system, we "paste" a polar plane onto a Cartesian xy plane, creating a *reference plane*. We call the positive Cartesian x axis the *reference axis*. Imagine a point P in three-space, along with its *projection point* P' onto the reference plane. In this context, the term "projection" means that P' is directly above or below P, so a line connecting the two points is perpendicular to the reference plane. We define three coordinates:

- The *direction angle*, which we call θ, is the angle in the reference plane as we turn counterclockwise from the reference axis to the ray that goes out from the origin through P'.
- The *radius*, which we call r, is the straight-line distance from the origin to P'.
- The *height*, which we call h, is the vertical displacement (positive, negative, or zero) from P' to P.

These three coordinates give us enough information to uniquely define the position of P as shown in Fig. 9-1. We express the cylindrical coordinates as an ordered triple

$$P = (\theta, r, h)$$

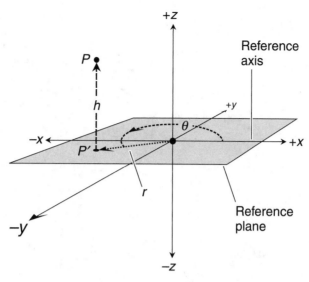

Figure 9-1   Cylindrical coordinates define points in
three dimensions according to an angle, a
radial distance, and a vertical
displacement.

Strange values

We can have nonstandard direction angles in cylindrical coordinates, but it's best to add or subtract whatever multiple of 2π to bring the angle into the preferred range of $0 \le \theta < 2\pi$. If $\theta \ge 2\pi$, then we're making at least one complete counterclockwise rotation from the reference axis. If $\theta < 0$, then we're rotating clockwise from the reference axis rather than counterclockwise.

We can have negative radii, but it's best to reverse the direction angle if necessary to keep the radius nonnegative. We can multiply a negative radius coordinate by -1 so it becomes positive, and then add or subtract π to or from the direction angle to ensure that $0 \le \theta < 2\pi$.

The height h can be any real number. We have $h > 0$ if and only if P is above the reference plane, $h < 0$ if and only if P is below the reference plane, and $h = 0$ if and only if P is in the reference plane.

An example

In the situation shown by Fig. 9-1, the direction angle θ appears to be somewhat more than π (half of a rotation from the reference axis) but less than $3\pi/2$ (three-quarters of a rotation). The radius r is positive, but we can't tell how large it is because there are no coordinate increments for reference. The height h is also positive, but again, we don't know its exact value because there are no reference increments.

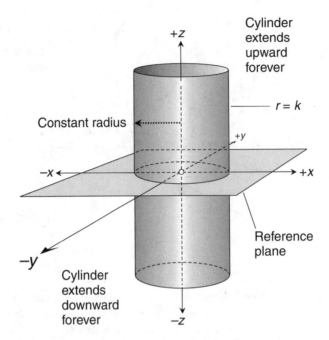

Figure 9-2   When we set the radius equal to a constant in cylindrical coordinates, we get an infinitely tall vertical cylinder whose axis corresponds to the vertical axis.

Another example

In Chap. 3, we learned that the equation of a circle in polar two-space is simple; all we have to do is specify a radius. If we do the same thing in cylindrical three-space, we get a vertical cylinder that's infinitely tall, with an axis that corresponds to the vertical coordinate axis. Figure 9-2 shows what we get when we graph the equation

$$r = k$$

in cylindrical three-space, where k is a nonzero constant.

Still another example

If we set the height equal to a nonzero constant in cylindrical coordinates, we get the set of all points at a specific distance either above or below the reference plane. That's always a plane parallel to the reference plane. Figure 9-3 is an example of the generic situation where

$$h = k$$

In this case, k is a positive real-number constant, but we don't know the exact value because the graph doesn't show us any reference increments for the height coordinate.

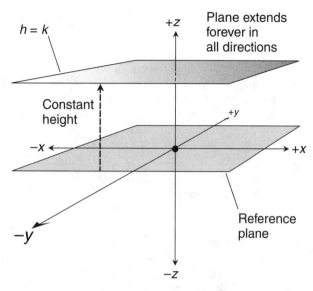

Figure 9-3   When we set the height equal to a constant in cylindrical coordinates, we get a plane parallel to the reference plane.

Are you confused?

Some texts will tell you that the cylindrical coordinates of a point are listed in an ordered triple with the radius first, then the angle, and finally the height, as

$$P = (r, \theta, h)$$

Don't let this notational inconsistency baffle you. For any particular set of coordinate values, we're talking about the same point, regardless of the order in which we list them. In this book, we indicate the angle before the radius to be consistent with the polar-coordinate system described in Chap. 3. When "traveling" from the origin out to some point P in space in the cylindrical system, most people find it easiest to think of the reference-plane angle θ first (as in "face northwest"), then the radius r (as in "walk 40 meters"), and finally the height h (as in "dig down 2 meters to find the treasure"). That's why, in this book, we use the form

$$P = (\theta, r, h)$$

Here's a challenge!

What do we get if we set the direction angle θ equal to a constant in cylindrical coordinates? As an example, draw a diagram showing the graph of the equation

$$\theta = \pi / 2$$

Solution

Let's think back again to Chap. 3. In polar coordinates, if we set the direction angle equal to a constant, we get a line passing through the origin. Cylindrical coordinates are simply a vertical extension of polar coordinates, going infinitely upward and infinitely downward. If we hold θ constant in cylindrical coordinates but allow the other coordinates to vary at will, we get a vertical plane, which is an infinite vertical extension of a horizontal line. If k is any real-number constant, then the graph of

$$\theta = k$$

is a plane that passes through the vertical axis. In the case where $\theta = \pi/2$, that plane also contains the ray for the direction angle $\pi/2$, as shown in Fig. 9-4.

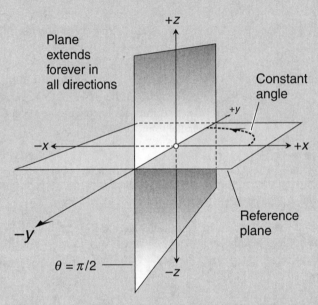

Figure 9-4   When we set the angle equal to a constant in cylindrical coordinates, we get a plane that contains the vertical axis.

Cylindrical Conversions

Conversion of coordinate values between cylindrical and Cartesian three-space is just as easy as conversion between polar and Cartesian two-space. The only difference is that in three-space, we add the vertical dimension. In *xyz* space, it's *z*; in cylindrical three-space, it's *h*.

Cylindrical to Cartesian

Let's look at the simplest conversions first. These transformations are like going down a river; we can simply "get into the boat" (sharpen our pencils) and make sure we don't "run aground" (make an arithmetic error). Suppose we have a point (θ, r, h) in cylindrical coordinates. We can find the Cartesian x value of this point using the formula

$$x = r \cos \theta$$

The Cartesian y value is

$$y = r \sin \theta$$

The Cartesian z value is

$$z = h$$

An example

Consider the point $(\theta, r, h) = (\pi, 2, -3)$ in cylindrical coordinates. Let's find the (x, y, z) representation in Cartesian three-space using the preceding formulas. Plugging in the numbers gives us

$$x = 2 \cos \pi = 2 \times (-1) = -2$$
$$y = 2 \sin \pi = 2 \times 0 = 0$$
$$z = h = -3$$

Therefore, we have the Cartesian equivalent point

$$(x, y, z) = (-2, 0, -3)$$

Cartesian to cylindrical: finding θ

Going from Cartesian to cylindrical coordinates is like navigating up a river. We not only have to "go against the current" (do some hard work), but we have to be sure we "take the right tributary" (use the correct angle values).

Cartesian-to-cylindrical angle conversion is the same as the Cartesian-to-polar angle conversion process that we learned in Chap. 3. That was messy, because we had to break the situation down into nine different ranges for θ. In the cylindrical context, the angle-conversion process works as follows:

$\theta = 0$ by default	When $x = 0$ and $y = 0$ that is, at the origin
$\theta = 0$	When $x > 0$ and $y = 0$
$\theta = \text{Arctan } (y/x)$	When $x > 0$ and $y > 0$
$\theta = \pi/2$	When $x = 0$ and $y > 0$
$\theta = \pi + \text{Arctan } (y/x)$	When $x < 0$ and $y > 0$
$\theta = \pi$	When $x < 0$ and $y = 0$
$\theta = \pi + \text{Arctan } (y/x)$	When $x < 0$ and $y < 0$
$\theta = 3\pi/2$	When $x = 0$ and $y < 0$
$\theta = 2\pi + \text{Arctan } (y/x)$	When $x > 0$ and $y < 0$

If you've forgotten what the Arctangent function is, and why we use a capital "A" to denote it, you can check in Chap. 3 to refresh your memory. Notice that the Cartesian z value is irrelevant when we want to find the direction angle in cylindrical coordinates.

Cartesian to cylindrical: finding r

When we want to calculate the r coordinate in cylindrical three-space on the basis of a point in Cartesian xyz space, we use the Cartesian two-space distance formula, exactly as we would in the polar plane. The radius depends only on the values of x and y; the z coordinate is irrelevant. The r coordinate is therefore equal to the distance between the projection point P' and the origin in the xy plane, which is

$$r = (x^2 + y^2)^{1/2}$$

Cartesian to cylindrical: finding h

When we want to change the Cartesian z value to the cylindrical h value in three-space, we can make the direct substitution

$$h = z$$

An example

Let's convert the Cartesian point $(x,y,z) = (1,1,1)$ to cylindrical three-space coordinates. In this situation, $x = 1$ and $y = 1$. To find the angle, we should use the formula

$$\theta = \mathrm{Arctan}\,(y/x)$$

because $x > 0$ and $y > 0$. When we plug in the values for x and y, we get

$$\theta = \mathrm{Arctan}\,(1/1) = \mathrm{Arctan}\,1 = \pi/4$$

When we input the values for x and y to the formula for r, we get

$$r = (1^2 + 1^2)^{1/2} = 2^{1/2}$$

Because $z = 1$, we know that

$$h = z = 1$$

We've just found that the cylindrical equivalent point is

$$(\theta,r,h) = (\pi/4, 2^{1/2}, 1)$$

Are you confused?

We must pay close attention to the meaning of the radius in cylindrical coordinates. The cylindrical radius goes from the origin to the *reference-plane projection* of the point whose coordinates we're interested in. It does *not* go straight through space to the point of interest, which is usually outside the reference plane.

Here's a challenge!

Convert the Cartesian point $(x,y,z) = (-5,-12,8)$ to cylindrical coordinates. Using a calculator, approximate all irrational values to four decimal places.

Solution

We have $x = -5$ and $y = -12$. To find the angle, we should use the formula

$$\theta = \pi + \text{Arctan}\,(y/x)$$

because $x < 0$ and $y < 0$. When we plug in $x = -5$ and $y = -12$, we get

$$\theta = \pi + \text{Arctan}\,[(-12)/(-5)] = \pi + \text{Arctan}\,(12/5)$$

That is a theoretically exact answer, but it's an irrational number. A calculator set to work in radians (not degrees) allows us to approximate this to four decimal places as

$$\theta \approx 4.3176$$

When we input $x = -5$ and $y = -12$ to the formula for r, we get

$$r = [(-5)^2 + (-12)^2]^{1/2} = (25 + 144)^{1/2} = 169^{1/2} = 13$$

Because $z = 8$, we know that

$$h = z = 8$$

We've found that the cylindrical equivalent point is

$$(\theta,r,h) \approx (4.3176,13,8)$$

The value of θ is approximate to four decimal places, while r and h are exact values.

Spherical Coordinates

Figure 9-5 illustrates a system of *spherical coordinates* for defining points in three-space. Instead of one angle and two displacements as in cylindrical coordinates, we now use two angles and one displacement.

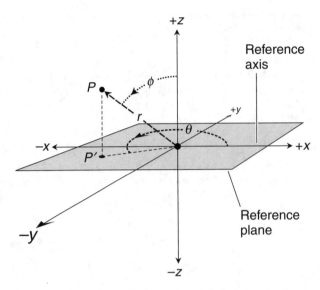

Figure 9-5 Spherical coordinates define points in three-space according to a horizontal angle, a vertical angle, and a radius.

How it works

In the spherical coordinate arrangement, we start with a horizontal Cartesian reference plane, just as we do when we set up cylindrical coordinates. The positive Cartesian x axis forms the reference axis. Suppose that we want to define the location of a point P. Consider its projection, P', onto the reference plane:

- The *horizontal angle*, which we call θ, turns counterclockwise in the reference plane from the reference axis to the ray that goes out from the origin through P'.
- The *vertical angle*, which we call ϕ, turns downward from the vertical axis to the ray that goes out from the origin through P.
- The *radius*, which we call r, is the straight-line distance from the origin to P.

These three coordinates, taken all together, provide us with sufficient information to uniquely define the location of P in three-space. We can express the spherical coordinates as an ordered triple

$$P = (\theta, \phi, r)$$

Strange values

In spherical three-space, we can have nonstandard horizontal direction angles, but it's always best to add or subtract whatever multiple of 2π will keep us within the preferred range of $0 \leq \theta < 2\pi$. If $\theta \geq 2\pi$, it represents at least one complete counterclockwise rotation from the reference axis. If $\theta < 0$, it represents clockwise rotation from the reference axis.

We can have nonstandard vertical angles, although things are simplest if we keep them nonnegative but no larger than π. Theoretically, all possible locations in space can be covered if we restrict the vertical angle to the range $0 \le \phi \le \pi$. If it's outside this range, such as $-\pi < \phi < 0$ or $\pi < \phi < 2\pi$, we can multiply the radius r by -1, and then add or subtract π to or from ϕ, and we'll end up at the point we want. But those are confusing ways to get there!

The radius r can be any real number, but things are simplest if we keep it nonnegative. If our horizontal and vertical direction angles put us on a ray that goes from the origin through P, then $r > 0$. If our direction angles put us on a ray that goes from the origin away from P, then $r < 0$. We have $r = 0$ if and only if P is at the origin. If we find ourselves working with a negative radius, we should reverse the direction by adding or subtracting π to or from both angles, keeping $0 \le \theta < 2\pi$ and $0 \le \phi \le \pi$. Then we can take the absolute value of the negative radius and use it as the radius coordinate.

An example

In the situation of Fig. 9-5, the horizontal direction angle θ appears to be somewhere between π and $3\pi/2$. The vertical direction angle ϕ appears to be roughly 1 radian. We can't be sure of the exact values of these angles, because we don't have any reference lines to compare them with. The radius r is positive, but we have no idea how large it is because there are no radial coordinate increments.

Another example

Imagine that we set the horizontal direction angle θ equal to a constant in spherical coordinates. For example, let's say that we have the equation

$$\theta = 7\pi/5$$

When we work in polar coordinates and set the direction angle equal to a constant, we get a line passing through the origin. In spherical coordinates, the horizontal angle in the reference plane is geometrically identical to the polar direction angle. Therefore, if k is any real-number constant, the graph of

$$\theta = k$$

is a plane that passes through the vertical axis. When $k = 7\pi/5$, that vertical plane also contains the ray for the direction angle $7\pi/5$, as shown in Fig. 9-6.

Still another example

If we set the radius equal to a constant in spherical coordinates, we get the set of all points at some fixed distance from the origin. That's a sphere centered at the origin. Figure 9-7 shows what happens when we graph the following equation:

$$r = k$$

in spherical three-space, where k is a nonzero constant.

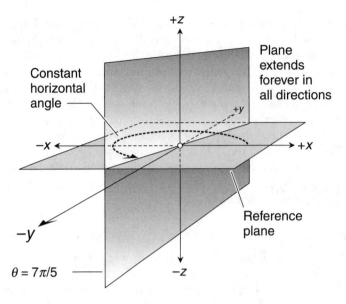

Figure 9-6   When we set the horizontal angle equal to a
constant in spherical coordinates, we get a plane
that contains the vertical axis.

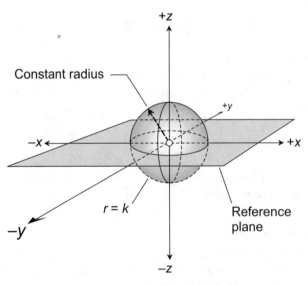

Figure 9-7   When we set the radius equal to a constant
in spherical coordinates, we get a sphere
centered at the origin.

Are you confused?

Don't get the wrong idea about the meaning of the radius in spherical coordinates. It's not the same as the cylindrical-coordinate radius! In spherical coordinates, the radius follows a straight-line path from the origin to the point whose coordinates we're interested in. This line almost never lies in the reference plane. In cylindrical coordinates, the radius goes from the origin to the projection of the point in the reference plane. You can see the difference if you compare Fig. 9-1 with Fig. 9-5.

Are you still confused?

If you've read a lot of other pre-calculus texts (and I recommend that you do), you might notice that the order in which we list spherical coordinates is different from the way it's done in some of those other texts. You might see the spherical coordinates of a point P go with the radius first, then the horizontal angle, and finally the vertical angle, as

$$P = (r, \theta, \phi)$$

Theoretically, it doesn't matter in which order we list the coordinates. For any particular values, we're always working with the same point. When we want to get from the origin to a point in spherical three-space, most people find it easiest to think of the horizontal angle θ first (as in "face southeast"), then the vertical angle ϕ (as in "fix your gaze at an angle that's $\pi/6$ radian from the zenith"), and finally the radius r (as in "follow the string for 150 meters to reach the kite"). That's why we use the form

$$P = (\theta, \phi, r)$$

Here's a challenge!

What sort of graph do we get if we set the vertical angle ϕ equal to a constant in spherical coordinates? As an example, draw a diagram showing the graph of the following equation:

$$\phi = \pi/4$$

Solution

This situation doesn't resemble anything we've seen so far in Cartesian, polar, or cylindrical coordinates. If we hold the vertical angle constant in a spherical coordinate system, we get the set of points formed by a line passing through the origin and rotated with respect to the vertical axis. If k is a real-number constant, then the graph of

$$\phi = k$$

is a double cone whose axis corresponds to the vertical axis and whose apex is at the origin, as shown in Fig. 9-8.

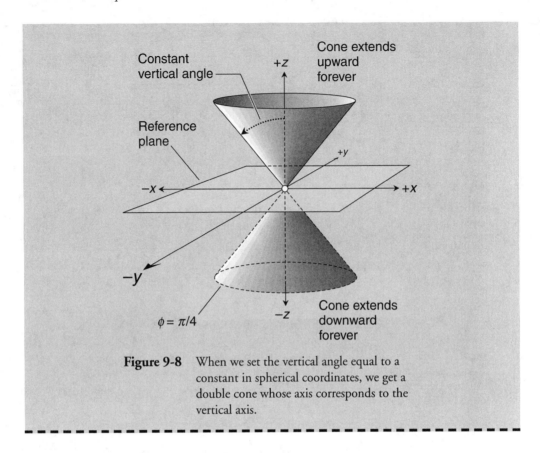

Figure 9-8 When we set the vertical angle equal to a constant in spherical coordinates, we get a double cone whose axis corresponds to the vertical axis.

Spherical Conversions

Converting coordinates between *xyz* space and spherical three-space is a little tricky, but not too difficult. Let's think about a point *P* whose spherical coordinates are (θ, ϕ, r) and whose Cartesian coordinates are (x, y, z).

Spherical to Cartesian: finding *x*

In spherical coordinates, the radius is usually outside of the reference plane, so we can't use it directly in the same formulas as the cylindrical radius. But we can construct a *projection radius* identical to the cylindrical radius: the distance from the origin to the projection point P' in the reference plane. In Fig. 9-9, the projection radius is called r'. From this geometry, we can see that r' is equal to the true spherical radius times the sine of the vertical angle. As an equation, we have

$$r' = r \sin \phi$$

The *x* value conversion formula from cylindrical coordinates, which we learned earlier in this chapter, tells us that

$$x = r' \cos \theta$$

where θ is the horizontal direction angle, which is the same in spherical and cylindrical coordinates. Substituting the quantity ($r \sin \phi$) for r' gives us

$$x = r \sin \phi \cos \theta$$

Spherical to Cartesian: finding *y*

When we found the cylindrical equivalent of the Cartesian *y* value, we took the radius in the reference plane and multiplied by the sine of the direction angle in that plane. In the spherical-coordinate situation of Fig. 9-9, that translates to

$$y = r' \sin \theta$$

where θ is the horizontal direction angle. We can substitute ($r \sin \phi$) for r' to get

$$y = r \sin \phi \sin \theta$$

Spherical to Cartesian: finding *z*

Let's look again at Fig. 9-9, and locate the projection point P^* on the *z* axis, such that the *z* values of P^* and *P* are equal. We can see that P^*, *P*, P', and the origin form the vertices of a

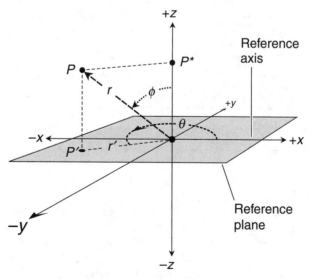

Figure 9-9   Conversion between spherical and Cartesian three-space coordinates involves several geometric variables.

rectangle perpendicular to the reference plane. It follows that P^*, P, and the origin are at the vertices of a right triangle. By trigonometry, the z value of P^* is equal to the spherical radius r times the cosine of the vertical angle ϕ. Because the z values of P and P^* are the same, we can deduce that the z value of P is given by

$$z = r \cos \phi$$

Cartesian to spherical: finding *r*

Now let's figure out how to get from Cartesian *xyz* space to spherical three-space. The radius is the easiest coordinate to find, so let's do it first. Recall that the spherical radius of a point is its distance from the origin. Therefore, when we want to find the spherical radius r for point P in terms of its *xyz* space coordinates, we can apply the Cartesian three-space distance formula to get

$$r = (x^2 + y^2 + z^2)^{1/2}$$

Cartesian to spherical: finding *θ*

The horizontal angle in spherical coordinates is identical to its counterpart in cylindrical coordinates, so we can use the conversion table from earlier in this chapter.

$\theta = 0$ by default	When $x = 0$ and $y = 0$ that is, at the origin
$\theta = 0$	When $x > 0$ and $y = 0$
$\theta = \text{Arctan}\,(y/x)$	When $x > 0$ and $y > 0$
$\theta = \pi/2$	When $x = 0$ and $y > 0$
$\theta = \pi + \text{Arctan}\,(y/x)$	When $x < 0$ and $y > 0$
$\theta = \pi$	When $x < 0$ and $y = 0$
$\theta = \pi + \text{Arctan}\,(y/x)$	When $x < 0$ and $y < 0$
$\theta = 3\pi/2$	When $x = 0$ and $y < 0$
$\theta = 2\pi + \text{Arctan}\,(y/x)$	When $x > 0$ and $y < 0$

The Arccosine

Before we can find the vertical spherical angle for a point that's given to us in Cartesian coordinates, we must be familiar with the *arccosine* relation. It's abbreviated arccos ($\cos^{-1}$ in some texts), and it "undoes" the work of the cosine function. For example, we know that

$$\cos\,(\pi/3) = 1/2$$

and

$$\cos\,\pi = -1$$

For things to work without ambiguity when we go the other way, we want the arccosine to be a true function. To do that, we must restrict its range (output) to an interval where we don't get into trouble with ambiguity. By convention, mathematicians specify the closed interval $[0,\pi]$ for this purpose. That happens to be the ideal range of values for our vertical angle ϕ in spherical coordinates. When we make this restriction, we capitalize the "A" and write Arccosine or Arccos to indicate that we're working with a true function. Then we can state the above facts "in reverse" using the Arccosine function, getting

$$\text{Arccos } 1/2 = \pi/3$$

and

$$\text{Arccos } (-1) = \pi$$

For any real number u, we can be sure that

$$\text{Arccos } (\cos u) = u$$

Going the other way, for any real number v such that $-1 \le v \le 1$, we know that

$$\cos (\text{Arccos } v) = v$$

We restrict v because the Arccosine function is not defined for input values less than -1 or larger than 1.

Cartesian to spherical: finding ϕ

We've learned how to find the vertical angle on the basis of the Cartesian coordinate z. That formula is

$$z = r \cos \phi$$

We can use algebra to rearrange this, getting

$$\cos \phi = z/r$$

provided $r \ne 0$. When we examine Fig. 9-9, we can see that for any given point P, the absolute value of z can never exceed r, so we can be sure that $-1 \le z/r \le 1$. Therefore, we can take the Arccosine of both sides of the preceding equation, getting

$$\text{Arccos } (\cos \phi) = \text{Arccos } (z/r)$$

Simplifying, we obtain

$$\phi = \text{Arccos } (z/r)$$

This formula works nicely if we know the value of r. But we sometimes want to find the vertical angle in terms of x, y, and z exclusively. We've found that

$$r = (x^2 + y^2 + z^2)^{1/2}$$

so we can substitute to obtain

$$\phi = \text{Arccos } [z \, / \, (x^2 + y^2 + z^2)^{1/2}]$$

An example

Consider a point P in spherical three-space whose coordinates are given by

$$P = (\theta, \phi, r) = (3\pi/2, \pi/2, 5)$$

Let's find the equivalent coordinates in Cartesian xyz space. We'll start by calculating the x value. The formula is

$$x = r \sin \phi \cos \theta$$

When we plug in the spherical values, we get

$$x = 5 \sin (\pi/2) \cos (3\pi/2) = 5 \times 1 \times 0 = 0$$

The formula for y is

$$y = r \sin \phi \sin \theta$$

Plugging in the spherical values yields

$$y = 5 \sin (\pi/2) \sin (3\pi/2) = 5 \times 1 \times -1 = -5$$

The formula for z is

$$z = r \cos \phi$$

When we put in the spherical values, we get

$$z = 5 \cos (\pi/2) = 5 \times 0 = 0$$

In xyz space, our point can be specified as

$$P = (0, -5, 0)$$

Another example

Let's convert the xyz space point $(-1, -1, 1)$ to spherical coordinates. To find the radius, we use the formula

$$r = (x^2 + y^2 + z^2)^{1/2}$$

Plugging in the values, we get

$$r = [(-1)^2 + (-1)^2 + 1^2]^{1/2} = (1 + 1 + 1)^{1/2} = 3^{1/2}$$

To find the horizontal angle, we use the formula

$$\theta = \pi + \text{Arctan}\ (y/x)$$

because $x < 0$ and $y < 0$. When we plug in the values for x and y, we get

$$\theta = \pi + \text{Arctan}\ [-1/(-1)] = \pi + \text{Arctan}\ 1 = \pi + \pi/4 = 5\pi/4$$

To find the vertical angle, we can use the formula

$$\phi = \text{Arccos}\ (z/r)$$

We already know that $r = 3^{1/2}$, so

$$\phi = \text{Arccos}\ (1/3^{1/2}) = \text{Arccos}\ 3^{-1/2}$$

Our spherical ordered triple, listing the coordinates in the order $P = (\theta, \phi, r)$, is

$$P = [5\pi/4, (\text{Arccos}\ 3^{-1/2}), 3^{1/2}]$$

Are you confused?

When you come across a messy ordered triple like this, you might ask, "Is there any way to make it look simpler?" Sometimes there is. In this case, there isn't. You can get rid of the grouping symbols if you're willing to use a calculator to approximate the values. But even if you do that, you'll have to remember that in spherical coordinates, the first two values represent angles in radians, and the third value represents a linear distance.

Here's a challenge!

Suppose we're given the coordinates of a point P in spherical three-space as

$$P = (\theta, \phi, r) = (3\pi/4, \pi/4, 3^{1/2})$$

Find the coordinates of P in cylindrical coordinates.

Solution

We haven't learned any formulas for direct conversion between spherical and cylindrical coordinates, so we must convert to Cartesian coordinates first, and then to cylindrical coordinates from there. The Cartesian x value is

$$x = r \sin \phi \cos \theta = 3^{1/2} \sin (\pi/4) \cos (3\pi/4)$$

$$= 3^{1/2} \times 2^{1/2}/2 \times (-2^{1/2}/2) = -3^{1/2}/2$$

The Cartesian y value is

$$y = r \sin \phi \sin \theta = 3^{1/2} \sin (\pi/4) \sin (3\pi/4)$$
$$= 3^{1/2} \times 2^{1/2}/2 \times 2^{1/2}/2 = 3^{1/2}/2$$

The Cartesian z value is

$$z = r \cos \phi = 3^{1/2} \cos (\pi/4) = 3^{1/2} \times 2^{1/2}/2 = 6^{1/2}/2$$

Our Cartesian ordered triple is therefore

$$P = (x,y,z) = (-3^{1/2}/2, 3^{1/2}/2, 6^{1/2}/2)$$

Now let's convert these coordinates to their cylindrical counterparts. We have

$$x = -3^{1/2}/2$$

and

$$y = 3^{1/2}/2$$

To find the cylindrical direction angle θ, we use the formula

$$\theta = \pi + \text{Arctan} \, (y/x)$$

because $x < 0$ and $y > 0$. When we plug in the values for x and y, we get

$$\theta = \pi + \text{Arctan} \, [(3^{1/2}/2) \, / \, (-3^{1/2}/2)] = \pi + \text{Arctan} \, (-1)$$
$$= \pi + (-\pi/4) = 3\pi/4$$

This is the same as the horizontal direction angle in the original set of spherical coordinates, as we should expect. (If things hadn't come out that way, we'd have made a mistake!) When we input the values for x and y to the formula for the cylindrical radius r, we get

$$r = [(-3^{1/2}/2)^2 + (3^{1/2}/2)^2]^{1/2} = (3/4 + 3/4)^{1/2}$$
$$= (6/4)^{1/2} = 6^{1/2}/2$$

We calculated that $z = 6^{1/2}/2$, so the cylindrical height h is

$$h = z = 6^{1/2}/2$$

We've found that the cylindrical equivalent point is

$$(\theta,r,h) = (3\pi/4, 6^{1/2}/2, 6^{1/2}/2)$$

Practice Exercises

This is an open-book quiz. You may (and should) refer to the text as you solve these problems. Don't hurry! You'll find worked-out answers in App. A. The solutions in the appendix may not represent the only way a problem can be figured out. If you think you can solve a particular problem in a quicker or better way than you see there, by all means try it!

1. Describe the graphs of the following equations in cylindrical coordinates. What would they look like in Cartesian *xyz* space?

$$\theta = 0$$
$$r = 0$$
$$h = 0$$

2. Plot the point $(\theta,r,h) = (3\pi/4,6,8)$ in the cylindrical coordinate system.

3. Consider the point $(\theta,r,h) = (\pi/4,0,1)$ in cylindrical coordinates. Find the equivalent of this point in Cartesian *xyz* space.

4. Consider the point $(-4,1,0)$ in *xyz* space. Find the equivalent of this point in cylindrical three-space. First, find the exact coordinates. Then, using a calculator, approximate the irrational coordinates to four decimal places.

5. In the chapter text, we used the conversion formulas to find that the cylindrical equivalent of $(x,y,z) = (1,1,1)$ is $(\theta,r,h) = (\pi/4,2^{1/2},1)$. Convert these coordinates back to Cartesian *xyz* coordinates to verify that the result we got was correct and unambiguous.

6. Describe the graphs of the following equations in spherical coordinates. What would they look like in Cartesian *xyz* space?

$$\theta = 0$$
$$\phi = 0$$
$$r = 0$$

7. Plot the point $(\theta,\phi,r) = (3\pi/4,\pi/4,8)$ in the spherical coordinate system.

8. Consider the point $(\theta,\phi,r) = (\pi/4,0,1)$ in spherical coordinates. Find the equivalent of this point in Cartesian *xyz* space.

9. Consider the point $(-4,1,0)$ in *xyz* space. Find the equivalent of this point in spherical three-space. First, find the exact coordinates. Then, using a calculator, approximate the irrational coordinates to four decimal places.

10. Work the final "challenge" backward to verify that we did our calculations correctly. Consider the point *P* in cylindrical three-space given by

$$P = (\theta,r,h) = [3\pi/4,6^{1/2}/2,6^{1/2}/2]$$

Find the coordinates of *P* in Cartesian coordinates, and from there, convert to spherical coordinates.

10

Review Questions and Answers

Part One

This is not a test! It's a review of important general concepts you learned in the previous nine chapters. Read it through slowly and let it sink in. If you're confused about anything here, or about anything in the section you've just finished, go back and study that material some more.

Chapter 1

Question 1-1

What's the difference between an open interval, a half-open interval, and a closed interval?

Answer 1-1

All three types of intervals are continuous spans of values that a variable can attain between a specific minimum and a specific maximum, which are called the extremes. But there are subtle differences between the three types as listed below:

- In an open interval, neither extreme is included.
- In a half-open interval, one extreme is included, but not the other.
- In a closed interval, both extremes are included.

Question 1-2

Imagine two real numbers a and b, such that $a < b$. These numbers can be the extremes of four different intervals: one open, two half-open, and one closed. How can we denote these four intervals for a variable x?

Answer 1-2

If we include neither a nor b, we have an open interval where $a < x < b$. We can write

$$x \in (a,b)$$

which means "*x* is an element of the open interval (*a*,*b*)." If we include *a* but not *b*, we have a half-open interval where $a \le x < b$. We can write

$$x \in [a,b)$$

which translates to "*x* is an element of the half-open interval [*a*,*b*)." If we include *b* but not *a*, we have an open interval where $a < x \le b$. We write

$$x \in (a,b]$$

which means "*x* is an element of the half-open interval (*a*,*b*]." If we include both *a* and *b*, we have a closed interval where $a \le x \le b$. We can write

$$x \in [a,b]$$

which means "*x* is an element of the closed interval [*a*,*b*]."

Question 1-3

What point of confusion must we avoid when working with interval notation?

Answer 1-3

We must never confuse an open interval with an ordered pair, which uses the same notation. If we pay close attention to the context in which the expression appears, we shouldn't have trouble.

Question 1-4

Relations and functions are operations that map specific values of a variable into specific values of another variable. There's an important distinction between a relation and a function. What is it?

Answer 1-4

In a relation, we can have more than one value of the dependent (or output) variable for a single value of the independent (or input) variable. In a function, we're allowed no more than one output for any given input. All functions are relations, but not all relations are functions.

Question 1-5

The Cartesian plane can be used for graphing relations and functions between an independent variable and a dependent variable. The plane is divided into four sections, called quadrants. How do we identify them?

Answer 1-5

In the first quadrant (usually the upper right), both variables are positive. In the second quadrant (usually the upper left), the independent variable is negative and the dependent variable is positive. In the third quadrant (usually the lower left), both variables are negative. In the fourth quadrant (usually the lower right), the independent variable is positive and the dependent variable is negative.

Question 1-6

Suppose we have a point S in Cartesian two-space that is represented by the ordered pair (x_s, y_s). We can write this as

$$S = (x_s, y_s)$$

What's the straight-line distance d_s between S and the coordinate origin? What's the minimum possible distance between S and the origin? Can the distance be negative? What's the maximum possible distance?

Answer 1-6

We can find the distance using the formula that we derived from the Pythagorean theorem in geometry. In this situation, the formula is

$$d_s = (x_s^2 + y_s^2)^{1/2}$$

The minimum possible distance between S and the origin is zero, which occurs if and only if $x_s = 0$ and $y_s = 0$, so that

$$S = (x_s, y_s) = (0,0)$$

We can never have a negative distance. There is no maximum possible distance between S and the origin. We can make it as large as we want by making x_s or y_s (or both) huge positively or huge negatively.

Question 1-7

Imagine two points in Cartesian two-space, called S and T, such that

$$S = (x_s, y_s)$$

and

$$T = (x_t, y_t)$$

What's the straight-line distance d_{st} going from S to T? What's the straight-line distance d_{ts} going from T to S? Does it make any difference which way we go?

Answer 1-7

If we go from S to T, the distance between the points is

$$d_{st} = [(x_t - x_s)^2 + (y_t - y_s)^2]^{1/2}$$

If we go from T to S, the distance is

$$d_{ts} = [(x_s - x_t)^2 + (y_s - y_t)^2]^{1/2}$$

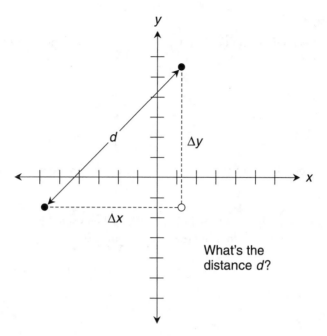

Figure 10-1   Illustration for Question and Answer 1-8.

It doesn't matter which way we go when we want to determine the straight-line distance between two points. Therefore, $d_{st} = d_{ts}$.

Question 1-8

In Fig. 10-1, what do the expressions Δx and Δy mean? What's the straight-line distance d between the two points, based on the values of Δx and Δy?

Answer 1-8

We read Δx as "delta x," which means "the difference in x." We read Δy as "delta y," which means "the difference in y." The straight-line distance d between the points can be found by squaring Δx and Δy individually, adding the squares, and then taking the nonnegative square root of the result, getting

$$d = (\Delta x^2 + \Delta y^2)^{1/2}$$

Question 1-9

Suppose we want to find the midpoint of a line segment connecting two known points in the Cartesian xy plane. How can we do this?

Answer 1-9

We average the x coordinates of the endpoints to get the x coordinate of the midpoint, and we average the y coordinates of the endpoints to get the y coordinate of the midpoint.

Question 1-10

Once again, imagine two points S and T in the Cartesian plane with the coordinates

$$S = (x_s, y_s)$$

and

$$T = (x_t, y_t)$$

What are the coordinates of the point B that bisects the line segment connecting S and T?

Answer 1-10

The point B is the midpoint of the line segment. When we follow the procedure described in Answer 1-9, we obtain the coordinates (x_b, y_b) of point B as

$$(x_b, y_b) = [(x_s + x_t)/2, (y_s + y_t)/2]$$

Chapter 2

Question 2-1

What is a radian?

Answer 2-1

A radian is the standard unit of angular measure in mathematics. If we have two rays pointing out from the center of a circle, and those rays intersect the circle at the endpoints of an arc whose length is equal to the circle's radius, then the smaller (acute) angle between the rays measures one radian (1 rad).

Question 2-2

How many radians are there in a full circle? In 1/4 of a circle? In 1/2 of a circle? In 3/4 of a circle?

Answer 2-2

There are 2π rad in a full circle. Therefore, 1/4 of a circle is $\pi/2$ rad, 1/2 of a circle is π rad, and 3/4 of a circle is $3\pi/2$ rad.

Question 2-3

Suppose we have an angle whose radian measure is $7\pi/6$. What fraction of a complete circular rotation does this represent?

Answer 2-3

Remember that an angle of 2π represents a full rotation. The quantity $\pi/6$ is 1/12 of 2π, so an angle of $\pi/6$ represents 1/12 of a rotation. Therefore, an angle of $7\pi/6$ represents 7/12 of a rotation.

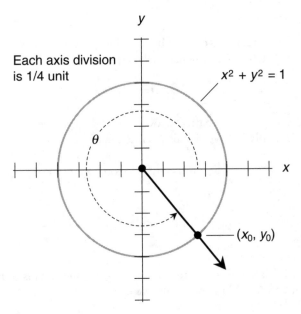

Each axis division is 1/4 unit

$x^2 + y^2 = 1$

θ

(x_0, y_0)

Figure 10-2 Illustration for Questions and Answers 2-4 through 2-9. Each axis division represents 1/4 unit.

Question 2-4

In Fig. 10-2, the gray circle is a graph of the equation $x^2 + y^2 = 1$. The point (x_0, y_0) lies on this circle. A ray from the origin through (x_0, y_0) subtends an angle θ going counterclockwise from the positive x axis. How can we define the sine of the angle θ?

Answer 2-4

The sine of θ as shown in Fig. 10-2 is equal to y_0. Mathematically, we write this as

$$\sin \theta = y_0$$

Question 2-5

How can we define the cosine of the angle θ in Fig. 10-2?

Answer 2-5

The cosine of θ is equal to x_0. Mathematically, we write this as

$$\cos \theta = x_0$$

Question 2-6

How can we define the tangent of the angle θ in Fig. 10-2?

Answer 2-6

The tangent of θ is equal to y_0 divided by x_0, as long as x_0 is nonzero. If $x_0 = 0$, then the tangent of the angle is not defined. Mathematically, we have

$$\tan \theta = y_0/x_0 \Leftrightarrow x_0 \neq 0$$

The double-headed, double-shafted arrow ($\Leftrightarrow$) is the *logical equivalence symbol*. It translates to the words "if and only if." We can also define the tangent as

$$\tan \theta = \sin \theta/\cos \theta \Leftrightarrow \cos \theta \neq 0$$

Question 2-7

How can we define the cosecant of the angle θ in Fig. 10-2?

Answer 2-7

The cosecant of θ is equal to the reciprocal of y_0, as long as y_0 is nonzero. If $y_0 = 0$, then the cosecant is not defined. Mathematically, we have

$$\csc \theta = 1/y_0 \Leftrightarrow y_0 \neq 0$$

We can also define the cosecant as

$$\csc \theta = 1/\sin \theta \Leftrightarrow \sin \theta \neq 0$$

Question 2-8

How can we define the secant of the angle θ in Fig. 10-2?

Answer 2-8

The secant of θ is equal to the reciprocal of x_0, as long as x_0 is nonzero. If $x_0 = 0$, then the secant is not defined. Mathematically, we have

$$\sec \theta = 1/x_0 \Leftrightarrow x_0 \neq 0$$

We can also define the secant as

$$\sec \theta = 1/\cos \theta \Leftrightarrow \cos \theta \neq 0$$

Question 2-9

How can we define the cotangent of the angle θ in Fig. 10-2?

Answer 2-9

The cotangent of θ is equal to x_0 divided by y_0, as long as y_0 is nonzero. If $y_0 = 0$, then the cotangent is not defined. Mathematically, we have

$$\cot \theta = x_0/y_0 \Leftrightarrow y_0 \neq 0$$

We can also define the cotangent as

$$\cot \theta = 1/\tan \theta \iff \tan \theta \neq 0$$

or as

$$\cot \theta = \cos \theta/\sin \theta \iff \sin \theta \neq 0$$

Question 2-10

What are the Pythagorean identities for trigonometric functions? Which, if any, of these should be memorized?

Answer 2-10

The Pythagorean identities are the three formulas

$$\sin^2 \theta + \cos^2 \theta = 1$$
$$\sec^2 \theta - \tan^2 \theta = 1$$
$$\csc^2 \theta - \cot^2 \theta = 1$$

The first of these is worth memorizing, because it comes up quite often in applied mathematics and engineering. The second and third identities can be derived from the first one.

Chapter 3

Question 3-1

How are variables and points portrayed on the polar-coordinate plane?

Answer 3-1

The independent variable is rendered as a direction angle θ, expressed counterclockwise from a reference axis. This reference axis normally goes outward from the origin toward the right (or "due east"), in the same direction as the positive x axis in the Cartesian xy plane. The dependent variable is rendered as a radius r, expressed as the straight-line distance from the origin. Points in the plane are expressed as ordered pairs of the form (θ,r), as shown in Fig. 10-3. In some texts, the ordered pair is written as (r,θ).

Question 3-2

Can a point in polar coordinates have a negative direction angle, or an angle that represents a full rotation or more?

Answer 3-2

Yes. If $\theta < 0$, it represents clockwise rotation from the reference axis. If $\theta \geq 2\pi$, it represents at least one complete counterclockwise rotation from the reference axis.

Question 3-3

Can a point in polar coordinates have a negative radius?

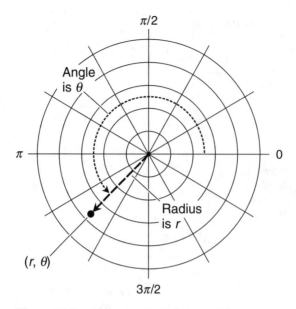

Figure 10-3   Illustration for Question and Answer 3-1.

Answer 3-3

Yes. If $r < 0$, we can multiply r by -1 so it becomes positive, and then add or subtract π to or from the direction angle, keeping it within the preferred range $0 \leq \theta < 2\pi$.

Question 3-4

How we can we portray a relation or function in polar coordinates when the independent variable is θ and the dependent variable is r?

Answer 3-4

We can write down an equation with r on the left-hand side and the name of the function followed by θ in parentheses on the right-hand side. For example, if our function is g, we write

$$r = g\,(\theta)$$

and read it as "r equals g of θ."

Question 3-5

If we set the polar-coordinate angle equal to a constant, say k, what graph do we get?

Answer 3-5

The graph is a straight line passing through the origin. The line appears at an angle of k radians with respect to the reference axis.

Question 3-6

If we set the polar-coordinate radius equal to a constant, say m, what graph do we get?

Answer 3-6

The graph a circle centered at the origin, so that every point on the circle is m units from the origin.

Question 3-7

Suppose we have a point (θ, r) in polar coordinates. How can we convert this to coordinates in the Cartesian xy plane?

Answer 3-7

We can convert the polar point (θ, r) to Cartesian (x, y) using the formulas

$$x = r \cos \theta$$

and

$$y = r \sin \theta$$

Question 3-8

Suppose we have a point (x, y) in the Cartesian plane. What's the polar radius r of this point?

Answer 3-8

The polar radius of a point is its distance from the origin. We can use the formula for the distance of a point from the origin to find that

$$r = (x^2 + y^2)^{1/2}$$

This gives us a positive value for the radius, which is preferred.

Question 3-9

Suppose we have a point (x, y) in the Cartesian plane. What's the polar angle θ of this point?

Answer 3-9

This problem breaks down into following nine cases, depending on where in the Cartesian plane our point (x, y) lies:

- If $x = 0$ and $y = 0$, then $\theta = 0$ by default.
- If $x > 0$ and $y = 0$, then $\theta = 0$.
- If $x > 0$ and $y > 0$, then $\theta = \text{Arctan}\,(y/x)$.
- If $x = 0$ and $y > 0$, then $\theta = \pi/2$.
- If $x < 0$ and $y > 0$, then $\theta = \pi + \text{Arctan}\,(y/x)$.
- If $x < 0$ and $y = 0$, then $\theta = \pi$.
- If $x < 0$ and $y < 0$, then $\theta = \pi + \text{Arctan}\,(y/x)$.
- If $x = 0$ and $y < 0$, then $\theta = 3\pi/2$.
- If $x > 0$ and $y < 0$, then $\theta = 2\pi + \text{Arctan}\,(y/x)$.

If we follow this process carefully, we always get an angle in the range $0 \le \theta < 2\pi$, which is preferred.

Question 3-10

What does "Arctan" mean in the conversions listed in Answer 3-9?

Answer 3-10

It stands for "Arctangent." That's the function that undoes the work of the trigonometric tangent function. The domain of the Arctangent function is the entire set of real numbers. The range is the open interval $(-\pi/2, \pi/2)$. For any real number u within this interval, we have

$$\text{Arctan}\,(\tan u) = u$$

Conversely, for any real number v, we have

$$\tan\,(\text{Arctan}\,v) = v$$

Chapter 4

Question 4-1

What is a vector?

Answer 4-1

A vector is a quantity with two independent properties: magnitude and direction. A vector can also be defined as a directed line segment having an originating point (beginning) and a terminating point (end).

Question 4-2

What's the standard form of a vector in the xy plane? What's the standard form of a vector in the polar plane? What's the advantage of putting a vector into its standard form?

Answer 4-2

In any coordinate system, a vector is in standard form if and only if its originating point is at the coordinate origin. The standard form allows us to uniquely define a vector as an ordered pair that represents the coordinates of its terminating point alone.

Question 4-3

How can we find the magnitude of a standard-form vector **b** in the xy plane whose terminating point has the coordinates (x_b, y_b)?

Answer 4-3

The magnitude of **b**, which we can write as r_b, is found by using the formula for the distance of the terminating point from the origin. In this case, we get

$$r_b = (x_b{}^2 + y_b{}^2)^{1/2}$$

In some texts, the magnitude of **b** would be denoted as $|b|$ or b.

Question 4-4

How can we find the direction of a standard-form vector **b** in the xy plane whose terminating point has the coordinates (x_b, y_b)?

Answer 4-4

We find the polar direction angle of the point (x_b, y_b). If we call this angle θ_b, the process can be broken down into the following nine possible cases:

- If $x_b = 0$ and $y_b = 0$, then $\theta_b = 0$ by default.
- If $x_b > 0$ and $y_b = 0$, then $\theta_b = 0$.
- If $x_b > 0$ and $y_b > 0$, then $\theta_b = \text{Arctan}\,(y_b/x_b)$.
- If $x_b = 0$ and $y_b > 0$, then $\theta_b = \pi/2$.
- If $x_b < 0$ and $y_b > 0$, then $\theta_b = \pi + \text{Arctan}\,(y_b/x_b)$.
- If $x_b < 0$ and $y_b = 0$, then $\theta_b = \pi$.
- If $x_b < 0$ and $y_b < 0$, then $\theta_b = \pi + \text{Arctan}\,(y_b/x_b)$.
- If $x_b = 0$ and $y_b < 0$, then $\theta_b = 3\pi/2$.
- If $x_b > 0$ and $y_b < 0$, then $\theta_b = 2\pi + \text{Arctan}\,(y_b/x_b)$.

In some texts, the direction of **b** is denoted as dir **b**.

Question 4-5

Imagine two vectors **a** and **b** in the xy plane, in standard form with terminating-point coordinates

$$\mathbf{a} = (x_a, y_a)$$

and

$$\mathbf{b} = (x_b, y_b)$$

How can we find the sum of these vectors?

Answer 4-5

We calculate the sum vector $\mathbf{a} + \mathbf{b}$ using the formula

$$\mathbf{a} + \mathbf{b} = [(x_a + x_b), (y_a + y_b)]$$

Question 4-6

How can we calculate the Cartesian negative of a vector that's in standard form? How does the Cartesian negative compare with the original vector?

Answer 4-6

We take the negatives of both coordinate values. For example, if we have

$$\mathbf{b} = (x_b, y_b)$$

then its Cartesian negative is

$$-\mathbf{b} = (-x_b, -y_b)$$

The Cartesian negative has the same magnitude as the original vector, but points in the opposite direction.

Question 4-7

Imagine two Cartesian vectors **a** and **b**, in standard form with terminating-point coordinates

$$\mathbf{a} = (x_a, y_a)$$

and

$$\mathbf{b} = (x_b, y_b)$$

How can we find $\mathbf{a} - \mathbf{b}$? How can we find $\mathbf{b} - \mathbf{a}$? How do these two vectors compare?

Answer 4-7

We calculate the difference vector $\mathbf{a} - \mathbf{b}$ using the formula

$$\mathbf{a} - \mathbf{b} = [(x_a - x_b),(y_a - y_b)]$$

We find difference vector $\mathbf{b} - \mathbf{a}$ by reversing the order of subtraction for each coordinate, getting

$$\mathbf{b} - \mathbf{a} = [(x_b - x_a),(y_b - y_a)]$$

In the Cartesian plane, the difference vector $\mathbf{b} - \mathbf{a}$ is always equal to the negative of the difference vector $\mathbf{a} - \mathbf{b}$.

Question 4-8

Suppose we have a vector expressed in polar form as

$$\mathbf{c} = (\theta_c, r_c)$$

where θ_c is the direction angle of **c**, and r_c is the magnitude of **c**. How can we convert **c** to a standard-form vector (x_c, y_c) in the Cartesian plane?

Answer 4-8

We use formulas adapted from the polar-to-Cartesian conversion. We get

$$(x_c, y_c) = [(r_c \cos \theta_c),(r_c \sin \theta_c)]$$

Question 4-9

What restrictions apply when we work with vectors in the polar-coordinate plane?

Answer 4-9

A polar vector is not allowed to have a negative radius, a negative direction angle, or a direction angle of 2π or more. These constraints prevent ambiguities, so we can be confident that the set of all polar-plane vectors can be paired off in a one-to-one correspondence with the set of all Cartesian-plane vectors.

Question 4-10

Suppose we're given two vectors in polar coordinates. What's the best way to find their sum and difference? What's the best way to find the negative of a vector in polar coordinates?

Answer 4-10

The best way to add or subtract polar vectors is to convert them to Cartesian vectors in standard form, then add or subtract those vectors, and finally convert the result back to polar form. The best way to find the negative of a polar vector is to reverse its direction and leave the magnitude the same. Suppose we have

$$\mathbf{a} = (\theta_a, r_a)$$

If $0 \leq \theta_a < \pi$, then the polar negative is

$$-\mathbf{a} = [(\theta_a + \pi), r_a]$$

If $\pi \leq \theta_a < 2\pi$, then the polar negative is

$$-\mathbf{a} = [(\theta_a - \pi), r_a]$$

Chapter 5

Question 5-1

What's the left-hand Cartesian product of a scalar and a vector? What's the right-hand Cartesian product of a vector and a scalar? How do they compare?

Answer 5-1

Consider a real-number constant k, along with a standard-form vector $\mathbf{a}$ defined in the xy plane as

$$\mathbf{a} = (x_a, y_a)$$

The left-hand Cartesian product of k and $\mathbf{a}$ is

$$k\mathbf{a} = (kx_a, ky_a)$$

The right-hand Cartesian product of $\mathbf{a}$ and k is

$$\mathbf{a}k = (x_a k, y_a k)$$

The left- and right-hand products of a scalar and a Cartesian vector are always the same. For all real numbers k and all Cartesian vectors $\mathbf{a}$, we can be sure that

$$k\mathbf{a} = \mathbf{a}k$$

Question 5-2

What's the left-hand polar product of a positive scalar and a vector? What's the right-hand polar product of a vector and a positive scalar? How do they compare?

Answer 5-2

Imagine a polar vector **a** with angle θ_a and radius r_a, such that

$$\mathbf{a} = (\theta_a, r_a)$$

When we multiply **a** on the left by a positive scalar k_+, we get

$$k_+\mathbf{a} = (\theta_a, k_+ r_a)$$

When we multiply **a** on the right by k_+, we get

$$\mathbf{a}k_+ = (\theta_a, r_a k_+)$$

The left- and right-hand polar products of a positive scalar and a polar vector are always the same. For all positive real numbers k_+ and all polar vectors **a**, we can be sure that

$$k_+\mathbf{a} = \mathbf{a}k_+$$

Question 5-3

What's the left-hand polar product of a negative scalar and a vector? What's the right-hand polar product of a vector and a negative scalar? How do they compare?

Answer 5-3

Once again, suppose we have a polar vector **a** with angle θ_a and radius r_a, such that

$$\mathbf{a} = (\theta_a, r_a)$$

When we multiply **a** on the left by a negative scalar k_-, we get

$$k_-\mathbf{a} = [(\theta_a + \pi), (-k_- r_a)]$$

if $0 \le \theta_a < \pi$, and

$$k_-\mathbf{a} = [(\theta_a - \pi), (-k_- r_a)]$$

if $\pi \le \theta_a < 2\pi$. Because k_- is negative, $-k_-$ is positive, so $-k_- r_a$ is positive, ensuring that we get a positive radius for the resultant vector. If we multiply **a** on the right by k_-, we get

$$\mathbf{a}k_- = [(\theta_a + \pi), r_a(-k_-)]$$

if $0 \le \theta_a < \pi$, and

$$\mathbf{a}k_- = [(\theta_a - \pi), r_a(-k_-)]$$

if $\pi \le \theta_a < 2\pi$. Because k_- is negative, $-k_-$ is positive, so $r_a(-k_-)$ is positive, ensuring that we get a positive radius for the resultant vector. For all negative real numbers k_- and all polar vectors **a**,

$$k_-\mathbf{a} = \mathbf{a}k_-$$

Question 5-4

Suppose we're given two standard-form vectors **a** and **b**, defined by the ordered pairs

$$\mathbf{a} = (x_a, y_a)$$

and

$$\mathbf{b} = (x_b, y_b)$$

What's the Cartesian dot product **a • b**? What's the Cartesian dot product **b • a**? How do they compare?

Answer 5-4

The Cartesian dot product **a • b** is a real number given by

$$\mathbf{a} \bullet \mathbf{b} = x_a x_b + y_a y_b$$

and the Cartesian dot product **b • a** is a real number given by

$$\mathbf{b} \bullet \mathbf{a} = x_b x_a + y_b y_a$$

The Cartesian dot product is commutative, so for any two vectors **a** and **b** in the xy plane, we can be confident that

$$\mathbf{a} \bullet \mathbf{b} = \mathbf{b} \bullet \mathbf{a}$$

Question 5-5

Imagine a polar vector **a** with angle θ_a and radius r_a, such that

$$\mathbf{a} = (\theta_a, r_a)$$

and a polar vector **b** with angle θ_b and radius r_b, such that

$$\mathbf{b} = (\theta_b, r_b)$$

What's the polar dot product **a • b**?

Answer 5-5

Let $\theta_b - \theta_a$ be the angle as we rotate from **a** to **b**. The polar dot product **a • b** is given by the formula

$$\mathbf{a} \bullet \mathbf{b} = r_a r_b \cos(\theta_b - \theta_a)$$

Question 5-6

Consider the same two polar vectors as we worked with in Question and Answer 5-5. What's the polar dot product $\mathbf{b} \cdot \mathbf{a}$?

Answer 5-6

We can define this dot product by reversing the roles of the vectors in the previous problem. Let $\theta_a - \theta_b$ be the angle going from $\mathbf{b}$ to $\mathbf{a}$. The polar dot product $\mathbf{b} \cdot \mathbf{a}$ is given by the formula

$$\mathbf{b} \cdot \mathbf{a} = r_b r_a \cos(\theta_a - \theta_b)$$

Question 5-7

How do the polar dot products $\mathbf{a} \cdot \mathbf{b}$ and $\mathbf{b} \cdot \mathbf{a}$, as defined in Answers 5-5 and 5-6, compare?

Answer 5-7

For any two vectors $\mathbf{a}$ and $\mathbf{b}$, the polar dot product is commutative. That is

$$\mathbf{a} \cdot \mathbf{b} = \mathbf{b} \cdot \mathbf{a}$$

Question 5-8

Imagine a polar vector $\mathbf{c}$ with angle θ_c and radius r_c, such that

$$\mathbf{c} = (\theta_c, r_c)$$

and a polar vector $\mathbf{d}$ with angle θ_d and radius r_d, such that

$$\mathbf{d} = (\theta_d, r_d)$$

What's the polar cross product $\mathbf{c} \times \mathbf{d}$?

Answer 5-8

Imagine that we start at vector $\mathbf{c}$ and rotate counterclockwise until we get to vector $\mathbf{d}$, so we turn through an angle of $\theta_d - \theta_c$. Suppose that $0 < \theta_d - \theta_c < \pi$. To calculate the magnitude $r_{c \times d}$ of the cross-product vector $\mathbf{c} \times \mathbf{d}$, we use the formula

$$r_{c \times d} = r_c r_d \sin(\theta_d - \theta_c)$$

In this situation, $\mathbf{c} \times \mathbf{d}$ points toward us. If $\pi < \theta_d - \theta_c < 2\pi$, we can consider the difference angle to be $2\pi + \theta_c - \theta_d$. Then the magnitude of $\mathbf{c} \times \mathbf{d}$ is

$$r_{c \times d} = r_c r_d \sin(2\pi + \theta_c - \theta_d)$$

and it points away from us.

Question 5-9

What's the right-hand rule for cross products?

Answer 5-9

Consider again the two vectors **c** and **d** that we defined in Question 5-8, and their difference angle $\theta_d - \theta_c$ that we defined in Answer 5-8. If $0 < \theta_d - \theta_c < \pi$, point your right thumb out, and curl your fingers counterclockwise from **c** to **d**. If $\pi < \theta_d - \theta_c < 2\pi$, point your right thumb out, and curl your right-hand fingers clockwise from **c** to **d**. Your thumb will then point in the general direction of **c** × **d**. The vector **c** × **d** is always perpendicular to the plane defined by **c** and **d**.

Question 5-10

How do the polar cross products of two vectors **c** × **d** and **d** × **c** compare?

Answer 5-10

They have identical magnitudes, but they point in opposite directions.

Chapter 6
Question 6-1

What's the unit imaginary number? What's the *j* operator?

Answer 6-1

These expressions both refer to the positive square root of -1. If we denote it as *j*, then

$$j = (-1)^{1/2}$$

and

$$j^2 = -1$$

Question 6-2

How is the set of imaginary numbers "built up"? How do we denote such numbers?

Answer 6-2

If we multiply *j* by a nonnegative real number *a*, we get a nonnegative imaginary number. If we multiply *j* by a negative real number $-a$, we get a negative imaginary number. We denote nonnegative imaginary numbers by writing *j* followed by the real-number coefficient. If $a \geq 0$, then

$$j \times a = a \times j = ja$$

We denote negative imaginary numbers as $-j$ followed by the absolute value of the real-number coefficient. If $-a < 0$, then

$$j \times (-a) = -a \times j = -ja$$

Question 6-3

How is the set of complex numbers "built up"? How do we denote such numbers?

Answer 6-3

A complex number is the sum of a real number and an imaginary number. If a is a real number and b is a nonnegative real number, then the general form for a complex number is

$$a + jb$$

If a is a real number and $-b$ is a negative real number, then we have

$$a + j(-b)$$

but it's customary to write the absolute value of $-b$ after j, and use a minus sign instead of a plus sign in the expression. That gives us the general form

$$a - jb$$

Question 6-4

How do the complex number $0 + j0$, the pure real number 0, and the pure imaginary number $j0$ compare?

Answer 6-4

They are all identical.

Question 6-5

How do we find the sum of two complex numbers $a + jb$ and $c + jd$? How do we find their difference? How do we find their product? How do we find their ratio?

Answer 6-5

When we want to add, we use the formula

$$(a + jb) + (c + jd) = (a + c) + j(b + d)$$

When we want to subtract, we use the formula

$$(a + jb) - (c + jd) = (a - c) + j(b - d)$$

When we want to multiply, we use the formula

$$(a + jb)(c + jd) = (ac - bd) + j(ad + bc)$$

When we want to find the ratio, we use the formula

$$(a + jb) / (c + jd) = [(ac + bd) / (c^2 + d^2)] + j [(bc - ad) / (c^2 + d^2)]$$

In a complex-number ratio, the denominator must not be equal to $0 + j0$.

Question 6-6

What are complex conjugates? What happens when we add a complex number to its conjugate? What happens when we multiply a complex number by its conjugate?

Answer 6-6

Complex conjugates have identical coefficients, but opposite signs between the real and imaginary parts, as in

$$a + jb$$

and

$$a - jb$$

When we add a complex number to its conjugate, we get

$$(a + jb) + (a - jb) = 2a$$

When we multiply a complex number by its conjugate, we get

$$(a + jb)(a - jb) = a^2 + b^2$$

Question 6-7

What's the Cartesian complex-number plane? What's the polar complex-number plane? How are complex vectors defined in these planes?

Answer 6-7

Figure 10-4 shows a Cartesian complex-number plane. The horizontal axis portrays the real-number part, and the vertical axis portrays the imaginary-number part. A Cartesian complex vector is rendered in standard form, going from the origin to the terminating point corresponding to the complex number. Figure 10-5 shows a polar complex-number plane. Polar complex vectors are defined in terms of their direction angle and magnitude, instead of their real and imaginary parts. Assuming that the axis divisions in Fig. 10-4 are the same size as the radial divisions in Fig. 10-5, the vectors in both drawings represent the same complex number.

Question 6-8

How can we convert a Cartesian complex vector to a polar complex vector?

Answer 6-8

Imagine a complex number $a + jb$ in the Cartesian complex plane, whose vector extends from the origin to the point (a,jb). We can derive the magnitude r of the equivalent polar vector by applying the distance formula to get

$$r = (a^2 + b^2)^{1/2}$$

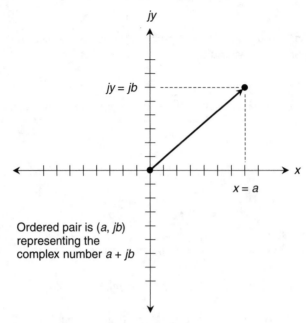

Figure 10-4 Illustration for Question and Answer 6-7.

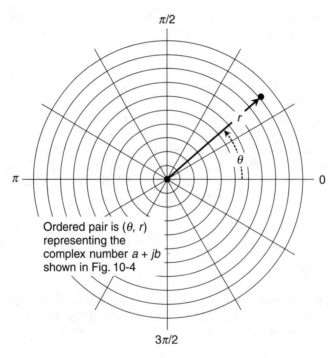

Figure 10-5 Another illustration for Question and Answer 6-7.

To determine the direction angle θ of the polar vector, we modify the polar-coordinate direction-finding system. Here's what happens:

- When $a = 0$ and $jb = j0$, we have $\theta = 0$ by default.
- When $a > 0$ and $jb = j0$, we have $\theta = 0$.
- When $a > 0$ and $jb > j0$, we have $\theta = \text{Arctan } (b/a)$.
- When $a = 0$ and $jb > j0$, we have $\theta = \pi/2$.
- When $a < 0$ and $jb > j0$, we have $\theta = \pi + \text{Arctan } (b/a)$.
- When $a < 0$ and $jb = j0$, we have $\theta = \pi$.
- When $a < 0$ and $jb < j0$, we have $\theta = \pi + \text{Arctan } (b/a)$.
- When $a = 0$ and $jb < j0$, we have $\theta = 3\pi/2$.
- When $a > 0$ and $jb < j0$, we have $\theta = 2\pi + \text{Arctan } (b/a)$.

Question 6-9

How can we convert a polar complex vector to a Cartesian complex vector?

Answer 6-9

Imagine a complex vector (θ, r) in the polar complex plane, whose direction angle is θ and whose radius is r. The Cartesian vector equivalent is

$$(a, jb) = [(r \cos \theta), j(r \sin \theta)]$$

which represents the complex number

$$a + jb = r \cos \theta + j(r \sin \theta)$$

Question 6-10

What are the two versions of De Moivre's theorem? How are they used?

Answer 6-10

The first, and more general, version of De Moivre's theorem involves products and ratios. Suppose we have two polar complex numbers c_1 and c_2, where

$$c_1 = r_1 \cos \theta_1 + j(r_1 \sin \theta_1)$$

and

$$c_2 = r_2 \cos \theta_2 + j(r_2 \sin \theta_2)$$

where r_1 and r_2 are real-number polar magnitudes, and θ_1 and θ_2 are real-number polar angles in radians. Then

$$c_1 c_2 = r_1 r_2 \cos (\theta_1 + \theta_2) + j \, [r_1 r_2 \sin (\theta_1 + \theta_2)]$$

and, as long as r_2 is nonzero,

$$c_1/c_2 = (r_1/r_2) \cos (\theta_1 - \theta_2) + j \, [(r_1/r_2) \sin (\theta_1 - \theta_2)]$$

The second version of De Moivre's theorem involves integer powers. Suppose that c is a complex number, where

$$c = r \cos \theta + j(r \sin \theta)$$

where r is the real-number polar magnitude and θ is the real-number polar angle. Also suppose that n is an integer. Then

$$c^n = r^n \cos (n\theta) + j[rn \sin (n\theta)]$$

Chapter 7
Question 7-1
How are the axes and variables defined in Cartesian xyz space?

Answer 7-1
We construct Cartesian xyz space by placing three real-number lines so that they all intersect at their zero points, and they're all mutually perpendicular. One number line represents the variable x, another represents the variable y, and the third represents the variable z. Figure 10-6 shows two perspective drawings of the typical system. Although the point of

Figure 10-6 Illustration for Question and Answer 7-1.

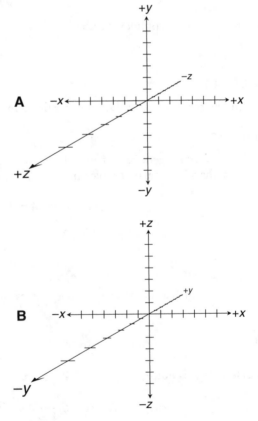

view differs between illustrations A and B, the relative axis orientation is the same in both cases. When we graph relations and functions having two independent variables in Cartesian xyz space, x and y are usually the independent variables, and z is usually the dependent variable.

Question 7-2

What's the difference between Cartesian xyz space and rectangular xyz space?

Answer 7-2

In Cartesian xyz space, the axes are all linear, and they're all graduated in increments of the same size. In rectangular xyz space, the divisions can differ in size between the axes, although each axis must be linear along its entire length.

Question 7-3

What's the "pool rule" for the relative axis orientation and coordinate values in Cartesian xyz space?

Answer 7-3

We can imagine that the origin of the coordinate grid rests on the surface of a swimming pool. We orient the positive x axis horizontally along the pool surface, pointing due east. Once we've done that, the coordinate values can be generalized as follows:

- Positive values of x are east of the origin.
- Negative values of x are west of the origin.
- Positive values of y are north of the origin.
- Negative values of y are south of the origin.
- Positive values of z are up in the air.
- Negative values of z are under the water.

Question 7-4

What are the biaxial planes in Cartesian xyz space?

Answer 7-4

The biaxial planes are the xy plane, the xz plane, and the yz plane. Each plane is perpendicular to the other two, and all three intersect at the origin. The biaxial planes are defined by pairs of axes as follows:

- The xy plane contains the axes for variables x and y.
- The xz plane contains the axes for variables x and z.
- The yz plane contains the axes for variables y and z.

Question 7-5

In Cartesian xyz space, a point can always be denoted as an ordered triple in the form (x,y,z). What do the x, y, and z coordinates represent geometrically?

Answer 7-5

We can think of this situation in two different ways. First, we can use the notion of a point's projection. We get the projection of a point onto an axis by drawing a line from the point to the axis, and making sure that the line intersects that axis at a right angle. That way, the coordinates and projection points are related as follows:

- The *x* coordinate represents the point's projection onto the *x* axis.
- The *y* coordinate represents the point's projection onto the *y* axis.
- The *z* coordinate represents the point's projection onto the *z* axis.

We can also think of the *x*, *y*, and *z* values for a particular point in terms of perpendicular displacements from the biaxial planes as follows:

- The *x* coordinate is the point's perpendicular displacement (positive, negative, or zero) from the *yz* plane.
- The *y* coordinate is the point's perpendicular displacement (positive, negative, or zero) from the *xz* plane.
- The *z* coordinate is the point's perpendicular displacement (positive, negative, or zero) from the *xy* plane.

Question 7-6

What semantical distinction should we keep in mind when we talk about points in terms of ordered triples?

Answer 7-6

An ordered triple represents the coordinates of a point in three-space, not the geometric point itself. Informally, the ordered triple is the name of the point. We can talk about the ordered triple as if it were the actual point, as long as we're aware of the technical difference between the object and its name.

Question 7-7

How can we find the distance of a point from the origin in Cartesian *xyz* space?

Answer 7-7

Suppose we name the point Q, and assign it the coordinates

$$Q = (x_q, y_q, z_q)$$

If we call the distance between Q and the origin by the name d_q, then

$$d_q = (x_q^2 + y_q^2 + z_q^2)^{1/2}$$

This distance is always defined, it's always unique (unambiguous), it's never negative, and it doesn't depend on whether we go from the origin to the point or from the point to the origin.

Question 7-8

How can we find the distance between two points in Cartesian *xyz* space?

Answer 7-8

Let's call the points and their coordinates

$$S = (x_s, y_s, z_s)$$

and

$$T = (x_t, y_t, z_t)$$

where each coordinate can range over the entire set of real numbers. If we go from S to T, the distance between the points is

$$d_{st} = [(x_t - x_s)^2 + (y_t - y_s)^2 + (z_t - z_s)^2]^{1/2}$$

If we go from T to S, the distance is

$$d_{ts} = [(x_s - x_t)^2 + (y_s - y_t)^2 + (z_s - z_t)^2]^{1/2}$$

This distance is always defined and unique. It's never negative, and it doesn't depend on which direction we go. Therefore

$$d_{st} = d_{ts}$$

Question 7-9

How can we find the midpoint of a line segment connecting two points in Cartesian *xyz* space?

Answer 7-9

Let's call the points and their coordinates

$$P = (x_p, y_p, z_p)$$

and

$$Q = (x_q, y_q, z_q)$$

We can call the midpoint M, and say that its coordinates are

$$M = (x_m, y_m, z_m)$$

Given this information, the coordinates of M in terms of the coordinates of P and Q are

$$(x_m, y_m, z_m) = [(x_p + x_q)/2, (y_p + y_q)/2, (z_p + z_q)/2]$$

This midpoint is always defined, it's always unique, and it doesn't depend on which direction we go.

Question 7-10

Suppose that we have two points in Cartesian *xyz* space where all three pairs of corresponding coordinates are negatives of each other. Where is the midpoint of a line segment connecting these two points?

Answer 7-10

It's always at the origin.

Chapter 8

Question 8-1

What's the Cartesian standard form for a vector in *xyz* space?

Answer 8-1

Any vector in *xyz* space, no matter where its originating and terminating points are located, has an equivalent standard-form vector whose originating point is at (0,0,0). Consider a vector $\mathbf{c}'$ whose originating point is Q_1 and whose terminating point is Q_2, such that

$$Q_1 = (x_1, y_1, z_1)$$

and

$$Q_2 = (x_2, y_2, z_2)$$

The standard form of $\mathbf{c}'$, denoted $\mathbf{c}$, has the originating point (0,0,0) and the terminating point Q_c such that

$$Q_c = (x_c, y_c, z_c) = [(x_2 - x_1), (y_2 - y_1), (z_2 - z_1)]$$

The two vectors $\mathbf{c}$ and $\mathbf{c}'$ have identical direction angles and identical magnitudes. That's why we say they're equivalent.

Question 8-2

What's the advantage of putting a three-space vector into its standard form?

Answer 8-2

The standard form allows us to uniquely define a vector as an ordered triple that represents the coordinates of its terminating point alone. We don't have to worry about the originating point.

Question 8-3

How can we find the magnitude r_b of a standard-form vector $\mathbf{b}$ in *xyz* space whose terminating point has the coordinates (x_b, y_b, z_b)?

Answer 8-3

We can do it by calculating the distance of the terminating point from the origin. In this case, the formula is

$$r_b = (x_b^2 + y_b^2 + z_b^2)^{1/2}$$

Question 8-4

How can we define the direction of a standard-form vector in *xyz* space whose terminating point has the coordinates (x_b, y_b, z_b)?

Answer 8-4

The *x*, *y*, and *z* coordinates implicitly contain all the information we need to define the direction of a standard-form vector in Cartesian three-space. But this information is "indirect." Alternatively, we can define the vector's direction if we know the measures of the angles θ_x, θ_y, and θ_z that the vector subtends relative to the +*x*, +*y*, and +*z* axes, respectively. These angles are never negative, and they're never larger than π. There is a one-to-one correspondence between all possible vector orientations and all possible values of the ordered triple $(\theta_x, \theta_y, \theta_z)$.

Question 8-5

Imagine two Cartesian *xyz* space vectors **a** and **b**, in standard form with terminating-point coordinates

$$\mathbf{a} = (x_a, y_a, z_a)$$

and

$$\mathbf{b} = (x_b, y_b, z_b)$$

How can we find the sum **a** + **b**? How can we find the difference **a** − **b**? How can we find the difference **b** − **a**? How can we calculate the Cartesian *xyz* space negative of a vector that's in standard form? How does the Cartesian negative compare with the original vector? How do the differences **a** − **b** and **b** − **a** compare?

Answer 8-5

We can calculate the sum vector **a** + **b** using the formula

$$\mathbf{a} + \mathbf{b} = [(x_a + x_b), (y_a + y_b), (z_a + z_b)]$$

We can calculate the difference vector **a** − **b** using the formula

$$\mathbf{a} - \mathbf{b} = [(x_a - x_b), (y_a - y_b), (z_a - z_b)]$$

We can find the difference vector **b** − **a** using the formula

$$\mathbf{b} - \mathbf{a} = [(x_b - x_a), (y_b - y_a), (z_b - z_a)]$$

To find the Cartesian *xyz* space negative of a vector that's in standard form, we take the negatives of all three terminating-point coordinate values. For example, if we have

$$\mathbf{b} = (x_b, y_b, z_b)$$

then its Cartesian negative is

$$-\mathbf{b} = (-x_b, -y_b, -z_b)$$

The Cartesian negative has the same magnitude as the original vector, but points in the opposite direction. In *xyz* space, the difference vector $\mathbf{b} - \mathbf{a}$ is always equal to the Cartesian negative of the difference vector $\mathbf{a} - \mathbf{b}$.

Question 8-6

What's the left-hand Cartesian product of a scalar and a vector in *xyz* space? What's the right-hand Cartesian product of a vector and a scalar in *xyz* space? How do they compare?

Answer 8-6

Consider a real-number constant k, along with a standard-form vector $\mathbf{a}$ defined in *xyz* space as

$$\mathbf{a} = (x_a, y_a, z_a)$$

The left-hand Cartesian product of k and $\mathbf{a}$ is

$$k\mathbf{a} = (kx_a, ky_a, kz_a)$$

The right-hand Cartesian product of $\mathbf{a}$ and k is

$$\mathbf{a}k = (x_a k, y_a k, z_a k)$$

For all real numbers k and all Cartesian *xyz* space vectors $\mathbf{a}$, we can be sure that

$$k\mathbf{a} = \mathbf{a}k$$

Question 8-7

What are the three standard unit vectors (SUVs) in Cartesian *xyz* space?

Answer 8-7

The three SUVs in Cartesian *xyz* space are defined as the standard-form vectors

$$\mathbf{i} = (1,0,0)$$
$$\mathbf{j} = (0,1,0)$$
$$\mathbf{k} = (0,0,1)$$

Any Cartesian *xyz* space vector in standard form can be split up into a sum of scalar multiples of the three SUVs. The scalar multiples are the coordinates of the ordered triple representing the vector. For example, suppose we have

$$\mathbf{a} = (x_a, y_a, z_a)$$

We can break the vector **a** up in the following manner:

$$\begin{aligned}
\mathbf{a} &= (x_a, y_a, z_a) \\
&= (x_a, 0, 0) + (0, y_a, 0) + (0, 0, z_a) \\
&= x_a(1, 0, 0) + y_a(0, 1, 0) + z_a(0, 0, 1) \\
&= x_a\mathbf{i} + y_a\mathbf{j} + z_a\mathbf{k}
\end{aligned}$$

Question 8-8

Suppose we have two standard-form vectors in Cartesian xyz space, defined as

$$\mathbf{a} = (x_a, y_a, z_a)$$

and

$$\mathbf{b} = (x_b, y_b, z_b)$$

How can we calculate the dot product $\mathbf{a} \bullet \mathbf{b}$? How can we calculate the dot product $\mathbf{b} \bullet \mathbf{a}$? How do they compare?

Answer 8-8

We can calculate $\mathbf{a} \bullet \mathbf{b}$ as a real number using the formula

$$a \bullet b = x_a x_b + y_a y_b + z_a z_b$$

Alternatively, it is

$$a \bullet b = r_a r_b \cos \theta_{ab}$$

where r_a is the magnitude of **a**, r_b is the magnitude of **b**, and θ_{ab} is the angle between the vectors as determined in the plane containing them both, rotating from **a** to **b**. In the same fashion, we can calculate $\mathbf{b} \bullet \mathbf{a}$ using the formula

$$\mathbf{b} \bullet \mathbf{a} = x_b x_a + y_b y_a + z_b z_a$$

Alternatively, it is

$$\mathbf{b} \bullet \mathbf{a} = r_b r_a \cos \theta_{ba}$$

where r_b is the magnitude of **b**, r_a is the magnitude of **a**, and θ_{ba} is the angle between the vectors as determined in the plane containing them both, rotating from **b** to **a**. The dot product is commutative. In other words, for all vectors **a** and **b** in Cartesian xyz space, we can be sure that

$$\mathbf{a} \bullet \mathbf{b} = \mathbf{b} \bullet \mathbf{a}$$

Question 8-9

How can we find the cross product of two standard-form vectors **a** and **b** in three-space if we know their magnitudes and the angle between them?

Answer 8-9

The cross product **a** × **b** is a vector perpendicular to the plane containing both **a** and **b**, and whose magnitude $r_{a \times b}$ is given by

$$r_{a \times b} = r_a r_b \sin \theta_{ab}$$

where r_a is the magnitude of **a**, r_b is the magnitude of **b**, and θ_{ab} is the angle between **a** and **b**, expressed in the rotational sense going from **a** to **b**. We should define the angle so that it's always within the range

$$0 \leq \theta_{ab} \leq \pi$$

If we look at **a** and **b** from some point far outside of the plane containing them, and if θ_{ab} turns through a half circle or less counterclockwise as we go from **a** to **b**, then the cross-product vector **a** × **b** points toward us. If θ_{ab} turns through a half circle or less clockwise as we go from **a** to **b**, then **a** × **b** points away from us.

Question 8-10

Imagine that we have two vectors in *xyz* space whose coordinates are

$$\mathbf{a} = (x_a, y_a, z_a)$$

and

$$\mathbf{b} = (x_b, y_b, z_b)$$

How can we express **a** × **b** as an ordered triple?

Answer 8-10

We can plug in the coordinate values directly into the formula

$$\mathbf{a} \times \mathbf{b} = [(y_a z_b - z_a y_b), (z_a x_b - x_a z_b), (x_a y_b - y_a x_b)]$$

Chapter 9
Question 9-1

How do we determine the cylindrical coordinates of a point in three-space?

Answer 9-1

We "paste" a polar plane onto a Cartesian *xy* plane, creating a reference plane. The positive Cartesian *x* axis is the reference axis. To determine the cylindrical coordinates of a point P, we first locate its projection point, P' on the reference plane:

- The direction angle θ is expressed counterclockwise from the reference axis to the ray that goes out from the origin through P'.
- The radius r is the distance from the origin to P'.
- The height h is the vertical displacement (positive, negative, or zero) from P' to P.

The basic scheme is shown in Fig. 10-7. We express the cylindrical coordinates of our point of interest as an ordered triple:

$$P = (\theta, r, h)$$

Question 9-2

Can we have nonstandard direction angles in cylindrical coordinates? Can we have negative radii? Are there any restrictions on the values of the height coordinate?

Answer 9-2

Theoretically, we can have a nonstandard direction angle. But if we come across that situation, it's best to add or subtract whatever multiple of 2π will bring the direction angle into the preferred range $0 \le \theta < 2\pi$. If $\theta \ge 2\pi$, it represents at least one complete counterclockwise rotation from the reference axis. If $\theta < 0$, it represents clockwise rotation from the reference axis.

We can have a negative radius in theoretical terms. However, if we come across that sort of situation, it's best to reverse the direction angle and then consider the radius positive. If $r < 0$, we can take the absolute value of the negative radius and use it as the radius coordinate. Then we must add or subtract π to or from θ to reverse the direction, while also making sure that the new angle is larger than 0 but less than 2π.

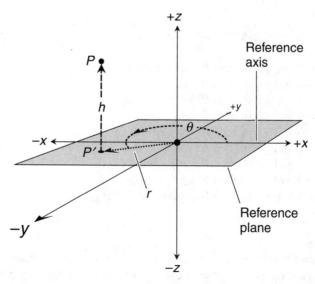

Figure 10-7 Illustration for Question and Answer 9-1.

The height h can be any real number. There are no restrictions on it whatsoever. We have $h > 0$ if and only if P is above the reference plane, $h < 0$ if and only if P is below the reference plane, and $h = 0$ if and only if P is in the reference plane.

Question 9-3

Consider a point $P = (\theta, r, h)$ in cylindrical coordinates. How can we determine the coordinates of P in Cartesian xyz space?

Answer 9-3

The Cartesian x value of P is

$$x = r \cos \theta$$

The Cartesian y value is

$$y = r \sin \theta$$

The Cartesian z value is

$$z = h$$

Question 9-4

Consider a point $P = (x, y, z)$ in Cartesian three-space. How can we find the direction angle θ of the point P in cylindrical coordinates?

Answer 9-4

Cartesian-to-cylindrical angle conversion is the same as Cartesian-to-polar angle conversion:

- If $x = 0$ and $y = 0$, then $\theta = 0$ by default.
- If $x > 0$ and $y = 0$, then $\theta = 0$.
- If $x > 0$ and $y > 0$, then $\theta = \text{Arctan}\,(y/x)$.
- If $x = 0$ and $y > 0$, then $\theta = \pi/2$.
- If $x < 0$ and $y > 0$, then $\theta = \pi + \text{Arctan}\,(y/x)$.
- If $x < 0$ and $y = 0$, then $\theta = \pi$.
- If $x < 0$ and $y < 0$, then $\theta = \pi + \text{Arctan}\,(y/x)$.
- If $x = 0$ and $y < 0$, then $\theta = 3\pi/2$.
- If $x > 0$ and $y < 0$, then $\theta = 2\pi + \text{Arctan}\,(y/x)$.

Question 9-5

Consider a point $P = (x, y, z)$ in Cartesian three-space. How can we find the radius r of the point P in cylindrical coordinates? How can we find the height h of the point P in cylindrical coordinates?

Answer 9-5

To find the cylindrical radius coordinate of P, we find the distance between its projection point P' and the origin in the xy plane. The z value is irrelevant, so the formula is

$$r = (x^2 + y^2)^{1/2}$$

The cylindrical height is simply equal to z. The x and y values are irrelevant, so the formula is

$$h = z$$

Question 9-6

How do we determine the spherical coordinates of a point in three-space?

Answer 9-6

We start with a Cartesian reference plane. The positive Cartesian x axis forms the reference axis. To determine the spherical coordinates of a point P, we first locate its projection point, P', on the reference plane:

- The horizontal angle θ turns counterclockwise in the reference plane from the reference axis to the ray that goes out from the origin through P'.
- The vertical angle ϕ turns downward from the vertical axis to the ray that goes out from the origin through P.
- The radius r is the straight-line distance from the origin to P.

The basic scheme is shown in Fig. 10-8. We express the spherical coordinates as an ordered triple

$$P = (\theta, \phi, r)$$

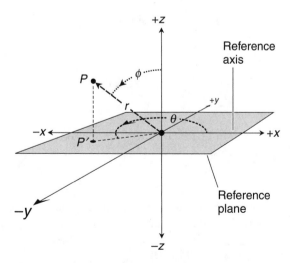

Figure 10-8 Illustration for Question and Answer 9-6.

Question 9-7

Are their any restrictions on the horizontal or vertical angles in spherical coordinates? Are there any restrictions on the radius?

Answer 9-7

Theoretically, we can have a nonstandard horizontal direction angle, but it's best to add or subtract whatever multiple of 2π will bring it into the preferred range $0 \leq \theta < 2\pi$. If $\theta \geq 2\pi$, it represents at least one complete counterclockwise rotation from the reference axis. If $\theta < 0$, it represents clockwise rotation from the reference axis.

Theoretically, we can have a nonstandard vertical angle, but it's best to restrict it to the range $0 \leq \phi \leq \pi$. We can do that by making sure that we traverse the smallest possible angle between the positive z axis and the ray connecting the origin with P.

The radius can be any real number, but things are simplest if we keep it nonnegative. If we find ourselves working with a negative radius, we should reverse the direction by adding or subtracting π to or from both angles, making sure that we end up with $0 \leq \theta < 2\pi$ and $0 \leq \phi \leq \pi$. Then we must take the absolute value of the negative radius and use it as the radius coordinate.

Question 9-8

Consider a point $P = (\theta, \phi, r)$ in spherical coordinates. How can we determine the coordinates of P in Cartesian xyz space?

Answer 9-8

The Cartesian x value of P is

$$x = r \sin \phi \cos \theta$$

The Cartesian y value is

$$y = r \sin \phi \sin \theta$$

The Cartesian z value is

$$z = r \cos \phi$$

Question 9-9

Consider a point $P = (x, y, z)$ in Cartesian three-space. How can we find the horizontal angle coordinate θ of the point P in spherical coordinates?

Answer 9-9

The Cartesian-to-spherical horizontal-angle conversion process is identical to the Cartesian-to-cylindrical direction-angle conversion process:

- If $x = 0$ and $y = 0$, then $\theta = 0$ by default.
- If $x > 0$ and $y = 0$, then $\theta = 0$.
- If $x > 0$ and $y > 0$, then $\theta = \text{Arctan}(y/x)$.

- If $x = 0$ and $y > 0$, then $\theta = \pi/2$.
- If $x < 0$ and $y > 0$, then $\theta = \pi + \mathrm{Arctan}\ (y/x)$.
- If $x < 0$ and $y = 0$, then $\theta = \pi$.
- If $x < 0$ and $y < 0$, then $\theta = \pi + \mathrm{Arctan}\ (y/x)$.
- If $x = 0$ and $y < 0$, then $\theta = 3\pi/2$.
- If $x > 0$ and $y < 0$, then $\theta = 2\pi + \mathrm{Arctan}\ (y/x)$.

Question 9-10

Consider a point $P = (x,y,z)$ in Cartesian three-space. How can we find the radius coordinate r of the point P in spherical coordinates? How can we find the vertical angle coordinate ϕ of the point P in spherical coordinates?

Answer 9-10

To find the spherical radius, we use the formula

$$r = (x^2 + y^2 + z^2)^{1/2}$$

To find the spherical vertical angle, we use the formula

$$\phi = \mathrm{Arccos}\ [z\ /\ (x^2 + y^2 + z^2)^{1/2}]$$

If we already know the radius r, then we have

$$\phi = \mathrm{Arccos}\ (z/r)$$

PART 2

Analytic Geometry

Relations in Two-Space

If you've taken the course *Algebra Know-It-All*, you've already had some basic training on relations and functions. They've been mentioned a few times in this book as well. Let's look more closely at how relations and functions behave in two-space.

What's a Two-Space Relation?

A relation is a special way of assigning, or *mapping*, the elements of a "source" set to the elements of a "destination" set. In two-space, both the source and destination sets usually consist of numbers. The sets might be identical, partially overlapping, or entirely disjoint. For example, we might have a relation between the set of negative integers and the set of positive integers, or the set of positive real numbers and the set of all real numbers, or the set of all real numbers and itself.

Ordered pairs

Any point in the Cartesian plane or the polar plane can be uniquely represented by an ordered pair in which a value of the independent variable (an element of the source set) is listed first, followed by a value of the dependent variable (an element of the destination set). The *domain* is the set of all values of the independent variable for which the relation produces defined values of the dependent variable. The *range* is the set of all values of the dependent variable that come from the elements of the domain. Here's an example of a relation written as a set of ordered pairs:

$$\{(3,2),(4,3),(5,4),(6,5)\}$$

The domain of this particular relation (let's call it set D) is the set of first numbers in the ordered pairs. Therefore

$$D = \{3,4,5,6\}$$

The range (let's call it set R) of the relation is the set of second numbers in the ordered pairs, so

$$R = \{2,3,4,5\}$$

Injection, surjection, and bijection

Imagine a relation between numbers x in a set X and numbers y in a set Y. Suppose that each number x in set X corresponds to one, but only one, number y in set Y. Also suppose that no number in Y has more than one "mate" in X. (There might be some numbers in Y without any "mate" in X.) A relation of this type is called an *injection*. In some older texts, it's called *one-to-one*.

Now imagine a relation that assigns the elements of set X to the elements of set Y so that every element of Y has at least one "mate" in X. This type of relation is called a *surjection*. Set Y is completely "spoken for." A surjection is sometimes called an *onto relation*, because it *maps* (assigns) the values from set X completely *onto* the entire set Y.

Finally, imagine a relation that is both an injection and a surjection. This type of relation is called a *bijection*. In older texts, you might see it referred to as a *one-to-one correspondence* (not to be confused with *one-to-one*, which means an injection). A bijection assigns every value of x in set X to a unique value of y in set Y. Conversely, every y in set Y corresponds to a unique value of x in set X. In this context, "a unique value" means "one and only one value" or "exactly one value."

Example 1

Relations are commonly represented by equations. Here's an example of a simple two-space relation that subtracts 1 from every value in the domain to generate values in the range:

$$y = x - 1$$

This relation could describe a one-to-one correspondence between the elements of the domain

$$X = \{3,4,5,6\}$$

and the elements of the range

$$Y = \{2,3,4,5\}$$

which we saw a few moments ago. If we allow the domain of the relation to extend over the entire set of real numbers, then the range also covers the entire set of real numbers. When we put specific values of x into the equation, we get results such as the following:

- If $x = -13$, then $(x,y) = (-13,-14)$.
- If $x = -1.6$, then $(x,y) = (-1.6,-2.6)$.
- If $x = 0$, then $(x,y) = (0,-1)$.
- If $x = 1$, then $(x,y) = (1,0)$.
- If $x = 3/2$, then $(x,y) = (3/2,1/2)$.
- If $x = 8^{1/2}$, then $(x,y) = [8^{1/2},(8^{1/2} - 1)]$.

For every value of *x*, the relation assigns one and only one value of *y*. The converse is also true; for every value of *y*, there is one and only one corresponding value of *x*.

Example 2

Next, let's consider a real-number relation that squares each element in the domain to produce values in the range. We can write this relation as the following equation:

$$y = x^2$$

In the set of real numbers, this relation is defined for all possible values of *x*, but we never get any negative values of *y*. The range is the set of all *y* such that $y \geq 0$. When we plug specific numbers into this equation, we get results such as the following:

- If $x = -4$, then $(x,y) = (-4,16)$.
- If $x = -1$, then $(x,y) = (-1,1)$.
- If $x = -1/2$, then $(x,y) = (-1/2,1/4)$.
- If $x = 0$, then $(x,y) = (0,0)$.
- If $x = 1/2$, then $(x,y) = (1/2,1/4)$.
- If $x = 1$, then $(x,y) = (1,1)$.
- If $x = 4$, then $(x,y) = (4,16)$.

For every value of *x*, the relation assigns a unique value of *y*, but for every assigned value of *y* except $y = 0$ in the range, the domain contains two values of *x*.

Example 3

Now let's look at a real-number relation that takes the positive or negative square root of elements in the domain to get elements in the range. We can write it as the equation

$$y = \pm(x^{1/2})$$

When we plug in some numbers here, we get results like the following:

- If $x = 1/9$, then $(x,y) = (1/9,1/3)$ or $(1/9,-1/3)$.
- If $x = 1/4$, then $(x,y) = (1/4,1/2)$ or $(1/4,-1/2)$.
- If $x = 1$, then $(x,y) = (1,1)$ or $(1,-1)$.
- If $x = 4$, then $(x,y) = (4,2)$ or $(4,-2)$.
- If $x = 9$, then $(x,y) = (9,3)$ or $(9,-3)$.
- If $x = 0$, then $(x,y) = (0,0)$.

In the set of real numbers, the domain of this relation is confined to nonnegative values of *x*. That is, the domain is the set of all *x* such that $x \geq 0$. For every positive value of *x* in the domain, there are two values of *y* in the range. If $x = 0$, then $y = 0$. The range encompasses all possible real-number values of *y*. For any value of *y* in the range, there exists one and only one corresponding value of *x* in the domain.

Example 4

Finally, let's examine a real-number relation that takes the nonnegative square root of values in the domain to get values in the range. We can denote it as

$$y = x^{1/2}$$

The domain is the set of all real numbers x such that $x \geq 0$, and the range is the set of all real numbers y such that $y \geq 0$. Following are a few examples of what happens when we input values of x into this equation:

- If $x = 1/9$, then $(x,y) = (1/9, 1/3)$.
- If $x = 1/4$, then $(x,y) = (1/4, 1/2)$.
- If $x = 1$, then $(x,y) = (1,1)$.
- If $x = 4$, then $(x,y) = (4,2)$.
- If $x = 9$, then $(x,y) = (9,3)$.
- If $x = 0$, then $(x,y) = (0,0)$.

For every x in the domain, there is one and only one y in the range. The converse is also true. For every y in the range, there is one and only one x in the domain.

Are you confused?

Sometimes a relation fails to take all of the elements of the source or destination sets into account. Figure 11-1 illustrates a generic example of a situation of this sort using a graphical scheme called a *Venn diagram*:

- The entire source set is called the *maximal domain*.
- The entire destination set is called the *co-domain*.
- The domain of a relation is a subset of its maximal domain.
- The range of a relation is a subset of its co-domain.

Here's a challenge!

Classify each of the relations in Examples 1 through 4 as an injection, a surjection, a bijection, or "none of them" from the set of real numbers to itself.

Solution

In each of these relations, our source set is the entire set of real numbers, and that's not necessarily the domain. Also, our destination set is the entire set of reals, and that's not necessarily the range:

- In Example 1, we subtract 1 from each value of the independent variable to get a value of the dependent variable. This operation produces a one-to-one correspondence between the set of real numbers and itself. For every value we input, we get a unique output. Also, every output value is the result of one and only one input value. It follows that this relation is a bijection.

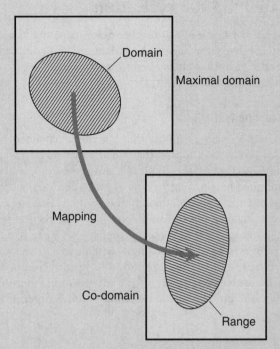

Figure 11-1   The domain of a relation is a
subset of the maximal domain.
The range is a subset of the
co-domain.

- In Example 2, we square each value of the independent variable to get a value of the dependent variable. For every value of the dependent variable except 0, two different values of the independent variable are assigned to it. The relation is not an injection, because it's not one-to-one. It can't be a bijection, then, either. The independent variable can attain any real value, but the dependent variable can never be negative, so this relation is not a surjection onto the set of real numbers. We must therefore classify this relation as "none of them."
- In Example 3, we take the positive or negative square root of each value of the independent variable to get a value of the dependent variable. The independent variable can't be negative, but the dependent variable can be any real number. The relation is therefore a surjection onto the set of real numbers. But it's not an injection, because most values of the independent variable map to two values of the dependent variable. It's not a bijection then, either.
- In Example 4, we take the nonnegative square root of each value of the independent variable to get a value of the dependent variable. As in Example 3, the independent variable can never be negative. Neither can the dependent variable. In this case we don't have an injection, because some real numbers in the source set don't have any counterparts in the destination set. We don't have a surjection either, because the range fails to cover the entire set of real numbers. We must categorize this as a "none of them" relation.

What's a Two-Space Function?

In two-space, a *function* is a relation that never maps any value of the independent variable to more than one value of the dependent variable. All functions are relations, but not all relations are functions. Figure 11-2 shows Venn diagrams of a "legal" assignment for a function (left) and an "illegal" assignment (right).

The vertical-line test

In the Cartesian xy plane, suppose that x is the independent variable, and we plot it against the horizontal axis. Also suppose that y is the dependent variable, and we plot it against the vertical axis. When we see the graph of a simple relation, it usually appears as a line or curve. More complicated relations may graph as groups of lines and/or curves.

We can test the graph of any relation in the Cartesian xy plane to see if it represents a function of x. Imagine an infinitely long, movable vertical line that's always parallel to the dependent-variable axis (the y axis). Suppose that we're free to move the line to the left or right, so it intersects the independent-variable axis (the x axis) wherever we want. If the graph is a function of x, then the movable vertical line never intersects the graph of our relation at more than one point. If, in any position, the vertical line intersects the graph at more than one point, then the relation is not a function of x. We call this exercise the *vertical-line test*.

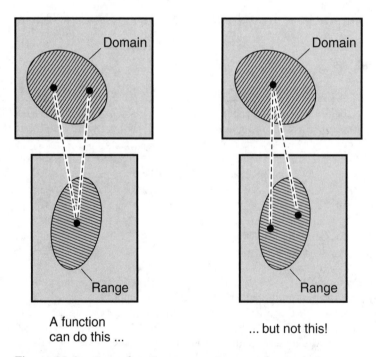

A function
can do this ...

... but not this!

Figure 11-2　A true function never assigns any element in its
domain to more than one element in its range.

Example 1 revisited

Let's take another look at the relation given by Example 1 in the previous section. We described it using the following equation:

$$y = x - 1$$

Figure 11-3 is a graph of this equation in the Cartesian *xy* plane. It's a straight line with a slope of 1 and a *y* intercept of −1. If we imagine an infinitely long, movable vertical line sweeping back and forth, it's easy to see that the vertical line never intersects our graph at more than one point. Therefore, the relation is a function.

Example 2 revisited

The relation in Example 2 in the previous section has a graph that's a parabola opening upward, as shown in Fig. 11-4. The equation is

$$y = x^2$$

The vertex of the parabola represents the *absolute minimum* value of the relation, and it coincides with the coordinate origin (0,0). The curve rises symmetrically on either side of the *y* axis. It's not difficult to see that a movable vertical line never intersects the parabola at more than one point. This fact tells us that the relation is a function of *x*.

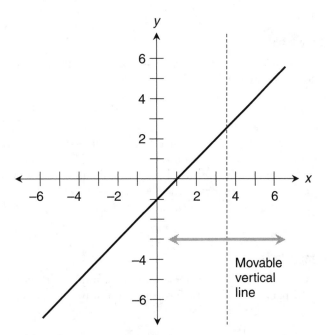

Figure 11-3 Cartesian graph of the relation $y = x - 1$.
The vertical-line test reveals that it's a
function of *x*.

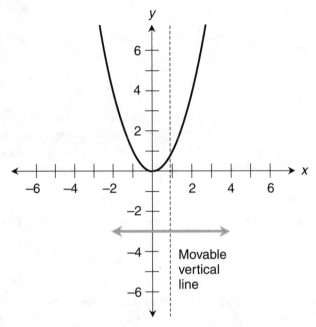

Figure 11-4   Cartesian graph of the relation $y = x^2$. The vertical-line test reveals that it's a function of x.

Example 3 revisited

Figure 11-5 is a graph of the relation we saw in Example 3 in the previous section. The equation for that relation was stated as

$$y = \pm(x^{1/2})$$

In this case, the graph is a parabola that opens to the right. The vertex coincides with the coordinate origin, but there is no absolute minimum or maximum for the dependent variable. When we construct a movable vertical line in this situation, we find that it doesn't intersect the graph when $x < 0$. When $x = 0$, the vertical line intersects the graph at the single point $(0,0)$. When $x > 0$, the vertical line intersects the graph at two points. Therefore, this relation is not a function of x.

Example 4 revisited

Figure 11-6 is a graph of the relation we saw in Example 4 in the previous section. It's the upper half of the parabola of Fig. 11-5, with the point $(0,0)$ included. The equation is

$$y = x^{1/2}$$

The vertical-line test tells us that this relation is a function of x. No matter where we position the vertical line, it never intersects the graph more than once.

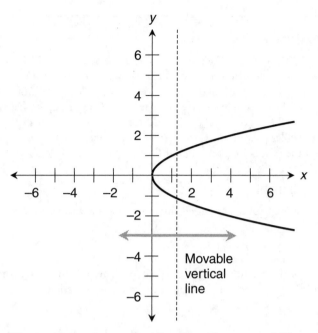

Figure 11-5 Cartesian graph of the relation $y = \pm(x^{1/2})$. The vertical-line test reveals that it isn't a function of x.

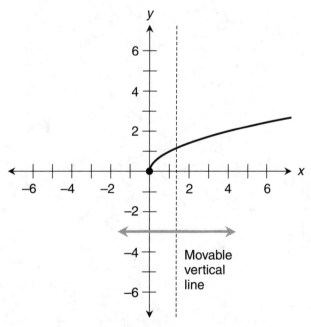

Figure 11-6 Cartesian graph of the relation $y = x^{1/2}$. The vertical-line test reveals that it's a function of x.

Are you confused?

By now you might wonder, "When we have a relation where the independent variable is represented by the polar angle θ and the dependent variable is represented by the polar radius r, how can we tell if the relation is a function of θ?" It's easy, but there's a little trick involved. We can draw the graph of the relation in a Cartesian plane with θ on the horizontal axis and r on the vertical axis. We must allow both θ and r to attain all possible real-number values. Once we've drawn the graph of the polar relation the Cartesian way, we can use the Cartesian vertical-line test to see whether or not the relation is a function of θ.

Here's a challenge!

Consider the relation between an independent variable x and a dependent variable y such that

$$x^2 - y^2 = 1$$

Sketch a graph of this relation in the Cartesian xy plane. Use the vertical-line test to determine, on the basis of the graph, whether or not this relation is a function of x.

Solution

Figure 11-7 is a graph of this relation. It's a geometric figure called a *hyperbola*. The vertical-line test tells us that the relation is not a function of x.

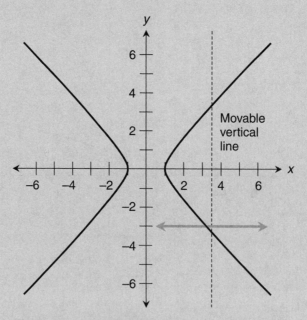

Figure 11-7   Cartesian graph of the relation $x^2 - y^2 = 1$. The vertical-line test reveals that it isn't a function of x.

Here's another challenge!

Consider the relation between an independent variable θ and a dependent variable r defined by the equation

$$r = 2\theta/\pi$$

Sketch a graph of this equation in the polar plane. Then redraw it in the Cartesian plane with θ on the horizontal axis and r on the vertical axis. Use the vertical-line test to determine, on the basis of the graph, whether or not the relation is a function of θ.

Solution

Figure 11-8 is a graph of the equation in the polar plane. It's a pair of "dueling spirals." When we draw the graph of the equation in a Cartesian plane with θ on the horizontal axis and r on the vertical axis, we get a straight line that passes through the origin with a slope of $2/\pi$, as shown in Fig. 11-9. The Cartesian vertical-line test indicates that the relation is a function of θ.

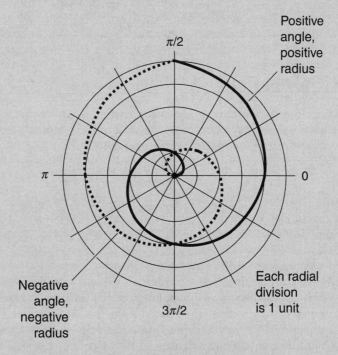

Figure 11-8   Polar graph of the relation $r = 2\theta/\pi$. Each radial division represents 1 unit.

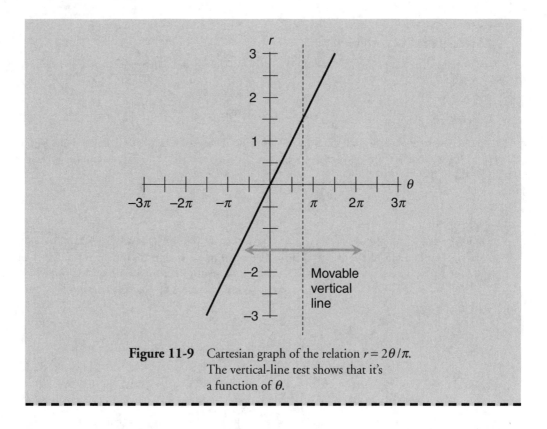

Figure 11-9   Cartesian graph of the relation $r = 2\theta/\pi$. The vertical-line test shows that it's a function of θ.

Algebra with Functions

Functions can always be written as equations. Therefore, when we want to add, subtract, multiply, or divide two functions, we can use ordinary algebra to add, subtract, multiply, or divide both sides of the equations representing the functions.

Cautions

There are three "catches" in the algebra of functions. Whenever we add, subtract, multiply, or divide one function by another, we must watch out for these potential pitfalls. Otherwise, we might get misleading or incorrect results:

- The independent variables of the two functions must match. That is, they must describe the same parameters or phenomena. We can't algebraically combine functions of two different variables in an attempt to get a new function in a single variable. If we try to do that, we won't know which variable the resultant function should operate on.
- The domain of the resultant function is the *intersection* of the domains of the two functions we combine. Any element in the domain of a sum, difference, product, or ratio function must belong to the domains of *both* of the constituent functions. The domain of a ratio function may, however, be restricted even further if the denominator function becomes 0 anywhere in its domain.

- If we divide a function by another function, the resultant function is undefined for any value of the independent variable where the denominator function becomes 0. This can, and often does, restrict the domain of the resultant function to a proper subset of the domain we would get if we were to add, subtract, or multiply the same two functions.

New names for old functions

So far in this chapter, we've encountered four different functions. Three of them are functions of x; the fourth is a function of θ. Following are the equations of the functions once again, for reference:

$$y = x - 1$$
$$y = x^2$$
$$y = x^{1/2}$$
$$r = 2\theta/\pi$$

Let's assign these functions specific names, so that we can write them in the conventional function notation. These are

$$f_1(x) = x - 1$$
$$f_2(x) = x^2$$
$$f_3(x) = x^{1/2}$$
$$f_4(\theta) = 2\theta/\pi$$

Sum of two functions

When we want to find the sum of two functions, we add both sides of their equations. This can be done in either order, producing identical results. For f_1 and f_2, we have

$$(f_1 + f_2)(x) = f_1(x) + f_2(x) = (x - 1) + x^2 = x^2 + x - 1$$

and

$$(f_2 + f_1)(x) = f_2(x) + f_1(x) = x^2 + (x - 1) = x^2 + x - 1$$

For f_1 and f_3, we have

$$(f_1 + f_3)(x) = f_1(x) + f_3(x) = (x - 1) + x^{1/2} = x + x^{1/2} - 1$$

and

$$(f_3 + f_1)(x) = f_3(x) + f_1(x) = x^{1/2} + (x - 1) = x + x^{1/2} - 1$$

For f_2 and f_3, we have

$$(f_2 + f_3)(x) = f_2(x) + f_3(x) = x^2 + x^{1/2}$$

and

$$(f_3 + f_2)(x) = f_3(x) + f_2(x) = x^{1/2} + x^2 = x^2 + x^{1/2}$$

It's customary to write polynomials (sums or differences of a variable raised to powers) with the largest power first, and then descending powers after that. That's why some of the sums and differences have been rearranged in the above examples.

What about f_4?

The independent variable in f_4 doesn't match the independent variable in any of the other three functions, so we can't combine and manipulate the equations as if the variables did match. We can add the equations straightaway, but that doesn't tell us much. For example, we can say that

$$f_2(x) + f_4(\theta) = x^2 + 2\theta/\pi$$

but that's all we can do with it. It's like trying to add minutes to millimeters. We can't get a resultant function of a single variable. The same problem occurs if we try to subtract, multiply, or find a ratio involving f_4 and any of the other three functions.

Difference between two functions

When we want to find the difference between two functions, we subtract both sides of their equations. This can be done in either order, usually producing different results. For f_1 and f_2, we have

$$(f_1 - f_2)(x) = f_1(x) - f_2(x) = (x - 1) - x^2 = -x^2 + x - 1$$

and

$$(f_2 - f_1)(x) = f_2(x) - f_1(x) = x^2 - (x - 1) = x^2 - x + 1$$

For f_1 and f_3, we have

$$(f_1 - f_3)(x) = f_1(x) - f_3(x) = (x - 1) - x^{1/2} = x - x^{1/2} - 1$$

and

$$(f_3 - f_1)(x) = f_3(x) - f_1(x) = x^{1/2} - (x - 1) = -x + x^{1/2} + 1$$

For f_2 and f_3, we have

$$(f_2 - f_3)(x) = f_2(x) - f_3(x) = x^2 - x^{1/2}$$

and

$$(f_3 - f_2)(x) = f_3(x) - f_2(x) = x^{1/2} - x^2 = -x^2 + x^{1/2}$$

Product of two functions

When we want to find the product of two functions, we multiply both sides of their equations. This can be done in either order, producing identical results. For f_1 and f_2, we have

$$(f_1 \times f_2)(x) = f_1(x) \times f_2(x) = (x-1)x^2 = x^3 - x^2$$

and

$$(f_2 \times f_1)(x) = f_2(x) \times f_1(x) = x^2(x-1) = x^3 - x^2$$

For f_1 and f_3, we have

$$(f_1 \times f_3)(x) = f_1(x) \times f_3(x) = (x-1)x^{1/2} = x^{3/2} - x^{1/2}$$

and

$$(f_3 \times f_1)(x) = f_3(x) \times f_1(x) = x^{1/2}(x-1) = x^{3/2} - x^{1/2}$$

For f_2 and f_3, we have

$$(f_2 \times f_3)(x) = f_2(x) \times f_3(x) = x^2 x^{1/2} = x^{5/2}$$

and

$$(f_3 \times f_2)(x) = f_3(x) \times f_2(x) = x^{1/2} x^2 = x^{5/2}$$

Ratio of two functions

When we want to find the ratio of two functions, we divide both sides of their equations. This can be done in either order, usually producing different results. For f_1 and f_2, we have

$$(f_1 / f_2)(x) = f_1(x) / f_2(x) = (x-1)/x^2 = x^{-1} - x^{-2}$$

and

$$(f_2 / f_1)(x) = f_2(x) / f_1(x) = x^2/(x-1) = x^2(x-1)^{-1}$$

For f_1 and f_3, we have

$$(f_1 / f_3)(x) = f_1(x) / f_3(x) = (x-1) / x^{1/2} = x^{1/2} - x^{-1/2}$$

and

$$(f_3 / f_1)(x) = f_3(x) / f_1(x) = x^{1/2}/(x-1) = x^{1/2}(x-1)^{-1}$$

For f_2 and f_3, we have

$$(f_2 / f_3)(x) = f_2(x) / f_3(x) = x^2/x^{1/2} = x^{3/2}$$

and

$$(f_3 / f_2)(x) = f_3(x) / f_2(x) = x^{1/2}/x^2 = x^{-3/2}$$

Are you confused?

You ask, "Do the commutative, associative, distributive, and other rules of arithmetic and algebra work with functions in the same ways as they do with numbers and variables?" The answer is a qualified yes. All the rules of addition, subtraction, multiplication, and division of functions are identical to the rules for arithmetic or algebra involving numbers or variables, as long as we heed the cautions outlined earlier in this section.

Here's a challenge!

Define the real-number domains of all the sum, difference, product, and ratio functions we've found in this section.

Solution

We found the real-number domains (which we can call the *real domains* for short) of the functions f_1, f_2, and f_3 earlier in this chapter. Here they are again, for reference:

- The real domain of $f_1(x)$, which subtracts 1 from x, is the set of all reals.
- The real domain of $f_2(x)$, which squares x, is the set of all reals.
- The real domain of $f_3(x)$, which takes the nonnegative square root of x, is the set of all nonnegative reals.

The real domains of the sum, difference, and product functions are the intersections of these. Let's list them:

- The real domains of $(f_1 + f_2)$, $(f_1 - f_2)$, and $(f_1 \times f_2)$ are the set of all real numbers.
- The real domains of $(f_2 + f_1)$, $(f_2 - f_1)$, and $(f_2 \times f_1)$ are the set of all real numbers.
- The real domains of $(f_1 + f_3)$, $(f_1 - f_3)$, and $(f_1 \times f_3)$ are the set of all nonnegative real numbers.
- The real domains of $(f_3 + f_1)$, $(f_3 - f_1)$, and $(f_3 \times f_1)$ are the set of all nonnegative real numbers.
- The real domains of $(f_2 + f_3)$, $(f_2 - f_3)$, and $(f_2 \times f_3)$ are the set of all nonnegative real numbers.
- The real domains of $(f_3 + f_2)$, $(f_3 - f_2)$, and $(f_3 \times f_2)$ are the set of all nonnegative real numbers.

The real domains of the ratio functions are subsets of the real domains for the sum, product, and difference functions. We have to look at each ratio function and check to see where the denominators are equal to 0:

- The denominator of (f_1 / f_2) becomes 0 when $x = 0$.
- The denominator of (f_2 / f_1) becomes 0 when $x = 1$.
- The denominator of (f_1 / f_3) becomes 0 when $x = 0$.
- The denominator of (f_3 / f_1) becomes 0 when $x = 1$.
- The denominator of (f_2 / f_3) becomes 0 when $x = 0$.
- The denominator of (f_3 / f_2) becomes 0 when $x = 0$.

On the basis of these observations, we can create one final list:

- The real domain of (f_1 / f_2) is the set of all real numbers except 0.
- The real domain of (f_2 / f_1) is the set of all real numbers except 1.

- The real domain of (f_1/f_3) is the set of all strictly positive real numbers.
- The real domain of (f_3/f_1) is the set of all nonnegative real numbers except 1.
- The real domain of (f_2/f_3) is the set of all strictly positive real numbers.
- The real domain of (f_3/f_2) is the set of all strictly positive real numbers.

Practice Exercises

This is an open-book quiz. You may (and should) refer to the text as you solve these problems. Don't hurry! You'll find worked-out answers in App. B. The solutions in the appendix may not represent the only way a problem can be figured out. If you think you can solve a particular problem in a quicker or better way than you see there, by all means try it!

1. Examine the relation illustrated in Fig. 11-10. Suppose that for every element x in set X, there exists *at most* one element y in set Y. Is this relation an injection? Is it a surjection? Is it a bijection? Note that the range of the relation is the entire co-domain. Is that true of all relations? As described here, is this relation a function? Explain each answer.

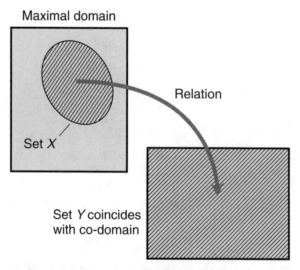

Figure 11-10   Illustration for Problem 1.

2. Imagine a relation in which the domain X is the set of all positive rational numbers, while the range Y is the set of all positive integers. Let's call the independent variable x and the dependent variable y. Suppose that for any x in set X, the relation rounds x up to the next larger integer to obtain the corresponding element y in set Y. Is this relation an injection? Is it a surjection? Is it a bijection? Is it a function of x? Explain each answer.

3. Suppose that we reverse the action of the relation described in Problem 2. Let the domain X be the set of all positive integers, while the range Y is the set of all positive rationals. Suppose that for any value of the independent variable x in set X, the relation

maps to the set of all rationals in the half open interval $(x - 1, x]$. Is this relation an injection? Is it a surjection? Is it a bijection? Is it a function of x? Explain each answer.

4. Can a relation whose graph is a circle or ellipse in the Cartesian xy plane ever be a function of x? Why or why not?

5. Can a relation whose graph is a circle in the polar θr plane ever be a function of θ? Why or why not?

6. Find all the sums, differences, products, and ratios of

$$f(x) = x + 2$$

and

$$g(x) = 3$$

7. Find all the sums, differences, products, and ratios of

$$f(x) = x + 1$$

and

$$g(x) = x - 1$$

8. Find all the sums, differences, products, and ratios of

$$f(x) = x^{-1}$$

and

$$g(x) = x^{-2}$$

9. Find all the sums, differences, products, and ratios of

$$f(x) = \sin^2 \theta$$

and

$$g(x) = \cos^2 \theta$$

10. What are the real-number domains of all the original and derived functions in Problems 6 through 9?

Inverse Relations in Two-Space

Any relation in two-space has a unique *inverse relation*, which can be called simply the *inverse* if we understand that we're dealing with a relation. We denote the fact that a relation is an inverse by writing a superscript -1 after its name. For example, if we have a relation $f(x)$, then its inverse is $f^{-1}(x)$.

Finding an Inverse Relation

A relation's inverse does the opposite of whatever the original relation does. To find the inverse of a relation, we can manipulate the equation so that the independent and dependent variables switch roles. We must therefore transpose the domain and range.

The algebraic way

Suppose we have a relation $f(x)$. The inverse of f, which we call f^{-1}, is another relation such that

$$f^{-1}[f(x)] = x$$

for all possible values of x in the domain of f, and

$$f[f^{-1}(y)] = y$$

for all possible values of y in the range of f. When we talk or write about an inverse relation, it's customary to swap the names of the variables so the inverse relation calls the independent and dependent variables by their original names. That means the preceding equation can be rewritten as

$$f[f^{-1}(x)] = x$$

for all possible values of x in the domain of f^{-1}.

An example

Let's find the inverse of the following relation:

$$f(x) = x + 2$$

If we call the dependent variable y, then we can rewrite our relation as

$$y = x + 2$$

Swapping the names of the variables, we get

$$x = y + 2$$

which can be manipulated with algebra to obtain

$$y = x - 2$$

If we replace the new variable y by the relation notation $f^{-1}(x)$, we get

$$f^{-1}(x) = x - 2$$

The domain and range of the original relation f both span the entire set of real numbers. Therefore, the domain and range of the inverse relation f^{-1} also both span the entire set of reals.

Another example

Let's find the inverse of the following relation:

$$g(x) = \pm(x^{1/2})$$

If we call the dependent variable y, then we can rewrite the relation as

$$y = \pm(x^{1/2})$$

When we switch the names of the variables, we get

$$x = \pm(y^{1/2})$$

Squaring both sides produces

$$x^2 = y$$

Reversing the left- and right-hand sides gives us

$$y = x^2$$

Replacing y by $g^{-1}(x)$, we get

$$g^{-1}(x) = x^2$$

The domain of the original relation g spans the set of nonnegative reals, and the range of g spans the set of all reals. Therefore, the domain of g^{-1} includes all reals, while the range of g^{-1} is confined to the set of nonnegative reals.

Still another example

Let's find the inverse of the relation

$$h(x) = x^{1/2}$$

When we write the 1/2 power of a quantity without including any sign, we mean the nonnegative square root of that quantity. If we call the dependent variable y, then we have

$$y = x^{1/2}$$

Swapping the names of the variables, we get

$$x = y^{1/2}$$

Squaring both sides, we obtain

$$x^2 = y$$

Reversing the left- and right-hand sides of this equation yields

$$y = x^2$$

Replacing y by $h^{-1}(x)$, we get

$$h^{-1}(x) = x^2$$

Is the inverse of h identical to the inverse of g we obtained a few moments ago? It looks that way "on the surface," but it's not so simple when we examine the situation more closely. The domain of h spans the set of nonnegative reals, just as the domain of g does. But the range of h spans the set of nonnegative reals only (not the set of all reals, as the range of g does). Transposing, we must conclude that the domain and range of h^{-1} are both confined to the set of nonnegative reals. The relations h and h^{-1} are therefore restricted versions of g and g^{-1}.

The graphical way

Imagine the line represented by the equation $y = x$ in the Cartesian xy plane as a "point reflector." For any point that's part of the graph of the original relation, we can locate its

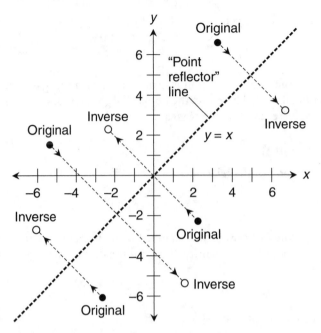

Figure 12-1 Any point on the graph of the inverse of a relation is the point's image on the opposite side of a point reflector line. The new coordinates are obtained by reversing the sequence of the ordered pair representing the original point.

counterpart in the graph of the inverse relation by going to the opposite side of the point reflector, exactly the same distance away. Figure 12-1 shows how this works. The line connecting a point in the original graph and its "mate" in the inverse graph is perpendicular to the point reflector. The point reflector is a *perpendicular bisector* of every point-connecting line.

Mathematically, we can do a point transformation of the sort shown in Fig. 12-1 by reversing the sequence of the ordered pair representing the point. For example, if (4,6) represents a point on the graph of a certain relation, then its counterpoint on the graph of the inverse relation is represented by (6,4).

When we want to graph the inverse of a relation, we flip the whole graph over along a "hinge" corresponding to the point reflector line $y = x$. That moves every point in the graph of the original relation to its new position in the graph of the inverse. Figures 12-2, 12-3, and 12-4 show how this process works with the three relations we dealt with a few moments ago. The positions of the x and y axes haven't changed, but the values of the variables, as well as the domain and range, have been reversed.

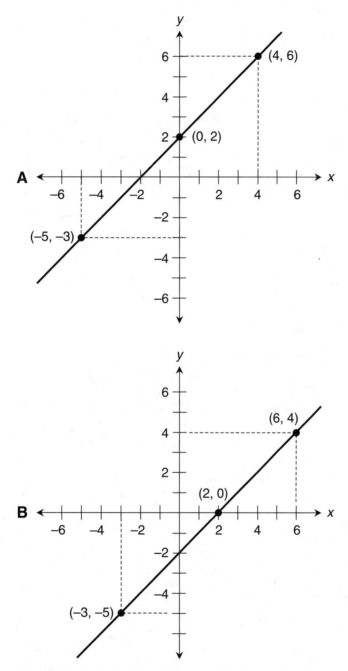

Figure 12-2   At A, Cartesian graph of the relation $y = x + 2$.
At B, Cartesian graph of the inverse relation.

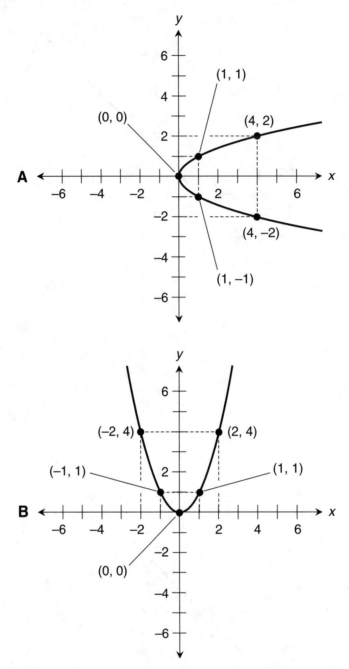

Figure 12-3   At A, Cartesian graph of the relation $y = \pm(x^{1/2})$.
At B, Cartesian graph of the inverse relation.

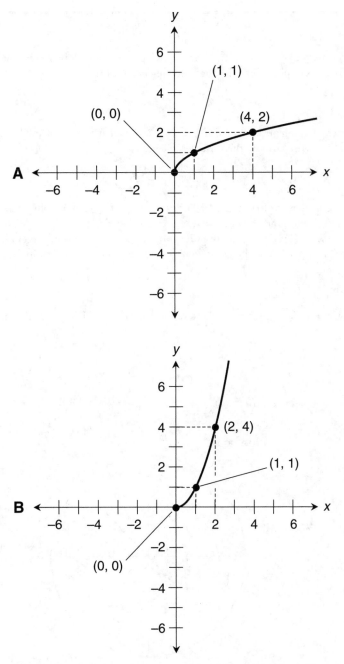

Figure 12-4 At A, Cartesian graph of the relation $y = x^{1/2}$.
At B, Cartesian graph of the inverse relation.

Are you confused?

It's reasonable for you to wonder, "Can any relation be its own inverse?" The answer is yes. There are plenty of examples. Consider the following equation:

$$x^2 + y^2 = 25$$

The Cartesian graph of this equation is a circle centered at the origin and having a radius of 5 units (Fig. 12-5). If we transpose the variables, we get

$$y^2 + x^2 = 25$$

which is equivalent to the original relation. If we perform the graphical transformation by mirroring the circle around the line $y = x$, we get another circle having the same radius and the same center. Theoretically, all but two of points on the new circle are in different places than the points on the original circle, but the graph looks the same as the one shown in Fig. 12-5.

Here's a challenge!

Consider the following relation where the independent variable is x and the dependent variable is y:

$$x^2/9 + y^2/25 = 1$$

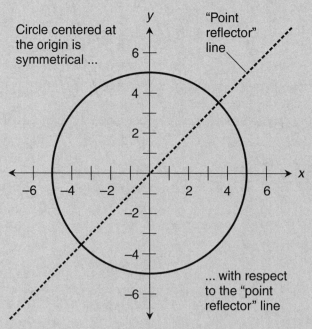

Figure 12-5   Cartesian graph of the relation $x^2 + y^2 = 25$. This relation is its own inverse.

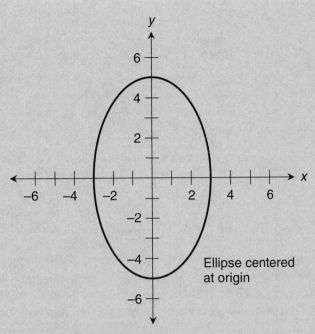

Figure 12-6 Cartesian graph of the relation $x^2/9 + y^2/25 = 1$.

Figure 12-6 is a graph of this relation in Cartesian coordinates. It's an ellipse centered at the origin. The distance from the center to the extreme right- or left-hand point on the ellipse measures 3 units, which is the square root of 9. The distance from the center to the uppermost or lowermost point on the ellipse measures 5 units, which is the square root of 25. Determine the inverse of this relation, and graph it.

Solution

We can obtain the inverse of this relation by swapping the variables. That gives us the equation

$$y^2/9 + x^2/25 = 1$$

which can be rewritten as

$$x^2/25 + y^2/9 = 1$$

Figure 12-7 illustrates the graphs of the original relation and its inverse in Cartesian coordinates. The new graph is another ellipse having the same shape as the original one, and centered at the origin just like the original one. But the horizontal and vertical axes of the ellipse have been transposed. The distance from the center to the extreme right- or left-hand point on the "inverse ellipse" measures 5 units, which is the square root of 25. The distance from the center to the uppermost or lowermost point on the "inverse ellipse" measures 3 units, which is the square root of 9.

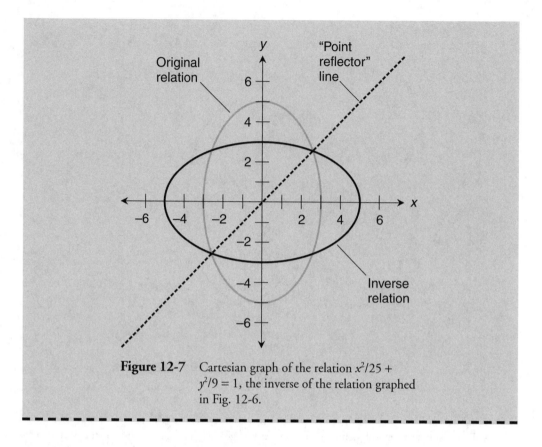

Figure 12-7   Cartesian graph of the relation $x^2/25 +$
$y^2/9 = 1$, the inverse of the relation graphed
in Fig. 12-6.

Finding an Inverse Function

If a function is a bijection (that is, a perfect one-to-one correspondence) over a certain domain and range, then we can transpose the domain and range, and the resulting inverse relation will always be a function. If a function is many-to-one, then its inverse relation is one-to-many, so it's not a function.

"Undoing" the work

Suppose that f and f^{-1} are both true functions that are inverses of each other. Then for all x in the domain of either function, we have

$$f^{-1}[f(x)] = x$$

and

$$f[f^{-1}(x)] = x$$

An inverse function undoes the work of the original function in an unambiguous manner when the domains and ranges are restricted so that the original function and the inverse are both bijections.

Sometimes we can simply turn f "inside-out" to get an inverse relation, and the inverse will be a true function for all the values in the domain and range of f. But often, when we seek the inverse of a function f, we get a relation that's not a true function, because some elements in the range of f map from more than one element in the domain of f. When this happens, we must restrict f to define an inverse f^{-1} that's a true function. We can usually (but not always) find a way to "force" f^{-1} to behave as a true function by excluding all values of either variable that map to more than one value of the other variable. Once we've done that, we get a bijection, ensuring that there is no ambiguity or redundancy either way.

Making a relation behave as a function

A little while ago, we looked at a relation whose graph is a circle with a radius of 5 units (Fig. 12-5). The equation of that relation, once again, is

$$x^2 + y^2 = 25$$

which can be rewritten as

$$y^2 = 25 - x^2$$

and then morphed to

$$y = \pm(25 - x^2)^{1/2}$$

If we use relation notation to express this equation and name the relation f, we have

$$f(x) = \pm(25 - x^2)^{1/2}$$

The vertical-line test tells us that f is not a true function of x. We can modify it so that it becomes a function of x if we restrict the range to nonnegative values. Graphically, that eliminates the lower half of the circle, so that for every input value in the domain, we get only one output in the range. Figure 12-8A is an illustration of this function, which we can call f_+ and define as

$$f_+(x) = (25 - x^2)^{1/2}$$

Once again, we mustn't forget that when we take the 1/2 power of a quantity without including any sign, we mean, by default, the nonnegative square root of that quantity. The solid dots indicate that the plotted points are part of the range of f_+.

Now suppose that we eliminate the top half of the circle including the points $(-5,0)$ and $(5,0)$, getting the graph shown in Fig. 12-8B. The vertical-line test indicates that this is a true function of x. If we call this function f_-, we can write

$$f_-(x) = -(25 - x^2)^{1/2}$$

The white dots (small open circles) tell us that the plotted points are *not* part of the range of f_-. We can restrict the range further, say to values strictly larger than 1 or values smaller than or

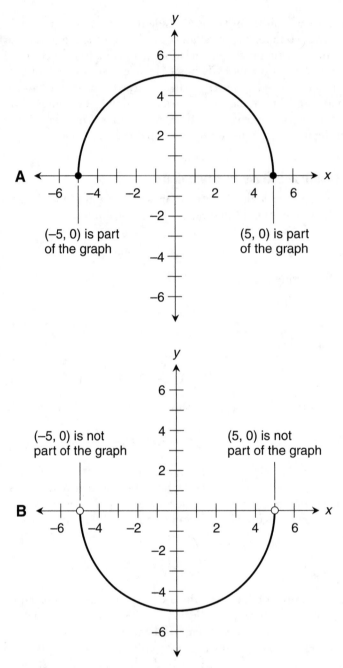

Figure 12-8   At A, Cartesian graph of the function $y = (25 - x^2)^{1/2}$. At B, Cartesian graph of the function $y = -(25 - x^2)^{1/2}$.

equal to −2, and we'll get more true functions. You can doubtless imagine other restrictions we can impose on the range of the original relation f to get true functions of x.

What about the inverse of f_+?

Let's manipulate f_+ algebraically to find its inverse. If we call the dependent variable y, then

$$y = (25 - x^2)^{1/2}$$

Swapping the names of the variables, we get

$$x = (25 - y^2)^{1/2}$$

Squaring both sides, we obtain

$$x^2 = 25 - y^2$$

Subtracting 25 from each side yields

$$x^2 - 25 = -y^2$$

When we multiply through by −1 and transpose the left-hand and right-hand sides of the equation, we obtain

$$y^2 = 25 - x^2$$

Taking the complete square root of both sides gives us

$$y = \pm(25 - x^2)^{1/2}$$

Replacing y by $f_+^{-1}(x)$ to indicate the inverse of f_+, we get

$$f_+^{-1}(x) = \pm(25 - x^2)^{1/2}$$

Does this look like the same thing as the inverse of the original relation f? Don't be fooled; it isn't the same! We haven't quite finished our work. We must transpose the domain and range of f_+ to get the domain and range of the inverse relation f_+^{-1}. The domain of f_+ is the closed interval [−5,5], and the range of f_+ is the closed interval [0,5]. Therefore, the domain of f_+^{-1} is the closed interval [0,5], and the range of f_+^{-1} is the closed interval [−5,5]. Figure 12-9A is a graph of this inverse relation. It's easy to see that f_+^{-1} fails the vertical-line test, so it's not a true function of x.

What about the inverse of f_-?

Now let's go through the algebra to figure out the inverse of the function f_-. This process is almost identical to the work we just finished, but it's a good practice to carry it out step by step anyway. If we call the dependent variable y, then

$$y = -(25 - x^2)^{1/2}$$

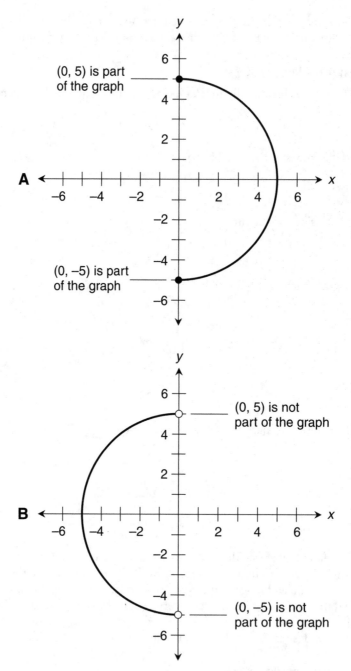

Figure 12-9 At A, Cartesian graph of the inverse of the function $y = (25 - x^2)^{1/2}$. At B, Cartesian graph of the inverse of the function $y = -(25 - x^2)^{1/2}$. Vertical-line tests indicate that neither of these inverse relations is a function.

Switching the names of the variables, we get

$$x = -(25 - y^2)^{1/2}$$

Squaring both sides, we obtain

$$x^2 = 25 - y^2$$

Subtracting 25 from each side gives us

$$x^2 - 25 = -y^2$$

When we multiply through by −1 and transpose the left- and right-hand sides of the equation, we get

$$y^2 = 25 - x^2$$

Taking the complete square root of both sides, we have

$$y = \pm(25 - x^2)^{1/2}$$

Replacing y by $f_-^{-1}(x)$ to indicate the inverse of f_-, we get

$$f_-^{-1}(x) = \pm(25 - x^2)^{1/2}$$

We transpose the domain and range of f_- to get the domain and range of f_-^{-1}. Things get a little tricky here. Refer again to Fig. 12-8B. The domain of f_- is the open interval (−5,5), and the range of f_- is the half-open interval [−5,0). Transposing, we can see that the domain of f_-^{-1} is the half-open interval [−5,0), and the range of f_-^{-1} is the open interval (−5,5). Figure 12-9B is a graph of f_-^{-1}. This inverse relation fails the vertical-line test, so it's not a true function of x.

Making an inverse behave as a function

Do you get the idea that we can't make the relation graphed in Fig. 12-5 behave as a function whose inverse is another function, no matter what limitations we impose on the domain and range? Don't give up. There are plenty of ways. For example, we can restrict both the domain and the range of the original relation

$$x^2 + y^2 = 25$$

to values that all show up in the first quadrant of the Cartesian plane. When we do that, the domain and range are both narrowed down to the open interval (0,5). The relation becomes a true function of x, and its inverse also becomes a true function. Similar things happen if we restrict both the domain and the range to values that show up entirely in the second quadrant, entirely in the third quadrant, or entirely in the fourth quadrant. Feel free to draw the graphs, put a point reflector line to work, and see for yourself.

Are you confused?

You might ask, "We've seen an example of a relation that's its own inverse. Can a function be its own inverse?" The answer is yes. The function $f(x) = x$ is its own inverse; the domain and range both span the entire set of real numbers. It's the ultimate in simplicity. The function's graph coincides with the point reflector line, so it's identical to its own reflection! We have

$$f^{-1}[f(x)] = f^{-1}(x) = x$$

so therefore

$$f[f^{-1}(x)] = f(x) = x$$

Another example

Consider $g(x) = 1/x$, with the restriction that the domain and range can attain any real-number value except zero. This function is its own inverse. We have

$$g^{-1}[g(x)] = g^{-1}(1/x) = 1/(1/x) = x$$

so therefore

$$g[g^{-1}(x)] = g(1/x) = 1/(1/x) = x$$

Still another example

Consider the function $h(x) = 3$ for all real numbers x. Figure 12-10 shows its graph. When we transpose the variables, domain, and range, we must set $x = 3$ for $h^{-1}(x)$ to mean anything. Then we end up with all the real numbers at the same time. This relation fails the vertical-line function test in the worst possible way, because the graph is itself a vertical line (Fig. 12-11).

Here's a challenge!

Consider the following three functions:

$$f(x) = x - 11$$

$$g(x) = x^2/4$$

$$h(x) = -32x^5$$

The inverse of one of these functions is not a function. Which one?

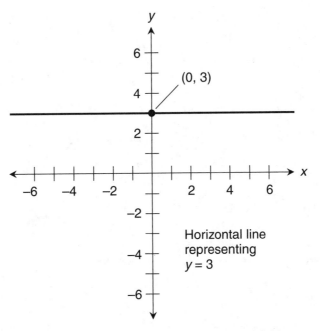

Figure 12-10   Cartesian graph of the function $h(x) = 3$.

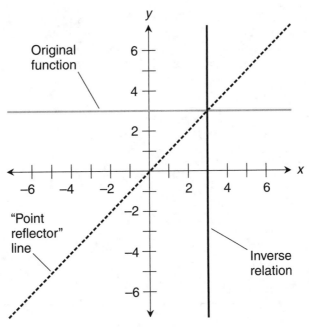

Figure 12-11   Cartesian graph of the inverse of $h(x) = 3$.
It's obviously not a function!

Solution

The inverse of g is not a function. If we call the dependent variable y, we get

$$y = x^2/4$$

The domain is the entire set of reals, and the range is the set of non-negative reals. If we swap the names of the independent and dependent variables, we get

$$x = y^2/4$$

which is the same as

$$y^2 = 4x$$

Taking the complete square root of both sides gives us

$$y = \pm 2x^{1/2}$$

The plus-or-minus symbol indicates that for every nonzero value of the independent variable x that we input to this relation, we get two values of the dependent variable y, one positive and the other negative. We can also write

$$g^{-1}(x) = \pm 2x^{1/2}$$

The original function g is two-to-one (except when $x = 0$). That's okay. But the inverse relation is one-to-two except when $x = 0$. That prevents g^{-1} from qualifying as a true function.

The other two functions, f and h, have inverses that are also functions. Both f and h are one-to-one, so their inverses are also one-to-one. We have

$$f(x) = x - 11$$

and

$$f^{-1}(x) = x + 11$$

We also have

$$h(x) = -32x^5$$

and

$$h^{-1}(x) = (-x)^{1/5}/2$$

Practice Exercises

This is an open-book quiz. You may (and should) refer to the text as you solve these problems. Don't hurry! You'll find worked-out answers in App. B. The solutions in the appendix may not represent the only way a problem can be figured out. If you think you can solve a particular problem in a quicker or better way than you see there, by all means try it!

1. Use algebra to find the inverse of the relation

$$f(x) = 2x + 4$$

2. Use algebra to find the inverse of the relation

$$g(x) = x^2 - 4x + 4$$

3. Use algebra to find the inverse of the relation

$$h(x) = x^3 - 5$$

4. Determine the real-number domains and ranges of the relations and inverses from the statements and solutions of Problems 1, 2, and 3.

5. Consider the two-space relation

$$x^2/4 - y^2/9 = 1$$

Figure 12-12 is a graph of this relation in Cartesian coordinates. It's a hyperbola centered at the origin, opening to the right and left, and crossing the x axis at (2,0) and (−2,0).

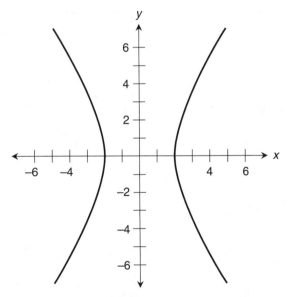

Figure 12-12   Illustration for Problem 5.

Call the independent variable x and the dependent variable y. Call the relation f. Determine $f(x)$ and $f^{-1}(x)$ mathematically. State them both using relation notation.

6. What is the real-number domain of the relation $f(x)$ that you determined when you solved Problem 5? What is its real-number range?

7. Sketch a graph of the inverse relation you found when you solved Problem 5. What is its real-number domain? What is its real-number range?

8. The relation described and graphed in Problem 5 can be modified by restricting its domain to the set of reals greater than or equal to 2. Show graphically, by means of the vertical-line test, that this restriction makes the inverse f^{-1} into a function.

9. The relation described and graphed in Problem 5 can be modified by restricting its domain to the set of reals smaller than or equal to -2. Show graphically, by means of the vertical-line test, that this restriction makes the inverse f^{-1} into a function.

10. The relation described and graphed in Problem 5 can be modified by restricting its range to the set of nonnegative reals. Show graphically, and by means of vertical-line tests, that this restriction makes f into a function, but does not make f^{-1} into a function.

13

Conic Sections

In this chapter, we'll learn the fundamental properties of curves called *conic sections*. These curves include the *circle*, the *ellipse*, the *parabola*, and the *hyperbola*. The conic sections can always be represented in the Cartesian plane as equations that contain the squares of one or both variables.

Geometry

Imagine a *double right circular cone* with a vertical axis that extends infinitely upward and downward. Also imagine a flat, infinitely large plane that can be moved around so that it slices through the double cone in various ways, as shown in Fig. 13-1. The intersection between the plane and the double cone is always a circle, an ellipse, a parabola, or a hyperbola, as long as the plane doesn't pass through the point where the apexes of the cones meet.

Geometry of a circle and an ellipse

Figure 13-1A shows what happens when the plane is perpendicular to the axis of the double cone. In that case, we get a circle. In Fig. 13-1B, the plane is not perpendicular to the axis of the cone, but it isn't tilted very much. The curve is closed, but it isn't a perfect circle. Instead, it's an "elongated circle" or ellipse.

Geometry of a parabola

As the plane tilts farther away from a right angle with respect to the double-cone axis, the ellipse becomes increasingly elongated. Eventually, we reach an angle of tilt where the curve is no longer closed. At precisely this threshold angle, the intersection between the plane and the cone is a parabola (Fig. 13-1C).

Geometry of a hyperbola

So far, the plane has only intersected one half of the double cone. If we tilt the plane beyond the angle at which the intersection curve is a parabola, the plane intersects both halves of the cone. In that case, we get a hyperbola. If we tilt the plane as far as possible so that it becomes parallel to the cone's axis, we still get a hyperbola (Fig. 13-1D).

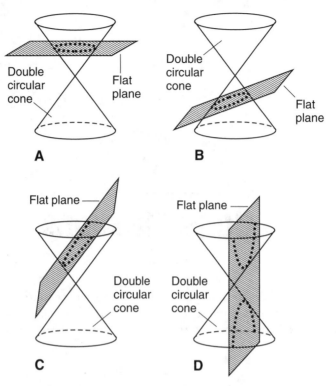

Figure 13-1   The conic sections can be defined by the intersection of a flat plane with a double right circular cone. At A, a circle. At B, an ellipse. At C, a parabola. At D, a hyperbola.

Are you confused?

You might ask, "We haven't mentioned the *flare angle* of the double cone (the measure of the angle between the axis of the cone and its surface). Does the size of this angle make any difference?" Quantitatively, it does. As the flare angle increases (the cones become "fatter"), we get ellipses less often and hyperbolas more often. As the flare angle decreases (the cones get "slimmer"), we obtain ellipses more often and hyperbolas less often. However, we can always get a circle, an ellipse, a parabola, or a hyperbola by manipulating the plane to the desired angle, regardless of the flare angle.

Here's a challenge!

Imagine that you're standing on a frozen lake at night, holding a flashlight that throws a cone-shaped beam with a flare angle of $\pi/10$; in other words, the outer face of the light cone subtends an angle of $\pi/10$ with respect to the beam center. How can you aim the flashlight so that the edge of the light cone forms a circle on the ice? An ellipse? A parabola?

Solution

The edge of the light cone is a circle if and only if the flashlight is pointed straight down, so the center of the beam is perpendicular to the surface of the ice (Fig. 13-2A). The edge of the region of light is an ellipse if and only if the beam axis subtends an angle of more than $\pi/10$ with respect to the ice, but the entire light cone still lands on the surface (Fig. 13-2B). The edge of the region of light is a parabola if and only if the beam axis subtends an angle of exactly $\pi/10$ with respect to the ice, so the top edge of the light cone is parallel to the surface (Fig. 13-2C).

Here's another challenge!

Imagine the scenario described above with the flashlight. How can you aim the flashlight so the edge of the light cone forms a half-hyperbola on the ice?

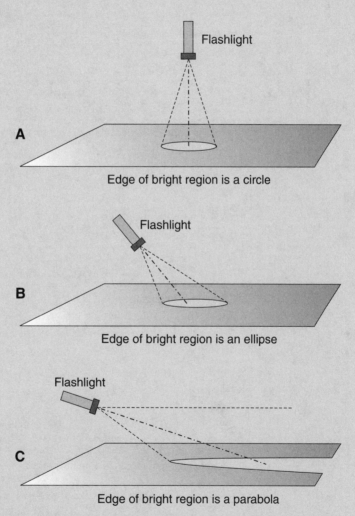

Figure 13-2   At A, the edge of the light cone creates a circle on the surface. At B, the edge of the light cone creates an ellipse on the surface. At C, the edge of the light cone creates a parabola on the surface. The dashed lines show the edges of the light cones. The dotted-and-dashed lines show the central axes of the light cones.

Solution

The edge of the region of light is a half-hyperbola if and only if one of the following conditions is met:

- The beam's central axis intersects the lake at an angle of less than $\pi/10$ with respect to the surface of the ice (Fig. 13-3A).
- The beam's central axis is aimed horizontally (Fig. 13-3B).
- The beam's central axis is aimed into the sky at an angle of less than $\pi/10$ above the horizon (Fig. 13-3C).

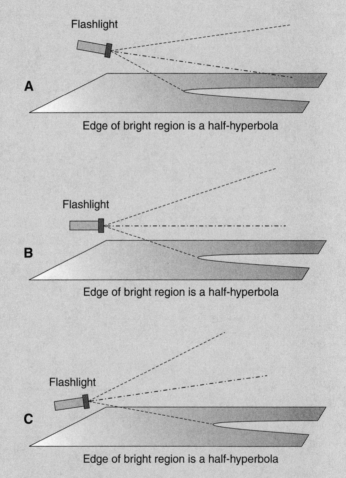

Figure 13-3   At A, B, and C, the edge of the light cone creates a half-hyperbola on the surface. The uppermost part of the light cone is above the horizon in all three cases. The dashed lines show the edges of the light cones. The dotted-and-dashed lines show the central axes of the light cones.

Basic Parameters

Figure 13-4 illustrates generic examples of a circle (at A), an ellipse (at B), and a parabola (at C) in the Cartesian xy plane. The circle and ellipse are closed curves, while the parabola is an open curve. In the circle, r is represents the radius. In the ellipse, a and b represent the *semi-axes*. The longer of the two is called the *major semi-axis*. The shorter of the two is called the *minor semi-axis*. In these examples, the circle and the ellipse are centered at the origin, and the parabola's *vertex* (the extreme point where the curvature is sharpest) is at the origin.

Specifications for a parabola

Suppose that we're traveling in a geometric plane along a course that has the contour of a parabola. At any given time, our location on the curve is defined by the ordered pair (x,y). To follow a parabolic path, we must always remain equidistant from a point called the *focus* and

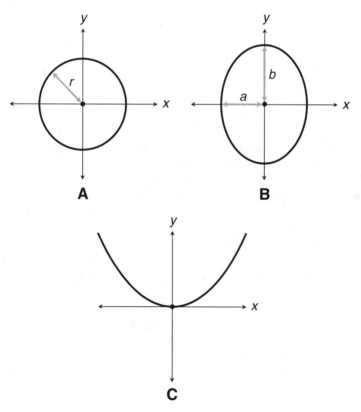

Figure 13-4 Three basic conic sections in the Cartesian xy plane. At A, a circle centered at the origin with radius r. At B, an ellipse centered at the origin with semi-axes a and b. At C, a parabola with the vertex at the origin.

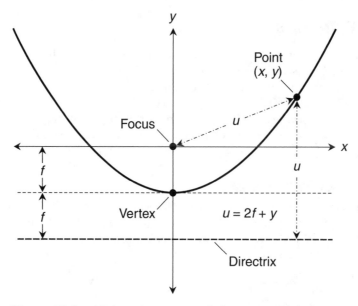

Figure 13-5 All the points on a parabola are at equal distances *u* from the focus and the directrix. The focus and the directrix are at equal distances *f* from the vertex of the curve.

a line called the *directrix* as shown in Fig. 13-5, where the focus and the directrix both lie in the same plane as the parabola. Let's call this distance *u*. In this illustration, the focus of the parabola is at the coordinate origin (0,0).

Now imagine a straight line passing through the focus and intersecting the directrix at a right angle. This line forms the *axis* of the parabola. In Fig. 13-5, the parabola's axis happens to coincide with the coordinate system's *y* axis. Along the axis line, the distance *u* is called the *focal length*, which mathematicians and scientists usually call *f*. (Be careful here! Don't confuse this *f* with the name of a relation or a function.) By drawing a line through the focus parallel to the directrix and perpendicular to the axis, we can divide *u*, measuring our distance from the directrix, into two line segments, one having length 2*f* and the other having length *y*. Therefore

$$u = 2f + y$$

The focus is at the point (0,0). Therefore, the distance *u* is the length of the hypotenuse of a right triangle whose base length is *x* and whose height is *y*. The Pythagorean theorem tells us that

$$x^2 + y^2 = u^2$$

If we divide the distance from the focus to point (*x,y*) on the curve by the distance from (*x,y*) to the directrix, we get a figure called the *eccentricity* of the curve. The eccentricity is symbolized *e*. (Don't confuse this with the exponential constant, which is also symbolized *e*.) In the case of a parabola, these distances are both equal to *u*, so

$$e = u/u = 1$$

Specifications for an ellipse and a circle

Suppose that we want to construct a curve in which the eccentricity is positive but less than 1. We can use a geometric arrangement similar to the one we used with the parabola, but the distance from the focus is *eu* instead of *u*, as shown in Fig. 13-6. In this situation we get an ellipse. The focus is at the origin (0,0). The ellipse has two vertices (points where the curvature is sharpest), both of which lie on the *y* axis, and the ellipse is taller than it is wide. When we draw an ellipse this way, its variables and parameters are related according to the equations

$$u = f + f/e + y$$

and

$$x^2 + y^2 = (eu)^2$$

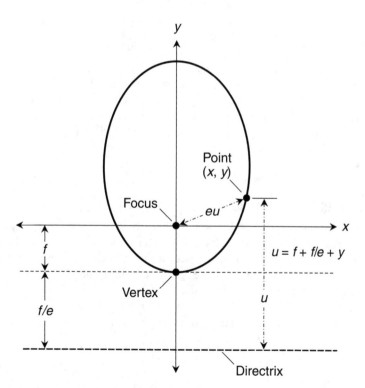

Figure 13-6 Construction of an ellipse based on a defined focus and directrix. The eccentricity *e* is an expression of the elongation of the ellipse.

As the eccentricity *e* approaches 0, the focus gets farther from the directrix, and the ellipse gets less elongated. When *e* reaches 0, then *f*/*e* becomes meaningless, the directrix vanishes ("runs away to infinity"), and we have a circle where *f* is equal to the radius *r*. A circle is actually an ellipse whose major and minor semi-axes are the same length. Going the other way, as the eccentricity *e* approaches 1, the focus gets closer to the directrix, and the ellipse gets more elongated. When *e* reaches 1, the ellipse "breaks open" at one end and becomes a parabola. Summarizing the above we can say

- For a circle, $e = 0$
- For an ellipse, $0 < e < 1$
- For a parabola, $e = 1$

The ellipse has another focus besides the one shown in Fig. 13-6. It's located at the same distance from the upper vertex of the curve as the coordinate origin is from the lower vertex. We can flip the ellipse in Fig. 13-6 upside-down, putting the upper focus in place of the lower focus and vice versa, and we'll get a diagram that looks exactly the same. The center of the ellipse is midway between the two foci.

How the foci, directrix, and eccentricity relate

Let's look at the circle, the ellipse, and the parabola in terms of the parameters we've just described. The circle has a single focus, which is at the center. The directrix is "at infinity." The ellipse has two foci separated by a finite distance. The curve is symmetrical with respect to a straight line that goes through the two foci. The curve is also symmetrical with respect to a straight line equidistant from the foci. The ellipse has two directrixes at finite distances from the vertices. We can think of a parabola as having two foci: one "within reach" and the other "at infinity." Its single directrix is at a distance from the vertex equal to the focal length.

There's an alternative way to define the eccentricity of an ellipse. Suppose we know the distance *d* between the foci, and we also know the length *s* of the major semi-axis. The eccentricity can be found by taking the ratio

$$e = d/(2s)$$

Specifications for a hyperbola

If we construct a conic section for which $e > 1$, we get a curve called a *hyperbola*. Figure 13-7 shows an example. The hyperbola looks like two parabolas back-to-back, but there's an important difference in the shape of a hyperbola compared with the shape of a double parabola. The parameters that help define hyperbolas are straight lines called *asymptotes*. Hyperbolas always have asymptotes, but parabolas never do.

In the scenario of Fig. 13-7, the hyperbola has two asymptotes that happen to pass through the origin. In this case, the equations of the asymptotes are

$$y = (b/a) \, x$$

and

$$y = -(b/a) \, x$$

The curve approaches the asymptotes as we move away from the center of the hyperbola, but the curves never quite reach the asymptotes, no matter how far from the center we go.

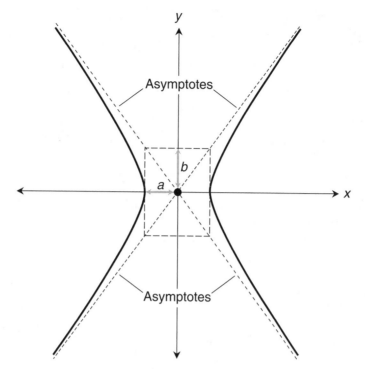

Figure 13-7   A basic hyperbola in the Cartesian *xy* plane. The
eccentricity is greater than 1. The distances *a* and *b*
are the semi-axes.

Are you confused?

You might ask, "Is it possible to have a conic section with negative eccentricity?" For our purposes
in this course, the answer is no. Negative eccentricity involves the notion of negative distances. If
we allow the eccentricity of a noncircular conic section to become negative, we get an "inside-out"
ellipse, parabola, or hyperbola. In ordinary geometry, such a curve is the same as a "real-world"
ellipse, parabola, or hyperbola.

That said, it's worth noting that in certain high-level engineering and physics applications, negative distances sometimes behave differently than positive distances. In those special situations, an
inside-out conic section might represent an entirely different phenomenon from a real-world conic
section. Keep that in the back of your mind if you plan on becoming an astronomer, cosmologist, or
high-energy physicist someday!

Here's a challenge!

Using the alternative formula for the eccentricity of an ellipse, show that if we have an ellipse in
which $e = 0$, then that ellipse is a circle.

Solution

First, we should realize that a circle is a special sort of ellipse in which the two semi-axes have identical length. With that in mind, let's plug $e = 0$ into the alternative equation for the eccentricity of an ellipse. That gives us

$$0 = d/(2s)$$

where d is the distance between the foci, and s is the length of the major semi-axis. We can multiply the above formula by $2s$ to obtain

$$0 = d$$

which tells us that the two foci are located at the same point, so there's really only one focus. A circle is the only type of ellipse that has a single focus.

Standard Equations

When we graph a conic section in the Cartesian xy plane, we can find a unique equation that represents that curve. These equations are always of the *second degree*, meaning that the equation must contain the square of one or both variables.

Equations for circles

We can write the standard-form general equation for a circle in terms of its center point and its radius as

$$(x - x_0)^2 + (y - y_0)^2 = r^2$$

where x_0 and y_0 are real constants that tell us the coordinates (x_0, y_0) of the center of the circle, and r is a positive real constant that tells us the radius (Fig. 13-8).

When a circle is centered at the origin, the equation is simpler because $x_0 = 0$ and $y_0 = 0$. Then we have

$$x^2 + y^2 = r^2$$

The simplest possible case is the *unit circle*, centered at the origin and having a radius equal to 1. Its equation is

$$x^2 + y^2 = 1$$

Equations for ellipses

The standard-form general equation of an ellipse in the Cartesian xy plane, as shown in Fig. 13-9, is

$$(x - x_0)^2/a^2 + (y - y_0)^2/b^2 = 1$$

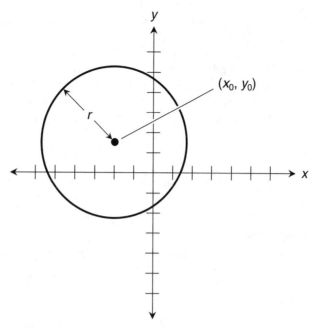

Figure 13-8   Graph of the circle for $(x - x_0)^2 + (y - y_0)^2 = r^2$.

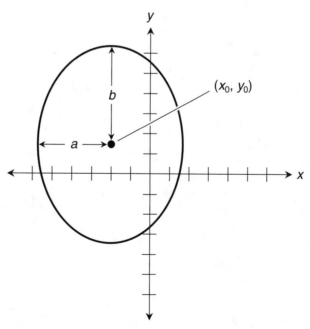

Figure 13-9   Graph of the ellipse for $(x - x_0)^2/a^2 + (y - y_0)^2/b^2 = 1$.

where x_0 and y_0 are real constants representing the coordinates (x_0,y_0) of the center of the ellipse, a is a positive real constant that represents the distance from (x_0,y_0) to the curve along a line parallel to the x axis, and b is a positive real constant that tells us the distance from (x_0,y_0) to the curve along a line parallel to the y axis. When we plot x on the horizontal axis and y on the vertical axis (the usual scheme), a is the length of the *horizontal semi-axis* or horizontal radius of the ellipse, and b is the length of the *vertical semi-axis* or vertical radius.

For ellipses centered at the origin, we have $x_0 = 0$ and $y_0 = 0$, so the general equation is

$$x^2/a^2 + y^2/b^2 = 1$$

If $a = b$, then the ellipse is a circle. Remember that a circle is an ellipse for which the eccentricity is 0.

Equations for parabolas

Figure 13-10 is an example of a parabola in the Cartesian xy plane. The standard-form general equation for this curve is

$$y = ax^2 + bx + c$$

The vertex is at the point (x_0,y_0). We can find these values according to the formulas

$$x_0 = -b/(2a)$$

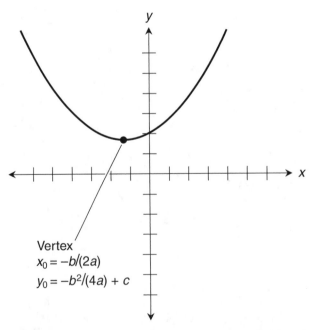

Vertex
$x_0 = -b/(2a)$
$y_0 = -b^2/(4a) + c$

Figure 13-10   Graph of the parabola for $y = ax^2 + bx + c$.

and

$$y_0 = ax_0^2 + bx_0 + c = -b^2/(4a) + c$$

If $a > 0$, the parabola opens upward, and the vertex represents the *absolute minimum* value of y. If $a < 0$, the parabola opens downward, and the vertex represents the *absolute maximum* value of y. In the graph of Fig. 13-10, the parabola opens upward, so we know that $a > 0$ in its equation.

Equations for hyperbolas

The standard-form general equation of a hyperbola in the Cartesian xy plane, as shown in Fig. 13-11, is

$$(x - x_0)^2/a^2 - (y - y_0)^2/b^2 = 1$$

where x_0 and y_0 are real constants that tell us the coordinates (x_0, y_0) of the center.

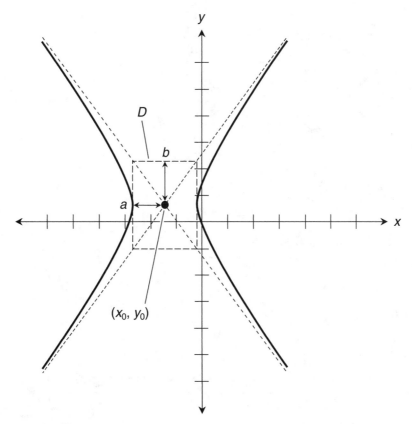

Figure 13-11 Graph of the hyperbola for $(x - x_0)^2/a^2 - (y - y_0)^2/b^2 = 1$.

The dimensions of a hyperbola are harder to define than the dimensions of a circle or an ellipse. Suppose that D is a rectangle whose center is at (x_0, y_0), whose vertical edges are tangent to the hyperbola, and whose corners lie on the asymptotes. When we define D this way, then a is the distance from (x_0, y_0) to D along a line parallel to the x axis, and b is the distance from (x_0, y_0) along a line parallel to the y axis. We call a the width of the horizontal semi-axis, and we call b the height of the vertical semi-axis.

For hyperbolas centered at the origin, we have $x_0 = 0$ and $y_0 = 0$, so the general equation becomes

$$x^2/a^2 - y^2/b^2 = 1$$

The simplest possible case is the *unit hyperbola* whose equation is

$$x^2 - y^2 = 1$$

- -

Are you astute?

You might imagine that the above-mentioned standard forms are not the only ways that the equations of conic sections can present themselves. If that's what you're thinking, you're right! However, you can always convert the equation of a conic section to its standard form. For example, suppose you encounter

$$49x^2 + 25y^2 = 1225$$

You say, "This looks like it might be the equation for an ellipse, but it's not in the standard form for any known conic section." Then you notice that 1225 is the product of 49 and 25. When you divide the whole equation through by 1225, you get

$$49x^2/1225 + 25y^2/1225 = 1225 / 1225$$

which simplifies to

$$x^2/25 + y^2/49 = 1$$

which can also be written as

$$x^2/5^2 + y^2/7^2 = 1$$

Now you know that the equation represents an ellipse centered at the origin whose horizontal semi-axis is 5 units wide, and whose vertical semi-axis is 7 units tall.

Here's a challenge!

Whenever we have an equation that can be reduced to the standard form

$$y = ax^2 + bx + c$$

we get a parabola that opens either upward or downward, and that represents a true function of x. How can we write the standard-form general equation of a parabola that opens to the right or the left? Does such a parabola represent a true function of x?

Solution

We can simply switch the variables to get

$$x = ay^2 + by + c$$

If $a > 0$, we have a parabola that opens to the right. If $a < 0$, we have a parabola that opens to the left. If we define x as the independent variable and y as the dependent variable as is usually done in Cartesian xy coordinates, then vertical-line tests reveal that these parabolas do not represent true functions of x.

Practice Exercises

This is an open-book quiz. You may (and should) refer to the text as you solve these problems. Don't hurry! You'll find worked-out answers in App. B. The solutions in the appendix may not represent the only way a problem can be figured out. If you think you can solve a particular problem in a quicker or better way than you see there, by all means try it!

1. At the beginning of this chapter, we learned that the intersection between a plane and a double right circular cone is always a circle, an ellipse, a parabola, or a hyperbola, as long as the plane doesn't pass through the point where the apexes of the two cones meet. What happens if the plane does pass through that point?

2. Figure 13-12 shows an ellipse in the Cartesian xy plane with some dimensions labeled. The lower focus is at the origin $(0,0)$. The lower vertex is at $(0,-2)$. Both foci and both vertices lie on the y axis. The ellipse is taller than it is wide. What is its eccentricity?

3. Recall the formulas relating the parameters of an ellipse when plotted in the manner of Fig. 13-12:

$$u = f + f/e + y$$

and

$$x^2 + y^2 = (eu)^2$$

Based on these formulas, the information provided in the figure, and the solution you worked out to Problem 2, determine a relation between x and y that describes our ellipse. The equation should include only the variables x and y, but it doesn't have to be in the standard form.

4. What are the coordinates of the upper vertex of the ellipse shown in Fig. 13-12? What are the coordinates of the upper focus of the ellipse shown in Fig. 13-12?

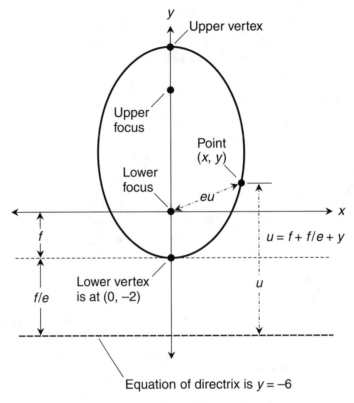

Figure 13-12   Illustration for Problems 2 through 5.

5. What are the coordinates of the center of the ellipse shown in Fig. 13-12? What is the length of the vertical semi-axis? What is the length of the horizontal semi-axis? Based on these results, write down the standard-form equation for the ellipse.

6. Determine the type of conic section the following equation represents, and then draw its graph:

$$x^2 + 9y^2 = 9$$

7. Determine the type of conic section the following equation represents, and then draw its graph:

$$x^2 + y^2 + 2x - 2y + 2 = 4$$

8. Determine the type of conic section the following equation represents, and then draw its graph:

$$x^2 - y^2 + 2x + 2y = 4$$

9. Determine the type of conic section the following equation represents, and then draw its graph:

$$x^2 - 3x - y + 3 = 1$$

10. Following is an equation in the standard form for a hyperbola:

$$(x - 1)^2/4 - (y + 2)^2/9 = 1$$

First, find the coordinates (x_0, y_0) of the center point. Then determine the length a of the horizontal semi-axis and the length b of the vertical semi-axis. Next, sketch a graph of the curve. Finally, work out the equations of the lines representing the asymptotes. Here's a hint: Use the point-slope form of the equation for a straight line in the xy plane. If it has slipped your memory, the general form is

$$y - y_0 = m(x - x_0)$$

where m is the slope of the line, and (x_0, y_0) represents a known point on the line.

Exponential and Logarithmic Curves

In your algebra courses, you learned about *exponential functions* and *logarithmic functions*. If you need a refresher, the basics are covered in Chap. 29 of *Algebra Know-It-All*. Let's look some graphs that involve these functions.

Graphs Involving Exponential Functions

An exponential function of a real variable x is the result of raising a positive real constant, called the *base*, to the xth power. The base is usually e (an irrational number called *Euler's constant* or the *exponential constant*) or 10. The value of e is approximately 2.71828.

Exponential: example 1

When we raise e to the xth power, we get the *natural exponential function* of x. When we raise 10 to the xth power, we get the *common exponential function* of x. Figure 14-1 shows graphs of these functions. At A, we see the graph of

$$y = e^x$$

over the portion of the domain between −2.5 and 2.5. At B, we see the graph of

$$y = 10^x$$

over the portion of the domain between −1 and 1. The curves have similar contours. When we "tailor" the axis scales in a certain relative way (as we do here), the two curves appear almost identical.

In the overall sense, both of these functions have domains that include all real numbers, because we can raise e or 10 to any real-number power and always get a real-number as the result. However, the ranges of both functions are confined to the set of positive reals. No matter what real-number exponent we attach to e or 10, we can never produce an output that's equal to 0, and we can never get an output that's negative.

Figure 14-1 Graphs of the natural exponential function (at A) and the common exponential function (at B).

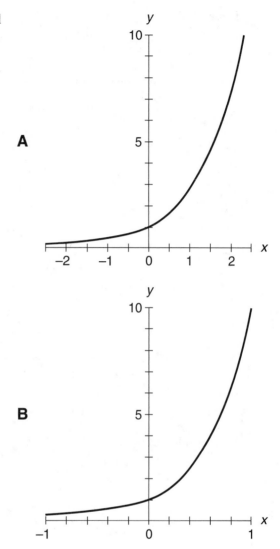

Exponential: example 2

Let's see what happens to the graphs of the foregoing functions when we take their reciprocals and then graph them over the same portions of their domains as we did before. Figure 14-2A is a graph of

$$y = 1/e^x$$

Figure 14-2B is a graph of

$$y = 1/10^x$$

Figure 14-2   Graphs of the reciprocals of the natural exponential (at A) and the common exponential (at B).

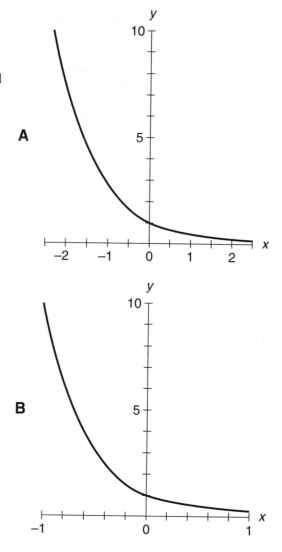

These curves are exactly reversed left-to-right from those in Fig. 14-1. The above reciprocal functions can be rewritten, respectively, as

$$y = e^{-x}$$

and

$$y = 10^{-x}$$

When we negate x before taking the power of the exponential base, we "horizontally mirror" all of the function values. The y axis acts as a "point reflector." The overall domains and ranges of these reciprocal functions are the same as the domains and ranges of the original functions.

Exponential: example 3

Now that we have two pairs of exponential functions, let's create two new functions by adding them, and see what their graphs look like. The solid black curve in Fig. 14-3A is a graph of

$$y = e^x + 1/e^x = e^x + e^{-x}$$

The solid black curve in Fig. 14-3B is a graph of

$$y = 10^x + 1/10^x = 10^x + 10^{-x}$$

The domains of these sum functions both encompass all the real numbers. The ranges are limited to the reals greater than or equal to 2.

Figure 14-3   Graphs of the natural exponential plus its reciprocal (solid black curve at A) and the common exponential plus its reciprocal (solid black curve at B). The dashed gray curves are the graphs of the original functions.

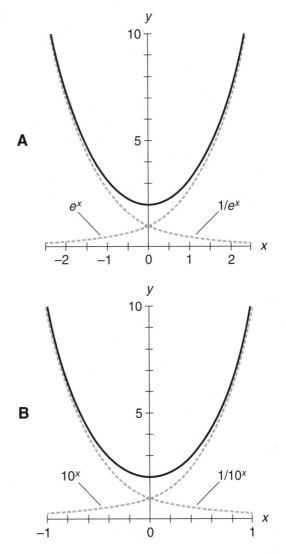

Here's a "heads up"!

In many of the graphs to come, you'll see two dashed gray curves representing functions to be combined in various ways, as is the case in Fig. 14-3. But the constituent functions won't be labeled as they are in Fig. 14-3. The absence of labels will keep the graphs from getting too cluttered, so you'll be able to see clearly how they relate. Also, the lack of labels will force you to think! Based on your knowledge of the way the functions behave, you should be able to tell which graph is which without having them labeled.

Exponential: example 4

Figure 14-4 shows what happens when we subtract the reciprocal of the natural exponential function from the original function and then graph the result. The solid black curve is the graph of

$$y = e^x - 1/e^x = e^x - e^{-x}$$

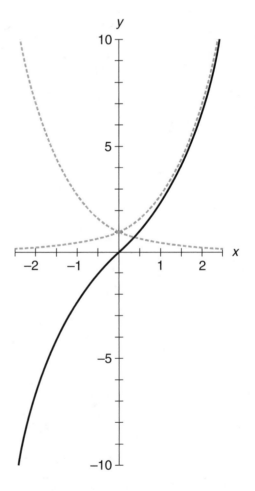

Figure 14-4 Graph of the natural exponential minus its reciprocal (solid black curve). The dashed gray curves are the graphs of the original functions.

Figure 14-5 Graph of the common exponential minus its reciprocal (solid black curve). The dashed gray curves are the graphs of the original functions.

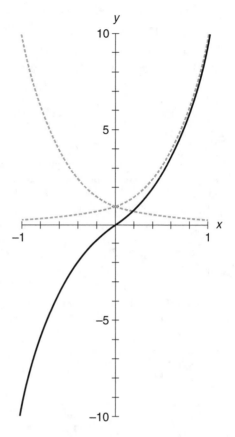

In Fig. 14-5, we do the same thing with the common exponential function and its reciprocal. Here, the solid black curve represents

$$y = 10^x - 1/10^x = 10^x - 10^{-x}$$

In both of these figures, the dashed gray curves represent the original functions. The domains and ranges of both difference functions include all real numbers.

Are you confused?

Do you wonder how we arrived at the graphs in Figs. 14-3 through 14-5? We can plot sum and difference functions in two ways. We can graph the original functions separately, and then add or subtract their values graphically (that is, geometrically) by moving vertically upward or downward at various points within the spans of the domains shown. Alternatively, we can, with the help of a calculator, plot several points for each sum or difference function after calculating the outputs for several different input values. Once we have enough points for the sum or difference function, we can draw an approximation of the graph for that function directly.

Here's a challenge!

Plot a graph of the function we get when we raise *e* to the power of 1/*x*. In rectangular *xy* coordinates, the curve is represented by the equation

$$y = e^{(1/x)}$$

What is the domain of this function? What is its range?

Solution

We can use a calculator to determine the values of *y* for various values of *x*. Figure 14-6 is the resulting graph for values of *x* ranging from −10 to 10. When we input *x* = 0, we get $e^{1/0}$, which is undefined. For any other real value of *x*, the output value *y* is a positive real number. Therefore, the domain of this function is the set of all nonzero reals. No matter how large we want *y* to be when *y* > 1, we can always find some value of *x* that will give it to us. Similarly, no matter how small we want *y* to be when 0 < *y* < 1, we can always find some value of *x* that will do the job. However, we can't find any value for *x* that will give us *y* = 1. For that to happen, we must raise *e* to the 0th power, meaning that we must find some *x* such that 1/*x* = 0. That's impossible! Therefore, the range of our function is the set of all positive reals except 1. The graph has a horizontal asymptote whose equation is *y* = 1, and a vertical asymptote corresponding to the *y* axis. The open circle at the point (0,0) indicates that it's not part of the graph.

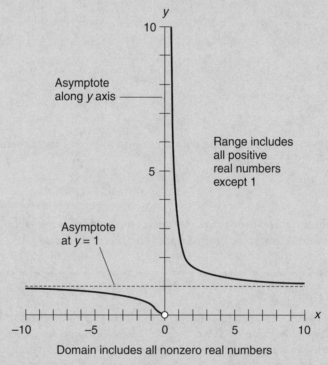

Figure 14-6   Graph of the function $y = e^{(1/x)}$. Note the "hole" in the domain at *x* = 0 and the "hole" in the range at *y* = 1.

Graphs Involving Logarithmic Functions

A logarithm (sometimes called a *log*) of a quantity is a power to which a positive real constant is raised to get that quantity. As with exponential functions, the constant is called the base, and it's almost always equal to either e or 10. The base-e log function, also called the *natural logarithm*, is usually symbolized by writing "ln" or "$\log_e$" followed by the argument (the quantity on which the function operates). The base-10 log function, also called the *common logarithm*, is usually symbolized by writing "$\log_{10}$" or "log" followed by the argument.

They're inverses!

A logarithmic function is the inverse of the exponential function having the same base. The natural logarithmic function "undoes" the work of the natural exponential function and vice versa, as long as we restrict the domains and ranges so that both functions are bijections. The common logarithmic and exponential functions also behave this way, so we can say that

$$\ln e^x = x = e^{(\ln x)}$$

and

$$\log 10^x = x = 10^{(\log x)}$$

For these formulas to work, we must restrict x to positive real-number values, because the logarithms of quantities less than or equal to 0 are not defined.

Logarithm: example 1

Figure 14-7 illustrates graphs of the two basic logarithmic functions operating on a variable x. At A, we see the graph of the base-e logarithmic function, over the portion of the domain from 0 to 10. The equation is

$$y = \ln x$$

At B in Fig. 14-7, we see the graph of the base-10 logarithmic function, over the portion of the domain from 0 to 10. The equation is

$$y = \log_{10} x$$

As with the exponential graphs, these curves have similar contours, and they look almost identical if we choose the axis scales as we've done here.

The domains of the natural and common log functions both span the entire set of positive reals. When we try to take a logarithm of 0 or a negative number, however, we get a meaningless quantity (or, at least, something outside the set of reals!). By inputting just the right positive real value to a log function, we can get any real-number output we want. The ranges of the log functions therefore include all real numbers.

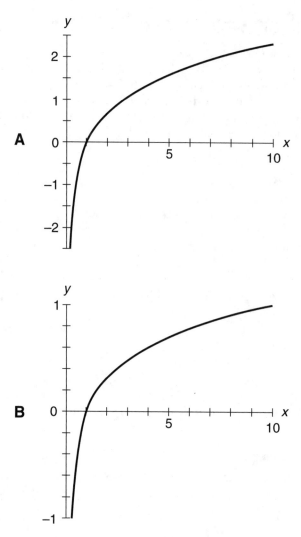

Figure 14-7   Graphs of the natural logarithmic function (at A) and the common logarithmic function (at B).

Logarithm: example 2

Let's take the reciprocal of the independent variable *x* before performing the natural or common log, and then plot the graphs. Figure 14-8 shows the results. At A, we see the graph of the function

$$y = \ln (1/x)$$

and at B, we see the graph of the function

$$y = \log_{10} (1/x)$$

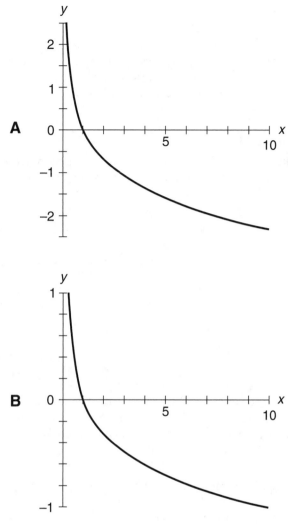

Figure 14-8 Graphs of the natural log of the reciprocal (at A) and the common log of the reciprocal (at B).

These functions can also be written as

$$y = \ln(x^{-1})$$

and

$$y = \log_{10}(x^{-1})$$

Based on our knowledge of logarithms from algebra, we can rewrite these functions, respectively, as

$$y = -1 \ln x = -\ln x$$

and

$$y = -1 \log_{10} x = -\log_{10} x$$

When we raise x to the -1 power before taking the logarithm, we negate all the function values, compared to what they'd be if we left x alone. The x axis acts as a point reflector. The domains and ranges of these reciprocal functions are the same as the domains and ranges of the original functions.

Logarithm: example 3

We can create two interesting functions by multiplying the functions defined in the previous two paragraphs. Let's do that, and see what the graphs look like. We want to graph the functions

$$y = (\ln x) [\ln (x^{-1})]$$

and

$$y = (\log_{10} x) [\log_{10} (x^{-1})]$$

Our knowledge of logarithms allows us to rewrite these functions, respectively, as

$$y = -(\ln x)^2$$

and

$$y = -(\log_{10} x)^2$$

The results are shown in Figs. 14-9 and 14-10. The domains of both product functions span the entire set of positive reals. The ranges of both functions are confined to the set of *nonpositive reals* (that is, the set of all reals less than or equal to 0).

Figure 14-9 Graph of the natural log times the natural log of the reciprocal (solid black curve). The dashed gray curves are the graphs of the original functions.

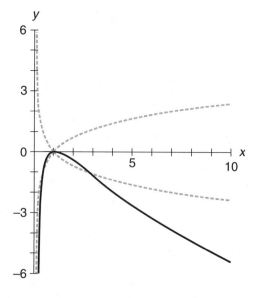

Figure 14-10 Graph of the common log times the common log of the reciprocal (solid black curve). The dashed gray curves are the graphs of the original functions.

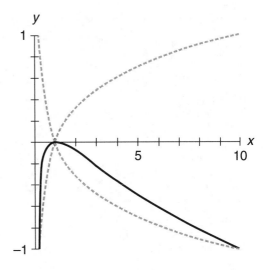

Logarithm: example 4

Finally, let's take the log functions we've been working with and find their ratios, as follows:

$$y = (\ln x) / [\ln (x^{-1})]$$

and

$$y = (\log_{10} x) / [\log_{10} (x^{-1})]$$

Our knowledge of logarithms allows us to simplify these, respectively, to

$$y = (\ln x) / (-\ln x) = -1$$

and

$$y = (\log_{10} x) / (-\log_{10} x) = -1$$

These functions are defined only if $0 < x < 1$ or $x > 1$. The domains have "holes" at $x = 1$ because when we input 1 to either quotient, we end up dividing by 0. (Try it and see!) The ranges are confined to the single value -1.

Don't let them confuse you!

In some texts, natural (base-*e*) logs are denoted by writing "log" without a subscript, followed by the argument. In other texts and in most calculators, "log" means the common (base-10) log.

To avoid confusion, you should include the base as a subscript whenever you write "log" followed by anything. For example, write "$\log_{10}$" or "$\log_e$," instead of "log" all by itself, unless it's impractical to write the subscript. You don't need a subscript when you write "ln" for the natural log.

If you aren't sure what the "log" key on a calculator does, you can do a test to find out. If your calculator says that the "log" of 10 is equal to 1, then it's the common log. If the "log" of 10 turns out to be an irrational number slightly larger than 2.3, then it's the natural log.

Here's a challenge!

Draw graphs of the ratio functions we found in "Logarithm: example 4." Be careful! They're a little tricky.

Solution

Figure 14-11 is a graph of the function

$$y = (\ln x) / [\ln (x^{-1})]$$

Figure 14-11   Graph of the natural log divided by the natural log of the reciprocal (solid black line with "holes"). The dashed gray curves are the graphs of the original functions.

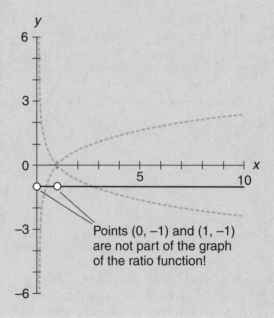

Points (0, −1) and (1, −1) are not part of the graph of the ratio function!

Figure 14-12 Graph of the common log divided by the common log of the reciprocal (solid black line with "holes"). The dashed gray curves are the graphs of the original functions.

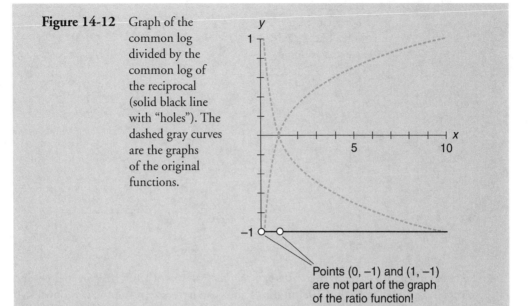

Points (0, –1) and (1, –1) are not part of the graph of the ratio function!

Figure 14-12 is a graph of the function

$$y = (\log_{10} x) \,/\, [\log_{10} (x^{-1})]$$

In both graphs, the original numerator and denominator functions are graphed as dashed gray curves. The ratio functions are graphed as solid black lines with "holes." The small open circles at the points (0,–1) and (1,–1) indicate that those points are *not* part of either graph. That's the trick I warned you about. Without the open circles, these graphs would be wrong.

Logarithmic Coordinate Planes

Engineers and scientists sometimes use coordinate systems in which one or both axes are graduated according to the common (base-10) logarithm of the displacement. Let's look at the three most common variants.

Semilog (*x*-linear) coordinates

Figure 14-13 shows *semilogarithmic* (*semilog*) *coordinates* in which the independent-variable axis is linear, and the dependent-variable axis is logarithmic. The values that can be depicted on the *y* axis are restricted to one sign or the other (positive or negative). The graphable intervals in this example are

$$-1 \leq x \leq 1$$

and

$$0.1 \leq y \leq 10$$

Figure 14-13 The semilog coordinate plane with a linear x axis and a logarithmic y axis.

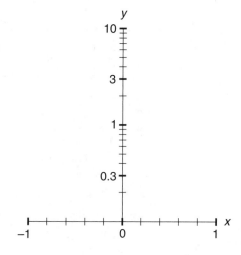

The y axis in Fig. 14-13 spans two *orders of magnitude* (powers of 10). The span could be increased to encompass more powers of 10, but the y values can never extend all the way down to 0.

Semilog (y-linear) coordinates

Figure 14-14 shows semilog coordinates in which the independent-variable axis is logarithmic, and the dependent-variable axis is linear. The values that can be depicted on the x axis are restricted to one sign or the other (positive or negative). The graphable intervals in this illustration are

$$0.1 \leq x \leq 10$$

and

$$-1 \leq y \leq 1$$

The x axis in Fig. 14-14 spans two orders of magnitude. The span could cover more powers of 10, but in any case the x values can't extend all the way down to 0.

Log-log coordinates

Figure 14-15 shows *log-log coordinates*. Both axes are logarithmic. The values that can be depicted on either axis are restricted to one sign or the other (positive or negative). In this example, the graphable intervals are

$$0.1 \leq x \leq 10$$

and

$$0.1 \leq y \leq 10$$

Both axes in Fig. 14-15 span two orders of magnitude. The span of either axis could cover more powers of 10, but neither axis can be made to show values down to 0.

Figure 14-14 The semilog coordinate plane with a logarithmic *x* axis and a linear *y* axis.

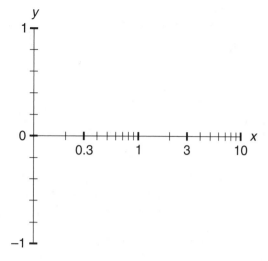

Figure 14-15 The log-log coordinate plane. The *x* and *y* axes are both logarithmic.

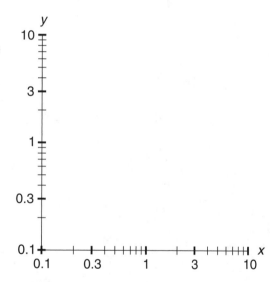

Are you confused?

Semilog and log-log coordinates distort the graphs of relations and functions because the axes aren't linear. Straight lines in Cartesian or rectangular coordinates usually show up as curves in semilog or log-log coordinates. Some functions whose graphs appear as curves in Cartesian or rectangular coordinates turn out to be straight lines in semilog or log-log coordinates. Try plotting some linear, logarithmic, and exponential functions in Cartesian, semilog, and log-log coordinates. See for yourself what happens! Use a calculator, plot numerous points, and then "connect the dots" for each function you want to graph.

Here's a challenge!

Plot graphs of each of the following three functions in *x*-linear semilog coordinates, *y*-linear semilog coordinates, and log-log coordinates (use the templates from Figs. 14-13 through 14-15):

$$y = x$$

$$y = \ln x$$

$$y = e^x$$

Solution

Use a scientific calculator and input various values of *x*. Plot several points for each function and then draw curves through them, interpolating as you go. Be sure that your calculator is set for the natural logarithmic and exponential functions (that is, base *e*), *not* common logarithm or common exponential functions (base 10). You should get graphs that look like those shown in Fig. 14-16. In two cases, only a single point of the function shows up in the coordinate spans portrayed here. You'd have to expand the linear axis (the *x* axis) at A beyond 1 to see any of the graph for $y = \ln x$. You'd have to expand the linear axis (the *y* axis) at B beyond 1 to see any of the graph for $y = e^x$.

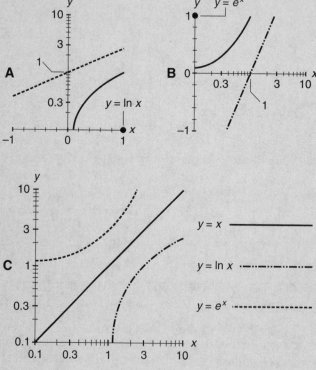

Figure 14-16 Simple functions in *x*-linear semilog
coordinates (at A), *y*-linear semilog coordinates
(at B), and log-log coordinates (at C).

Practice Exercises

This is an open-book quiz. You may (and should) refer to the text as you solve these problems. Don't hurry! You'll find worked-out answers in App. B. The solutions in the appendix may not represent the only way a problem can be figured out. If you think you can solve a particular problem in a quicker or better way than you see there, by all means try it! When plotting graphs here, feel free to use a calculator, locate numerous points, and "connect the dots."

1. When we discussed the range of the natural exponential function, we claimed that no real-number power of e is equal to 0. Prove it. Here's a hint: Use the technique of *reductio ad absurdum*, in which we assume the truth of a statement and then derive something obviously false or contradictory from that assumption.

2. Set up a rectangular coordinate system like the one in Fig. 14-1A, where the values of x are portrayed from -2.5 to 2.5, and the values of y are portrayed from 0 to 10. Sketch the graphs of the following functions on this coordinate grid:

$$y = e^x$$
$$y = e^{-x}$$
$$y = e^x e^{-x}$$

3. Set up a rectangular coordinate system like the one in Fig. 14-1B, where the values of x are portrayed from -1 to 1, and the values of y are portrayed from 0 to 10. Sketch the graphs of the following functions on this coordinate grid:

$$y = 10^x$$
$$y = 10^{-x}$$
$$y = 10^x/10^{-x}$$

4. Draw the graphs of the three functions from Problem 3 on an x-linear semilog coordinate grid, where the values of x are portrayed from -1 to 1, and the values of y are portrayed over the three orders of magnitude from 0.1 to 100.

5. Plot a rectangular-coordinate graph of the function we get when we raise 10 to the power of $1/x$. The curve is represented by the following equation:

$$y = 10^{(1/x)}$$

What is the domain of this function? What is its range? Include all values of the domain from -10 to 10.

6. We've claimed that the natural log of 0 isn't a real number. Prove it. Here's a hint: Use *reductio ad absurdum*, and use the solution to Problem 1 as a *lemma* (a theorem that helps in the proof of another theorem).

7. Plot a rectangular-coordinate graph of the sum of the natural log function and the common log function. The curve is represented by the following equation:

$$y = \ln x + \log_{10} x$$

What is the domain of this function? What is its range? Include all values of the domain from 0 to 10. Include all values of the range from -5 to 5.

8. Plot a rectangular-coordinate graph of the product of the natural log function and the common log function. The curve is represented by the following equation:

$$y = (\ln x)(\log_{10} x)$$

What is the domain of this function? What is its range? Include all values of the domain from 0 to 10, and all values of the range from 0 to 5.

9. Draw the graphs of the three functions from Fig. B-18 (the illustration for the solution to Problem 7) on a y-linear semilog coordinate grid. Portray values of x over the two orders of magnitude from 0.1 to 10. Portray values of y from -5 to 5.

10. Draw the graphs of the three functions from Fig. B-19 (the illustration for the solution to Problem 8) on a y-linear semilog coordinate grid. Portray values of x over the single order of magnitude from 1 to 10. Portray values of y from 0 to 2.5.

Trigonometric Curves

If you've taken basic algebra and geometry, you're familiar with the trigonometric functions. You also got some experience with them in Chap. 2 of this book. Now we'll graph some algebraic combinations of these functions.

Graphs Involving the Sine and Cosine

Let's find out what happens when we add, multiply, square, and divide the sine and cosine functions.

Sine and cosine: example 1

Figure 15-1 shows superimposed graphs of the sine and cosine functions along with a graph of their sum. You can follow along by inputting numerous values of θ into your calculator, determining the output values, plotting the points corresponding to the input/output ordered pairs, and then filling in the curve by "connecting the dots." In Fig. 15-1, the dashed gray curves are the individual sine and cosine waves. The solid black curve is the graph of the sum function:

$$f(\theta) = \sin \theta + \cos \theta$$

The sum-function wave has the same *period* (distance between the corresponding points on any two adjacent waves) as the sine and cosine waves. In this situation, that period is 2π. The new wave also has the same *frequency* as the originals. The frequency of any regular, repeating wave is always equal to the reciprocal of its period.

The *peaks* (recurring maxima and minima) of the sine and cosine waves attain values of ± 1. The peaks of the new wave attain values of $\pm 2^{1/2}$, which occur at values of θ where the graphs of the sine and cosine cross each other. By definition, the *peak amplitude* of the new function is $2^{1/2}$ times the peak amplitude of either original function. The new wave appears "sine-like," but we can't be sure that it's a true sinusoid on the basis of its appearance in this graph. The domain of our function f includes all real numbers. The range of f is the set of all reals in the closed interval $[-2^{1/2}, 2^{1/2}]$.

Figure 15-1   Graphs of the sine and cosine functions (dashed gray curves) and the graph of their sum (solid black curve). Each division on the horizontal axis represents $\pi/2$ units. Each division on the vertical axis represents 1/2 unit.

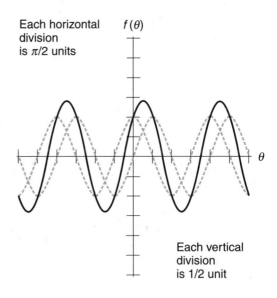

Each horizontal division is $\pi/2$ units

$f(\theta)$

θ

Each vertical division is 1/2 unit

Sine and cosine: example 2

In Fig. 15-2, we see graphs of the sine and cosine functions along with a graph of their product. The dashed gray curves are the superimposed sine and cosine waves. The solid black curve is the graph of the product function:

$$f(\theta) = \sin \theta \cos \theta$$

The new function's graph has a period of π, which is half the period of the sine wave, and half the period of the cosine wave. The peaks of the new wave are $\pm 1/2$, which occur at values of θ where the graphs of the sine and cosine intersect. As in the previous example, the new wave looks like a sinusoid, but we can't be sure about that by merely looking at it. The domain of the product function spans the entire set of reals. The range is the set of all reals in the closed interval $[-1/2, 1/2]$.

Sine and cosine: example 3

Figure 15-3 shows the graphs of the sine function (at A) and the cosine function (at B) along with their squares. The dashed gray curve at A is the sine wave; the dashed gray curve at B is the cosine wave. In illustration A, the solid black curve is the graph of

$$f(\theta) = \sin^2 \theta$$

In illustration B, the solid black curve is the graph of

$$g(\theta) = \cos^2 \theta$$

Figure 15-2 Graphs of the sine and cosine functions (dashed gray curves) along with the graph of their product (solid black curve). Each horizontal division represents $\pi/2$ units. Each vertical division represents 1/4 unit.

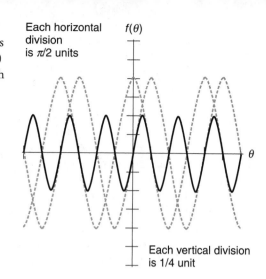

Each horizontal division is $\pi/2$ units

$f(\theta)$

θ

Each vertical division is 1/4 unit

Figure 15-3 The dashed gray curves are graphs of the sine function (at A) and the cosine function (at B). The solid black curves are graphs of the square of the sine function (at A) and the square of the cosine function (at B). Each division on the horizontal axes represents $\pi/2$ units. Each division on the vertical axes represents 1/4 unit.

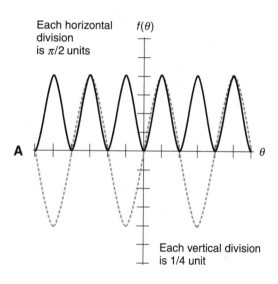

Each horizontal division is $\pi/2$ units

$f(\theta)$

A

θ

Each vertical division is 1/4 unit

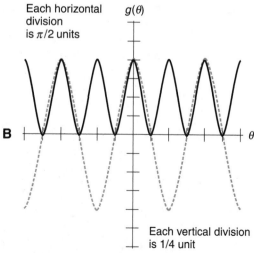

Each horizontal division is $\pi/2$ units

$g(\theta)$

B

θ

Each vertical division is 1/4 unit

The squared-function waves have periods of π, half the periods of the original functions. Therefore, the frequencies of the squared functions are twice those of the original functions.

The waves for the squared functions are displaced upward relative to the waves for the original functions. The squared functions attain repeated minima of 0 and repeated maxima of 1. In other words, the positive peak amplitudes of the squared-function waves are equal to 1, while the minimum peak amplitudes are equal to 0. We define the *peak-to-peak amplitude* of a regular, repeating wave as the difference between the positive peak value and the negative peak value. In this example, the original waves have peak-to-peak amplitudes of $1 - (-1) = 2$, while the squared-function waves have peak-to-peak amplitudes of $1 - 0 = 1$.

The waves representing the squared functions f and g look like sinusoids, but we can't be certain about that on the basis of their appearance alone. The domains of f and g include all real numbers. The ranges of f and g are confined to the set of all reals in the closed interval $[0,1]$.

Sine and cosine: example 4

Let's add the squared functions from the previous example and graph the result. The solid black line in Fig. 15-4 is a graph of the sum of the squares of the sine and the cosine functions, which are shown as superimposed dashed gray curves. We have

$$f(\theta) = \sin^2 \theta + \cos^2 \theta$$

In this case, the function has a constant value. The domain includes all of the real numbers. The range is the set containing the single real number 1.

Figure 15-4   Graph of the sum of the squares of the sine and cosine functions (solid black line). The dashed gray curves are the graphs of the original squared functions. Each horizontal division represents $\pi/2$ units. Each vertical division represents 1/4 unit.

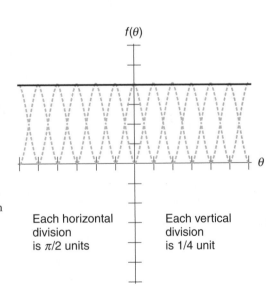

$f(\theta)$

θ

Each horizontal division is $\pi/2$ units

Each vertical division is 1/4 unit

Are you confused?

You might wonder, "How can we be certain that the graph in the previous example is actually a straight, horizontal line? When I input values into my calculator, I always get an output of 1, but I've learned that even a million examples can't prove a general truth in mathematics." Your skepticism shows that you're thinking! But let's remember one of the basic *trigonometric identities* that we learned in Chap. 2. For all real numbers θ, the following equation holds true:

$$\sin^2 \theta + \cos^2 \theta = 1$$

This fact assures us that in the previous example, we have

$$f(\theta) = 1$$

so the graph of the sum-of-squares function is indeed the horizontal, solid black line portrayed in Fig. 15-4.

Here's a challenge!

Sketch a graph of the ratio of the square of the sine function to the square of the cosine function. That is, graph

$$f(\theta) = (\sin^2 \theta)/(\cos^2 \theta)$$

Solution

The solid black complex of curves in Fig. 15-5 is the graph of the ratio of the square of the sine to the square of the cosine. The superimposed gray curves are graphs of the original sine-squared and cosine-squared functions.

Figure 15-5   Graph of the ratio of the square of the sine function to the square of the cosine function (solid black curves). Each horizontal division represents $\pi/2$ units. Each vertical division represents 1/2 unit. The dashed gray curves are the graphs of the original sine-squared and cosine-squared functions. The vertical dashed lines are asymptotes of f.

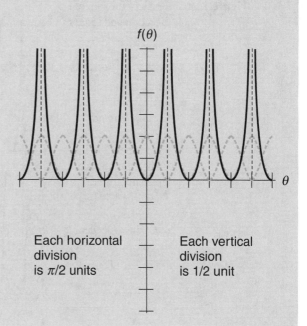

$f(\theta)$

θ

Each horizontal division is $\pi/2$ units

Each vertical division is 1/2 unit

This ratio function f is singular (that is, it "blows up") when θ is any odd-integer multiple of $\pi/2$. That's because $\cos^2 \theta$ (the denominator) equals 0 at those points, while $\sin^2 \theta$ (the numerator) equals 1. The function attains values of 0 at all integer multiples of π because at those points, $\sin^2 \theta$ (the numerator) equals 0, while $\cos^2 \theta$ (the denominator) equals 1.

The period of f is π, the distance between the asymptotes; the graph repeats itself completely between each adjacent pair of asymptotes. The peak amplitude and the peak-to-peak amplitude are both undefined. (It's tempting to call them "infinite," but let's not go there!) The domain includes all reals except the odd-integer multiples of $\pi/2$. The range is the set of all nonnegative reals.

Graphs Involving the Secant and Cosecant

In Chap. 2, we saw graphs of the basic secant and cosecant functions, which are the reciprocals of the cosine and sine, respectively. Let's combine these two functions after the fashion of the previous section, and see what the resulting graphs look like.

Secant and cosecant: example 1

The dashed gray curves in Fig. 15-6 are the superimposed graphs of the secant and cosecant functions. The complex of solid black curves is a graph of their sum. As always, you can reproduce this graph by inputting a sufficient number of values into your calculator, plotting the

Figure 15-6   Graph of the sum of the secant and cosecant functions (solid black curves). The dashed gray curves are the graphs of the original functions. Each division on the horizontal axis represents $\pi/2$ units. Each vertical division represents 1 unit. The vertical dashed lines are asymptotes of f. The dependent-variable axis is also an asymptote of f.

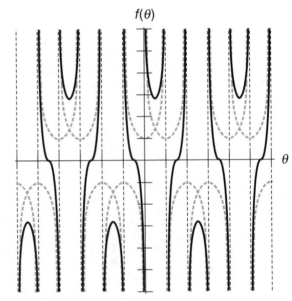

Each horizontal division is $\pi/2$ units
Each vertical division is 1 unit

output points, and then "connecting the dots." You'll have to take some time to "investigate" this function before you can accurately plot this graph, but be patient! We have

$$f(\theta) = \sec\theta + \csc\theta$$

The graph of f has asymptotes that pass through every point where the independent variable is an integer multiple of $\pi/2$. If you examine Fig. 15-6 closely, you'll see that the graph is regular and it repeats with a period of 2π, but we can't call it a wave. The domain includes all real numbers except the integer multiples of $\pi/2$ because, whenever θ attains one of those values, either the secant or the cosecant is undefined. The range spans the set of all real numbers.

Secant and cosecant: example 2

Figure 15-7 shows graphs of the secant and cosecant functions along with their product. The dashed gray curves are graphs of the original functions superimposed on each other; the solid black curves show the graph of

$$f(\theta) = \sec\theta\csc\theta$$

This function f has a period of π, which is half that of the secant and cosecant functions. Like the sum-function graph, this graph has asymptotes that pass through every point where the independent variable is an integer multiple of $\pi/2$. The domain is the set of all reals except the integer multiples of $\pi/2$. The range spans the set of all real numbers except those in the open interval $(-2,2)$. Alternatively, we can say that the range includes all reals y such that $y \geq 2$ or $y \leq -2$.

Figure 15-7 Graph of the product of the secant and cosecant functions (solid black curves). The dashed gray curves are the graphs of the original functions. Each division on the horizontal axis represents $\pi/2$ units. Each vertical division represents 1 unit. The vertical dashed lines are asymptotes of f. The dependent-variable axis is also an asymptote of f.

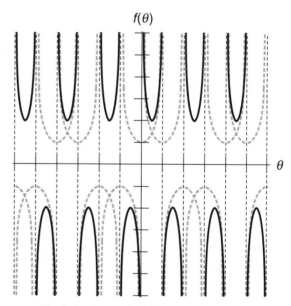

Each horizontal division is $\pi/2$ units

Each vertical division is 1 unit

Secant and cosecant: example 3

The dashed gray curves in Fig. 15-8 are the graphs of the secant function (at A) and the cosecant function (at B). At A, the solid black curve is the graph of

$$f(\theta) = \sec^2 \theta$$

At B, the solid black curve is the graph of

$$g(\theta) = \csc^2 \theta$$

Figure 15-8 The solid black curves are the graphs of the squares of the secant function (at A) and the cosecant function (at B). The dashed gray curves are the graphs of the original functions. Each division on the horizontal axes represents $\pi/2$ units. Each division on the vertical axes represents 1 unit. The vertical dashed lines are asymptotes of f and g. At B, the dependent-variable axis is also an asymptote of g.

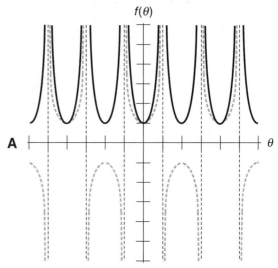

Each horizontal division is $\pi/2$ units
Each vertical division is 1 unit

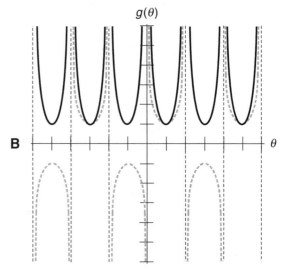

The squared functions have periods of π, which are half the periods of the original functions. Therefore, the frequencies of the squared functions are double those of the originals. Singularities occur at the same points on the independent-variable axes as they do for the original functions. The domain of the secant-squared function is the set of all reals except odd-integer multiples of $\pi/2$. The domain of the cosecant-squared function is the set of all reals except integer multiples of π. The ranges in both cases are confined to the set of reals y such that $y \geq 1$.

Secant and cosecant: example 4

Figure 15-9 shows what happens when we add the secant-squared function to the cosecant-squared function. The solid black curves compose the graph of

$$f(\theta) = \sec^2 \theta + \csc^2 \theta$$

The dashed gray curves are superimposed graphs of the original functions. This sum function has a period equal to half that of the original functions, or $\pi/2$. The domain includes all reals except the integer multiples of $\pi/2$. The range is the set of reals y such that $y \geq 4$.

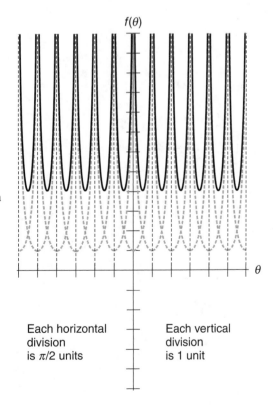

Figure 15-9   Graph of the sum of the squares of the secant and cosecant functions (solid black curves). The dashed gray curves are the graphs of the original squared functions. Each horizontal division represents $\pi/2$ units. Each vertical division represents 1 unit. The vertical dashed lines are asymptotes of f. The positive dependent-variable axis is also an asymptote of f.

Each horizontal division is $\pi/2$ units

Each vertical division is 1 unit

- -

Are you confused?

You're bound to wonder, "How do we know that the range of the sum-of-squares function in the previous example is the set of all reals greater than or equal to 4?" Another way of stating this fact is that the minima of the solid black curves in Fig. 15-9 have dependent-variable values equal to 4. These minima occur at values of θ where the graphs of the secant-squared and cosecant-squared functions (dashed gray curves) intersect. Every one of those points occurs where θ is an odd-integer multiple of $\pi/4$. With the help of your calculator, you can determine that whenever θ is an odd-integer multiple of $\pi/4$, the secant squared and cosecant squared are both 2, so their sum is 4. If you move slightly to the right or left of any of these points, the value of the sum-of-squares function increases (a fact that you can, again, check out with your calculator). It follows that the sum-of-squares function can never attain any real-number value less than 4. However, there's no limit to how large the value of the function can get. One or the other of the original functions "blows up positively" at every point where θ attains an integer multiple of $\pi/2$.

Here's a challenge!

Sketch a graph of the ratio of the square of the secant function to the square of the cosecant function. That is, graph

$$f(\theta) = (\sec^2 \theta)/(\csc^2 \theta)$$

Determine the domain and range of f. Be careful! Both the domain and the range have some tricky restrictions.

Solution

We can simplify this problem by remembering a few basic facts in trigonometry, and by applying a little algebra. First, let's remember that the cosecant is equal to the reciprocal of the sine, so the converse is also true. We have

$$1/(\csc \theta) = \sin \theta$$

When we square both sides, we get

$$1/(\csc^2 \theta) = \sin^2 \theta$$

Substituting in the equation for our function gives us

$$f(\theta) = (\sec^2 \theta)(\sin^2 \theta)$$

We've learned that the secant is equal to the reciprocal of the cosine. We have

$$\sec \theta = 1/(\cos \theta)$$

so we can square both sides to get

$$\sec^2 \theta = 1/(\cos^2 \theta)$$

Substituting again in the equation for our original function, we obtain

$$f(\theta) = (\sin^2 \theta)/(\cos^2 \theta) = [(\sin \theta)/(\cos \theta)]^2$$

The sine over the cosine is equal to the tangent, so we can substitute again to conclude that our original function is

$$f(\theta) = \tan^2 \theta$$

with the restriction that we can't define it for any input value where either the secant or the cosecant become singular.

The solid black curves in Fig. 15-10 show the result of squaring all the values of the tangent function, noting the additional undefined values as open circles. At the points shown by the open circles, the cosecant function is singular so its square is undefined. That means we can't define our ratio function f at any such point. At the asymptotes (dashed vertical lines), the secant function is singular so its square is undefined, making it impossible to define the ratio function f for those values of θ. Our function f has a period of π. The domain of f includes all real numbers except integer multiples of $\pi/2$, where one or the other of the original squared functions is singular. The range is the set of all positive reals.

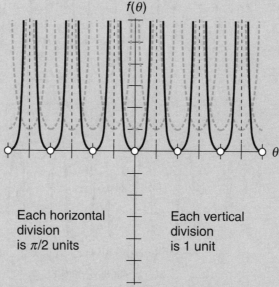

Each horizontal division is $\pi/2$ units

Each vertical division is 1 unit

Figure 15-10   Graph of the ratio of the square of the secant function to the square of the cosecant function (solid black curves). The dashed gray curves are the graphs of the original squared functions. Each horizontal division represents $\pi/2$ units. Each vertical division represents 1 unit. The vertical dashed lines are asymptotes of f.

Graphs Involving the Tangent and Cotangent

You were introduced to graphs of the basic tangent and cotangent functions in Chap. 2. The tangent is the ratio of the sine to the cosine, and the cotangent is the ratio of the cosine to the sine. Now we'll see what happens when we alter or combine these functions in a few different ways.

Tangent and cotangent: example 1

In Fig. 15-11, the dashed gray curves are superimposed graphs of the tangent and cotangent functions. The solid black curves portray the graph of

$$f(\theta) = \tan \theta + \cot \theta$$

The graph of f has asymptotes that pass through every point where the independent variable attains an integer multiple of $\pi/2$. The period is π. The domain is the set of all real numbers except the integer multiples of $\pi/2$. The range is the set of reals larger than or equal

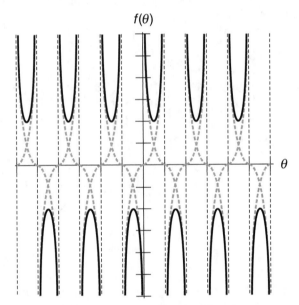

Each horizontal division is $\pi/2$ units

Each vertical division is 1 unit

Figure 15-11   Graph of the sum of the tangent and cotangent functions (solid black curves). The dashed gray curves are the graphs of the original functions. Each division on the horizontal axis represents $\pi/2$ units. Each vertical division represents 1 unit. The vertical dashed lines are asymptotes of f. The dependent-variable axis is also an asymptote of f.

to 2 or smaller than or equal to −2. We can also say that the range spans the set of all reals except those in the open interval (−2,2).

Tangent and cotangent: example 2

Figure 15-12 shows superimposed graphs of the tangent and cotangent functions (dashed gray curves) along with their product (black line with "holes" in it). We have

$$f(\theta) = \tan\theta\cot\theta$$

We can simplify the calculations to graph this function when we recall that the cotangent and the tangent are reciprocals of each other, so we have

$$\cot\theta = 1/(\tan\theta)$$

This equation is valid as long as both functions are defined and $\tan\theta \neq 0$. By substitution, the equation for our function f becomes

$$f(\theta) = (\tan\theta)/(\tan\theta) = 1$$

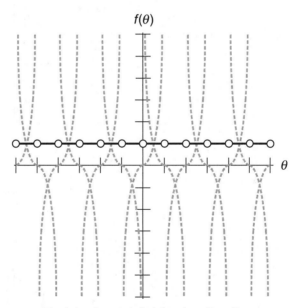

Each horizontal division is $\pi/2$ units

Each vertical division is 1 unit

Figure 15-12   Graph of the product of the tangent and cotangent functions (solid black curve). The dashed gray curves are the graphs of the original functions. Each division on the horizontal axis represents $\pi/2$ units. Each vertical division represents 1 unit.

The graph of f is a horizontal, straight line with infinitely many "holes," with each hole located at a point where θ is an integer multiple of $\pi/2$. If we want to get creative with our terminology, we can say that the graph of f consists of infinitely many open-ended line segments, each of length $\pi/2$, placed end-to-end in a collinear arrangement. The domain of f spans the set of all reals except the integer multiples of $\pi/2$. The range is the set containing the single real number 1.

Tangent and cotangent: example 3

The dashed gray curves in Fig. 15-13 are the graphs of the tangent function (at A) and the cotangent function (at B). The solid black curves in drawing A compose the graph of

$$f(\theta) = \tan^2 \theta$$

Figure 15-13   The solid black curves are graphs of the squares of the tangent function (at A) and the cotangent function (at B). The dashed gray curves are graphs of the original functions. Each division on the horizontal axes represents $\pi/2$ units. Each division on the vertical axes represents 1 unit. The vertical dashed lines are asymptotes of f and g. At B, the positive dependent-variable axis is also an asymptote of g.

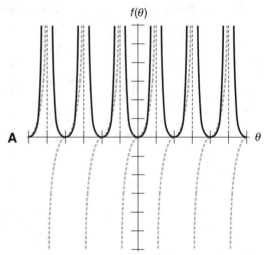

Each horizontal division is $\pi/2$ units
Each vertical division is 1 unit

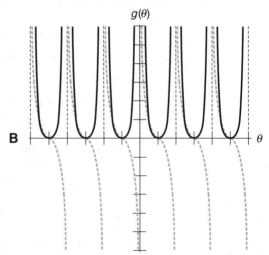

The solid black curves in drawing B compose the graph of

$$g(\theta) = \cot^2 \theta$$

Both f and g have periods of π, the same as the periods of the tangent and cotangent functions. Therefore, the frequencies of the squared functions are the same as those of the originals. Singularities occur in f and g at the same points on the independent-variable axis as they do for the original functions. The domain of f is the set of all reals except odd-integer multiples of $\pi/2$. The domain of g is the set of all reals except integer multiples of π. The ranges of both f and g span the set of nonnegative real numbers.

Tangent and cotangent: example 4

Figure 15-14 is a graph of the sum of the tangent-squared function and the cotangent-squared function. The solid black curves compose the graph of

$$f(\theta) = \tan^2 \theta + \cot^2 \theta$$

The dashed gray curves are superimposed graphs of the original squared functions. This sum function f has a period equal to half that of the original functions, or $\pi/2$. The domain includes all reals except the integer multiples of $\pi/2$. The range is the set of reals y such that $y \geq 2$.

Figure 15-14 Graph of the sum of the squares of the tangent and cotangent functions (solid black curves). The dashed gray curves are the graphs of the original squared functions. Each division on the horizontal axis represents $\pi/2$ units. Each vertical division represents 1 unit. The vertical dashed lines are asymptotes of f. The positive dependent-variable axis is also an asymptote of f.

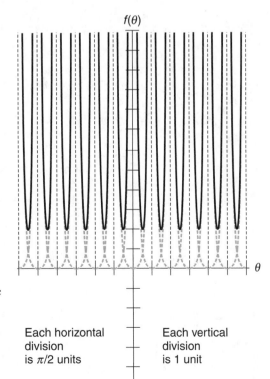

Are you confused?

You might wonder how we can be sure that the range of the sum-of-squares function graphed in Fig. 5-14 is the set of all reals greater than or equal to 2. To understand this, we can use the same reasoning as we did when we added the squares of the secant and the cosecant functions. All the minima on the solid black curves in Fig. 15-14 correspond to dependent-variable values of 2, because they occur where the graphs of the dashed gray curves intersect. At all such points, the tangent squared and cotangent squared are both 1, so their sum is 2. If you move slightly on either side of any such point, the value of the sum-of-squares function increases.

Here's a challenge!

Sketch a graph of the ratio of the square of the tangent function to the square of the cotangent function. That is, graph

$$f(\theta) = (\tan^2 \theta)/(\cot^2 \theta)$$

State the domain and range of f. Be careful! There are some tricky restrictions in the domain.

Solution

Let's use our knowledge of trigonometry to break this ratio down into sines and cosines. We recall that

$$\tan \theta = (\sin \theta)/(\cos \theta)$$

as long as θ isn't an odd-integer multiple of $\pi/2$, and

$$\cot \theta = (\cos \theta)/(\sin \theta)$$

provided θ isn't an integer multiple of π. Therefore,

$$\tan^2 \theta = (\sin^2 \theta)/(\cos^2 \theta)$$

and

$$\cot^2 \theta = (\cos^2 \theta)/(\sin^2 \theta)$$

with the same restrictions. By substitution, our ratio function becomes

$$f(\theta) = [(\sin^2 \theta)/(\cos^2 \theta)]/[(\cos^2 \theta)/(\sin^2 \theta)]$$

as long as θ isn't an integer multiple of $\pi/2$. The above equation can be rewritten as

$$f(\theta) = [(\sin^2 \theta)/(\cos^2 \theta)] \; [(\sin^2 \theta)/(\cos^2 \theta)]$$

which simplifies to

$$f(\theta) = (\sin^4 \theta)/(\cos^4 \theta)]$$

and finally to

$$f(\theta) = \tan^4 \theta$$

with, once again, the important restriction that θ cannot be any integer multiple of $\pi/2$. If we input any integer multiple of $\pi/2$ to the original function, we can't define the output because either the numerator or the denominator function encounters a singularity.

The black curves with the holes in Fig. 15-15 show the result of raising all the values of the tangent function to the fourth power, noting the additional undefined values as open circles. The dashed gray curves are the original tangent-squared and cotangent-squared functions. Our function f has a period of π. The domain includes all real numbers except integer multiples of $\pi/2$. The range includes all positive real numbers.

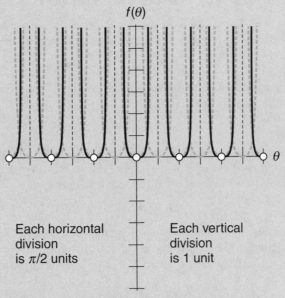

Each horizontal
division
is $\pi/2$ units

Each vertical
division
is 1 unit

Figure 15-15   Graph of the ratio of the square of the tangent function to the square of the cotangent function (solid black curves). The dashed gray curves are the graphs of the original squared functions. Each division on the horizontal axis represents $\pi/2$ units. Each vertical division represents 1 unit. The vertical dashed lines represent asymptotes of f.

Practice Exercises

This is an open-book quiz. You may (and should) refer to the text as you solve these problems. Don't hurry! You'll find worked-out answers in App. B. The solutions in the appendix may not represent the only way a problem can be figured out. If you think you can solve a particular problem in a quicker or better way than you see there, by all means try it!

1. Look back at Fig. 15-1, which shows the graphs of the sine and cosine waves (dashed gray curves) along with their sum (solid black curve). Sketch a graph, and state the domain and the range, of the difference function:

$$h(\theta) = \sin\theta - \cos\theta$$

2. Look back at Fig. 15-4, which shows the sine-squared and cosine-squared waves (dashed gray curves) along with their sum (solid black horizontal line). Sketch a graph, and state the domain and the range, of the difference-of-squares function:

$$h(\theta) = \sin^2\theta - \cos^2\theta$$

3. Look back at Fig. 15-5, which shows the graphs of the sine-squared and cosine-squared waves (dashed gray curves) along with the graph of the ratio-of-squares function:

$$f(\theta) = (\sin^2\theta)/(\cos^2\theta)$$

   Sketch a graph, and state the domain and the range, of the ratio-of-squares function going the other way. That function is

$$h(\theta) = (\cos^2\theta)/(\sin^2\theta)$$

4. Look back at Fig. 15-6, which shows the graphs of the secant and cosecant functions (dashed gray curves) along with their sum (solid black curves). Sketch a graph, and state the domain and the range, of the difference function:

$$h(\theta) = \sec\theta - \csc\theta$$

5. Look back at Fig. 15-9, which shows the graphs of the secant-squared and cosecant-squared functions (dashed gray curves) along with their sum (solid black curves). Sketch a graph, and state the domain and the range, of the difference function:

$$h(\theta) = \sec^2\theta - \csc^2\theta$$

6. Look back at Fig. 15-10, which shows the graphs of the secant-squared and cosecant-squared functions (dashed gray curves) along with the graph of

$$f(\theta) = (\sec^2\theta)/(\csc^2\theta)$$

   Sketch a graph, and state the domain and the range, of the ratio-of-squares function going the other way. That function is

$$h(\theta) = (\csc^2\theta)/(\sec^2\theta)$$

7. Look back at Fig. 15-11, which shows the graphs of the tangent and cotangent functions (dashed gray curves) along with the graph of their sum (solid black curves). Sketch a graph, and state the domain and the range, of the difference function:

$$h(\theta) = \tan\theta - \cot\theta$$

8. Look back at Fig. 15-14, which shows the graphs of the tangent-squared and cotangent-squared functions (dashed gray curves) along with the graph of their sum (solid black curves). Sketch a graph, and state the domain and the range, of the difference-of-squares function:

$$h\,(\theta) = \tan^2 \theta - \cot^2 \theta$$

9. In Chap. 2, we learned that square of the secant of an angle minus the square of the tangent of the same angle is always equal to 1, as long as the angle is not an odd-integer multiple of $\pi/2$. That is,

$$\sec^2 \theta - \tan^2 \theta = 1$$

Sketch a graph that illustrates this principle, which is sometimes called the Pythagorean theorem for the secant and tangent.

10. In Chap. 2, we learned that the square of the cosecant of an angle minus the square of the cotangent of the same angle is always equal to 1, as long as the angle is not an integer multiple of π. That is,

$$\csc^2 \theta - \cot^2 \theta = 1$$

Sketch a graph that illustrates this principle, which is sometimes called the Pythagorean theorem for the cosecant and cotangent.

16

Parametric Equations in Two-Space

In the two-space relations and functions we've seen so far, the value of one variable depends on the value of the other variable. In this chapter, we'll learn how to express two-space relations and functions in which both variables depend on an external factor called a *parameter*.

What's a Parameter?

In a two-space relation or function, a parameter acts as a "master controller" for one or both variables. When there exists a relation between x and y, for example, we don't have to say that x depends on y or vice versa. Instead, we can say that a parameter, which we usually call t, independently governs the values of x and y. We use *parametric equations* to describe how this happens.

A rectangular-coordinate example

Here's an example of a pair of parametric equations that produce a straight line in the Cartesian xy plane. Consider

$$x = 2t$$

and

$$y = 3t$$

To generate the graph of this system, we can input various values of t to both of the parametric equations, and then plot the ordered pairs (x,y) that come out. Following are some examples:

- When $t = -2$, we have $x = 2 \times (-2) = -4$ and $y = 3 \times (-2) = -6$.
- When $t = -1$, we have $x = 2 \times (-1) = -2$ and $y = 3 \times (-1) = -3$.
- When $t = 0$, we have $x = 2 \times 0 = 0$ and $y = 3 \times 0 = 0$.
- When $t = 1$, we have $x = 2 \times 1 = 2$ and $y = 3 \times 1 = 3$.
- When $t = 2$, we have $x = 2 \times 2 = 4$ and $y = 3 \times 2 = 6$.

When we plot the (x,y) ordered pairs based on the above list as points on a Cartesian plane and then "connect the dots," we get a line passing through the origin with a slope of 3/2, as shown in Fig. 16-1. From our knowledge of the slope-intercept form of a line in the xy plane, we can write down the equation in that form as

$$y = (3/2)x$$

Alternatively, we can use algebra to derive the equation of our system in terms of x and y alone, without t. Let's take the first parametric equation

$$x = 2t$$

and multiply it through by 3/2 to get

$$(3/2)x = (3/2)(2t) = 3t$$

Deleting the middle portion in the above three-way equation gives us

$$(3/2)x = 3t$$

The second parametric equation tells us that $3t = y$, so we can substitute directly in the above equation to obtain

$$(3/2)x = y$$

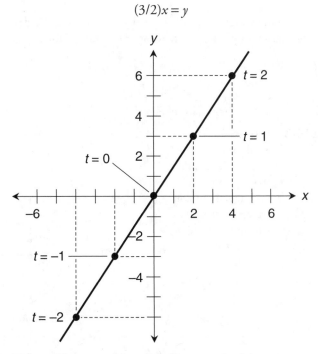

Figure 16-1   Cartesian-coordinate graph of the parametric equations $x = 2t$ and $y = 3t$.

which is identical to the slope-intercept equation

$$y = (3/2)x$$

The polar-coordinate counterpart

Let's see what happens if we change our pair of parametric equations to polar form. We'll put θ in place of x, and put r in place of y. Now we have

$$\theta = 2t$$

and

$$r = 3t$$

To create the polar graph, we can input various values of t, just as we did before. To keep things from getting messy, let's restrict ourselves to values of t such that we see only the part of the graph corresponding to the first full counterclockwise rotation of the direction angle, so $0 \le \theta \le 2\pi$. Consider the following cases:

- When $t = 0$, we have $\theta = 2 \times 0 = 0$ and $r = 3 \times 0 = 0$.
- When $t = \pi/4$, we have $\theta = 2 \times \pi/4 = \pi/2$ and $r = 3\pi/4$.
- When $t = \pi/2$, we have $\theta = 2 \times \pi/2 = \pi$ and $r = 3\pi/2$.
- When $t = 3\pi/4$, we have $\theta = 2 \times 3\pi/4 = 3\pi/2$ and $r = 3 \times 3\pi/4 = 9\pi/4$.
- When $t = \pi$, we have $\theta = 2\pi$ and $r = 3\pi$.

Our graph is a spiral, as shown in Fig. 16-2. Its equation can be derived with algebra exactly as we did in the Cartesian plane, substituting θ for x and r for y to get

$$r = (3/2)\theta$$

Figure 16-2   Polar-coordinate graph of the parametric equations $\theta = 2t$ and $r = 3t$. Each radial division represents π units. For simplicity, we restrict θ to the interval $[0, 2\pi]$.

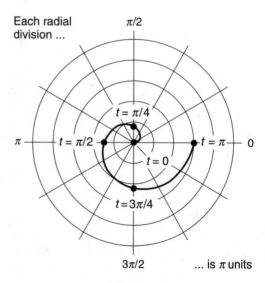

Are you confused?

If you're having trouble understanding the concept of a parameter, imagine the passage of time. In science and engineering, elapsed time t is the parameter on which many things depend. In the Cartesian situation described above, as time flows from the past ($t < 0$) through the present moment ($t = 0$) and into the future ($t > 0$), a point moves along the line in Fig. 16-1, going from the third quadrant (lower left, in the past) through the origin (right now) and into the first quadrant (upper right, in the future). In the polar case, as time flows from the present ($t = 0$) into the future ($t > 0$), a point travels along the spiral in Fig. 16-2, starting at the center (right now) and going counterclockwise, arriving at the outer end when $t = \pi$ (a little while from now).

Here's a challenge!

Find a pair of parametric equations that represent the line shown in Fig. 16-3.

Solution

We're given two points on the line. One of them, (0,3), tells us that the y-intercept is 3. We can deduce the slope from the coordinates of the other point. When we move 4 units to the right from (0,3), we must go downward by 3 units (or upward by -3 units) to reach (4,0). The "rise over run" ratio is -3 to 4, so the slope is $-3/4$. The slope-intercept form of the equation for our line is

$$y = (-3/4)x + 3$$

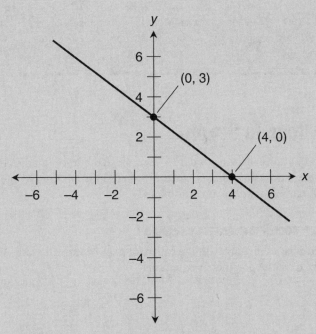

Figure 16-3   How can we represent this line as a pair of parametric equations?

We can let x vary directly with the parameter t. We describe that relation simply as

$$x = t$$

That's one of our two parametric equations. We can substitute t for x into the point-slope equation to get

$$y = (-3/4)t + 3$$

That's the other parametric equation.

Here's an experiment!

Do you suspect that the pair of equations

$$x = t$$

and

$$y = (-3/4)t + 3$$

isn't the only parametric way we can represent the line in Fig. 16-3? If so, maybe you're right. Let $x = -2t$, or $x = t + 1$, or $x = -2t + 1$, and see what happens when you generate the equation for y in terms of t on that basis. When you put the two parametric equations together, do you get the same line as the one shown in Fig. 16-3?

From Equations to Graph

Parametric equations allow us to define complicated curves in an elegant, and often simpler, way than we can do with ordinary equations. Let's look at a couple of examples, and plot their graphs in rectangular and polar coordinates.

Rectangular-coordinate graph: example 1

Suppose that x varies directly with the square of t, and y varies directly with the cube of t. In this situation, we have the parametric equations

$$x = t^2$$

and

$$y = t^3$$

Let's construct a graph of this equation by inputting several values of t to the system and then plotting the points. We can break the situation down as follows:

- When $t = -4$, we have $x = (-4)^2 = 16$ and $y = (-4)^3 = -64$.
- When $t = -3$, we have $x = (-3)^2 = 9$ and $y = (-3)^3 = -27$.
- When $t = -2$, we have $x = (-2)^2 = 4$ and $y = (-2)^3 = -8$.
- When $t = -1$, we have $x = (-1)^2 = 1$ and $y = (-1)^3 = -1$.
- When $t = 0$, we have $x = 0^2 = 0$ and $y = 0^3 = 0$.
- When $t = 1$, we have $x = 1^2 = 1$ and $y = 1^3 = 1$.
- When $t = 2$, we have $x = 2^2 = 4$ and $y = 2^3 = 8$.
- When $t = 3$, we have $x = 3^2 = 9$ and $y = 3^3 = 27$.
- When $t = 4$, we have $x = 4^2 = 16$ and $y = 4^3 = 64$.

We can plot the points for these nine xy-plane coordinates and then connect them by curve fitting to get the graph of Fig. 16-4. To keep the picture clean, the points aren't labeled. In this illustration, we have a rectangular-coordinate graph, but not a true Cartesian graph. That's because the divisions on the y axis represent different increments than those on the x axis. The result is a curve that's "vertically squashed" compared to the way it would look if plotted on a true Cartesian coordinate grid, but we can fit more of the curve into the available space.

Polar-coordinate graph: example 1

Figure 16-5 illustrates what happens when we substitute θ for x and r for y in the above example, and then graph the result in polar coordinates. For simplicity, let's restrict the graph

Figure 16-4   Rectangular-coordinate graph of the parametric equations $x = t^2$ and $y = t^3$.

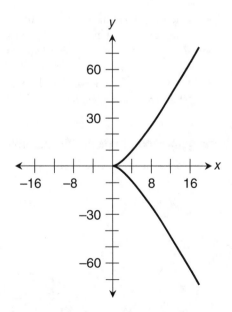

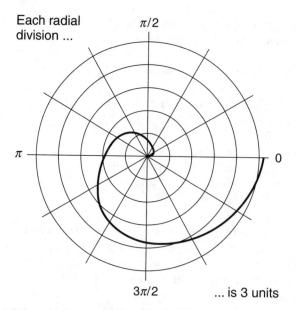

Each radial division ...

... is 3 units

Figure 16-5 Polar-coordinate graph of the parametric equations $\theta = t^2$ and $r = t^3$. Each radial division represents 3 units.

to values of t such that $0 \le t \le (2\pi)^{1/2}$. To keep the picture clean, we won't label any of the points. The situation breaks down as follows:

- When $t = 0$, we have $\theta = 0^2 = 0$ and $r = 0^3 = 0$.
- When $t = 1$, we have $\theta = 1^2 = 1$ and $r = 1^3 = 1$.
- When $t = \pi^{1/2}$, we have $\theta = (\pi^{1/2})^2 = \pi$ and $r = (\pi^{1/2})^3 = \pi^{3/2} \approx 5.57$.
- When $t = 2$, we have $\theta = 2^2 = 4$ and $r = 2^3 = 8$.
- When $t = 5^{1/2}$, we have $\theta = (5^{1/2})^2 = 5$ and $r = (5^{1/2})^3 = 5^{3/2} \approx 11.18$.
- When $t = (2\pi)^{1/2}$, we have $\theta = [(2\pi)^{1/2}]^2 = 2\pi$ and $r = [(2\pi)^{1/2}]^3 = (2\pi)^{3/2} \approx 15.75$.

Rectangular-coordinate graph: example 2

Suppose that x varies inversely with t, and y varies directly with $\ln t$. The parametric equations are

$$x = t^{-1}$$

and

$$y = \ln t$$

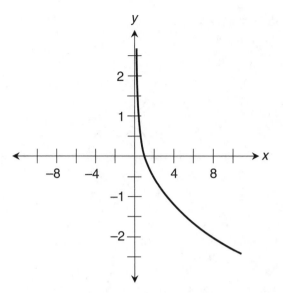

Figure 16-6   Rectangular-coordinate graph of the parametric equations $x = t^{-1}$ and $y = \ln t$.

We can construct a rectangular-coordinate graph of the relation between x and y by tabulating the values for several points, based on various values of t. Let's break things down into the following cases:

- When $t \leq 0$, $\ln t$ is undefined, so there are no points to plot.
- When $t = e^{-2} \approx 0.14$, we have $x = (e^{-2})^{-1} = e^{2} \approx 7.39$ and $y = \ln (e^{-2}) = -2$.
- When $t = e^{-1} \approx 0.37$, we have $x = (e^{-1})^{-1} = e \approx 2.72$ and $y = \ln (e^{-1}) = -1$.
- When $t = 1$, we have $x = 1^{-1} = 1$ and $y = \ln 1 = 0$.
- When $t = 2$, we have $x = 2^{-1} = 1/2$ and $y = \ln 2 \approx 0.69$.
- When $t = e \approx 2.72$, we have $x = e^{-1} \approx 0.37$ and $y = \ln e = 1$.
- When $t = e^{2} \approx 7.39$, we have $x = (e^{2})^{-1} = e^{-2} \approx 0.14$ and $y = \ln (e^{2}) = 2$.

Figure 16-6 shows the curve we obtain when we plot these points and "connect the dots." To keep the picture clean, the points aren't labeled. As in Fig. 16-4, we use distorted rectangular coordinates to help us fit more of the curve on the page than we could with a true Cartesian grid.

Polar-coordinate graph: example 2

We can directly substitute θ for x and r for y in the above example, tabulate some values, graph the results, and get the curve shown in Fig. 16-7. Let's restrict t to keep θ within the closed interval $[0, 2\pi]$ so we see only the first full positive revolution. The situation breaks down as follows:

- When $t = e^{5} \approx 148$, we have $\theta = (e^{5})^{-1} = e^{-5} \approx 0.0067$ and $r = \ln (e^{5}) = 5$.
- When $t = e^{2} \approx 7.39$, we have $\theta = (e^{2})^{-1} = e^{-2} \approx 0.14$ and $r = \ln (e^{2}) = 2$.

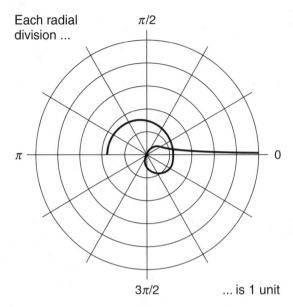

Each radial division ...

... is 1 unit

Figure 16-7   Polar-coordinate graph of the parametric equations $\theta = t^{-1}$ and $r = \ln t$. Each radial division represents 1 unit.

- When $t = 2$, we have $\theta = 2^{-1} = 1/2$ and $r = \ln 2 \approx 0.69$.
- When $t = 1$, we have $\theta = 1^{-1} = 1$ and $r = \ln 1 = 0$.
- When $t = 1/2$, we have $\theta = (1/2)^{-1} = 2$ and $r = \ln (1/2) \approx -0.69$.
- When $t = \pi^{-1} \approx 0.32$, we have $\theta = (\pi^{-1})^{-1} = \pi$ and $r = \ln (\pi^{-1}) \approx -1.14$.
- When $t = (2\pi)^{-1} \approx 0.16$, we have $\theta = [(2\pi)^{-1}]^{-1} = 2\pi$ and $r = \ln [(2\pi)^{-1}] \approx -1.84$.

As we plot the points to obtain this graph, we must remember that when we have a negative radius in polar coordinates, we go outward from the origin by a distance equal to $|r|$, but in the opposite direction from that indicated by θ.

Are you confused?

The polar graph in Fig. 16-7 can be baffling. Imagine that we start out facing east, in the direction $\theta = 0$. Our graph is infinitely far away in this direction. As we turn counterclockwise, the curve approaches us as r becomes finite and decreases. When we have turned counterclockwise through an angle of 1 rad (approximately 57°), the graph has come all the way in and reached the origin. As we continue to turn counterclockwise, the radius becomes negative, so the graph is behind us. As we rotate farther counterclockwise, r increases negatively. When we've rotated all the way around through a complete circle, the graph is approximately 1.84 units to our rear.

Here's a challenge!

Plot a rectangular-coordinate graph of the pair of parametric equations where x varies directly with e^t and y varies directly with t^2. Then plot a polar-coordinate graph of the pair of parametric equations where θ varies directly with e^t and r varies directly with t^2. For simplicity, restrict the polar graph to values of t such that $0 \le \theta \le 2\pi$.

Solution

The parametric equations for plotting the system in the rectangular xy plane are

$$x = e^t$$

and

$$y = t^2$$

Let's tabulate the x and y values for several points, based on various values of t:

- When $t = -2$, we have $x = e^{-2} \approx 0.14$ and $y = (-2)^2 = 4$.
- When $t = -1$, we have $x = e^{-1} \approx 0.37$ and $y = (-1)^2 = 1$.
- When $t = 0$, we have $x = e^0 = 1$ and $y = 0^2 = 0$.
- When $t = 1$, we have $x = e^1 = e \approx 2.72$ and $y = 1^2 = 1$.
- When $t = 3/2$, we have $x = e^{3/2} \approx 4.48$ and $y = (3/2)^2 = 9/4 = 2.25$.
- When $t = 2$, we have $x = e^2 \approx 7.39$ and $y = 2^2 = 4$.

Figure 16-8 shows the graph we obtain by plotting the points in the xy plane. This is a true Cartesian graph; the divisions on the x and y axes are the same size.

Figure 16-8   Cartesian-coordinate graph of the parametric equations $x = e^t$ and $y = t^2$.

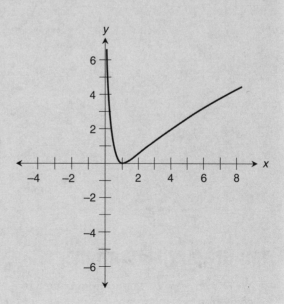

Now let's tabulate some values for a polar graph. We substitute θ for x and r for y, input values of t to get a good sampling of polar angles in the output, and restrict t to keep θ within the closed interval $[0, 2\pi]$. The situation breaks down into the following cases:

- When $t = -2$, we have $\theta = e^{-2} \approx 0.14$ and $r = (-2)^2 = 4$.
- When $t = -1$, we have $\theta = e^{-1} \approx 0.37$ and $r = (-1)^2 = 1$.
- When $t = 0$, we have $\theta = e^0 = 1$ and $r = 0^2 = 0$.
- When $t = 1$, we have $\theta = e^1 = e \approx 2.72$ and $r = 1^2 = 1$.
- When $t = 3/2$, we have $\theta = e^{3/2} \approx 4.48$ and $r = (3/2)^2 = 9/4 = 2.25$.
- When $t = \ln 2\pi$, we have $\theta = e^{(\ln 2\pi)} = 2\pi$ and $r = (\ln 2\pi)^2 \approx 3.38$.

Figure 16-9 shows the resulting curve in the polar plane. If the above tabulation doesn't generate enough points to satisfy you, feel free to work out a few more. As you gain experience in plotting graphs like this, you'll learn to get a sense of where the curves go without having to calculate very many discrete values.

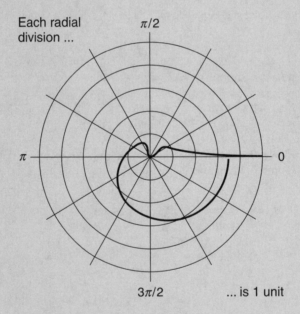

Figure 16-9 Polar-coordinate graph of the parametric equations $\theta = e^t$ and $r = t^2$. Each radial division represents 1 unit.

From Graph to Equations

We've seen how we can go from parametric equations to graphs. Now we'll do an exercise going from a graph to a pair of parametric equations.

Cartesian-coordinate graph to equations

Consider a circle of radius a, centered at the origin in the Cartesian xy plane as shown in Fig. 16-10. From trigonometry, we remember that

$$x = a \cos \phi$$

and

$$y = a \sin \phi$$

where ϕ is the angle going counterclockwise from the positive x axis. Both x and y depend on the value of ϕ. Let's rename ϕ and call it t, so our equations become

$$x = a \cos t$$

and

$$y = a \sin t$$

This is a pair of parametric equations representing a circle of radius a, centered at the origin in the Cartesian xy plane. For any particular circle, a is a constant (not a variable), so the parameter t is the only variable on the right-hand side of either equation.

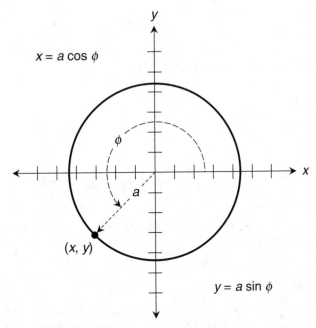

Figure 16-10   Cartesian-coordinate graph of a circle with radius a, centered at the origin. We can let $\phi = t$ to describe this circle as a pair of parametric equations.

Figure 16-11 Polar-coordinate graph of a circle with radius *a*, centered at the origin. We can let $\phi = t$ to describe this circle as a pair of parametric equations.

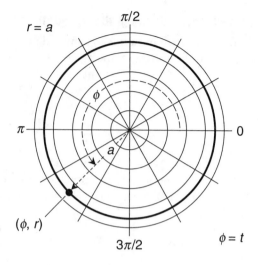

Polar-coordinate graph to equations

Now let's convert the circle in the previous example to a pair of polar-form parametric equations. Suppose the polar direction angle is ϕ, and the polar radius is r. The equation of a circle having radius *a* as shown in Fig. 16-11 is

$$r = a$$

Let's call the angle ϕ our parameter *t*, just as we did in the *xy*-plane situation. Then we can write the parametric equations of our circle as

$$\phi = t$$

and

$$r = a$$

Are you confused?

Does the above pair of parametric equations seem strange to you? The second equation doesn't contain the parameter! That's not a problem in this situation. The parameter has no effect because the polar radius *r* is always the same.

Here's a challenge!

Suppose that we come across a pair of parametric equations similar to the one in the Cartesian-coordinate example above, except that the cosine and sine of the parameter are multiplied by different nonzero real-number constants *a* and *b*, like this:

$$x = a \cos t$$

and

$$y = b \sin t$$

What sort of curve should we expect to get if we graph the relation defined by this pair of parametric equations?

Solution

We've been told that a and b are both nonzero real numbers. Therefore, we can divide the equations through by their respective constants to get

$$x/a = \cos t$$

and

$$y/b = \sin t$$

If we square both sides of both equations, we obtain

$$(x/a)^2 = \cos^2 t$$

and

$$(y/b)^2 = \sin^2 t$$

When we add these two equations, left-to-left and right-to-right, we obtain the new equation

$$(x/a)^2 + (y/b)^2 = \cos^2 t + \sin^2 t$$

From trigonometry, we remember that for any real number t, it's always true that

$$\cos^2 t + \sin^2 t = 1$$

Therefore, the preceding equation can be rewritten as

$$(x/a)^2 + (y/b)^2 = 1$$

Expanding the squared ratios on the left-hand side gives us

$$x^2/a^2 + y^2/b^2 = 1$$

which is the equation of an ellipse centered at the origin. The horizontal (x-coordinate) semi-axis is a units wide, and the vertical (y-coordinate) semi-axis is b units high.

Practice Exercises

This is an open-book quiz. You may (and should) refer to the text as you solve these problems. Don't hurry! You'll find worked-out answers in App. B. The solutions in the appendix may not represent the only way a problem can be figured out. If you think you can solve a particular problem in a quicker or better way than you see there, by all means try it!

1. In "Rectangular-coordinate graph: example 1" (Fig. 16-4), the parametric equations are

$$x = t^2$$

and

$$y = t^3$$

Find an equation for this relation that expresses x in terms of y without the parameter t. Then find an equation that expresses y in terms of x without the parameter t.

2. Is the relation defined in your second answer to Problem 1 a function of x?

3. In "Polar-coordinate graph: example 2" (Fig. 16-7), the parametric equations are

$$\theta = t^{-1}$$

and

$$r = \ln t$$

Find an equation for this relation that expresses θ in terms of r without the parameter t. Then find an equation that expresses r in terms of θ without the parameter t.

4. Is the relation defined in your second answer to Problem 3 a function of θ?

5. In "Cartesian-coordinate graph to equations" (Fig. 16-10), the parametric equations are

$$x = a \cos t$$

and

$$y = a \sin t$$

where a is a nonzero constant. Find an equation for this relation in terms of x and y only, without the parameter t.

6. Express the solution to Problem 5 as a relation in which x is the independent variable and y is the dependent variable. You should end up with y alone on the left-hand side of the equals sign, and an expression containing x (but not y) on the right-hand side. Is this relation a function of x?

7. Suppose that we come across the pair of parametric equations

$$x = \sec t$$

and

$$y = \tan t$$

Find an equation for this relation in terms of x and y only, without the parameter t. What sort of curve should we expect to get if we graph this relation in the Cartesian xy plane?

8. Consider the pair of parametric equations

$$x = a \csc t$$

and

$$y = b \cot t$$

where a and b are nonzero real-number constants. Find an equation for this relation in terms of x and y only, without the parameter t. What sort of curve should expect to get if we graph this relation in the Cartesian xy plane?

9. Express the relation

$$x = \sin (\cos y)$$

as a pair of parametric equations.

10. Manipulate the equation stated in Problem 9 so that y appears all by itself on the left-hand side of the equals sign, and operations involving x appear on the right-hand side. Then manipulate your answer to Problem 9 to get the same equation.

17

Surfaces in Three-Space

Three-space can contain an infinite variety of surfaces, all of which can be defined as equations in terms of three variables. In this chapter, we'll examine a few basic surfaces and their equations in Cartesian three-space.

Planes

An intuitive way to express the equation for a plane in Cartesian xyz space is to define the direction of a vector normal (perpendicular) to the plane, and then to identify the coordinates of a point in the plane. We don't have to know the magnitude of the vector, and the point in the plane doesn't have to be the one where the vector originates.

General equation of plane

Figure 17-1 shows a plane W in Cartesian three-space, a point $P = (x_0, y_0, z_0)$ in the plane W, and a vector $(a,b,c) = a\mathbf{i} + b\mathbf{j} + c\mathbf{k}$ that's normal to plane W. The vector (a,b,c) originates at a point Q that differs from P, and which is also located away from the coordinate origin. The values $x = a$, $y = b$, and $z = c$ for the vector are nevertheless based on the vector's standard form, as if it originated at $(0,0,0)$. The point and the vector give us enough information to uniquely define the plane and write its equation in standard form as

$$a(x - x_0) + b(y - y_0) + c(z - z_0) = 0$$

This equation can also be written as

$$ax + by + cz + d = 0$$

where d is a stand-alone constant. With a little algebra, we can work out its value in terms of the other constants and coefficients as

$$d = -ax_0 - by_0 - cz_0$$

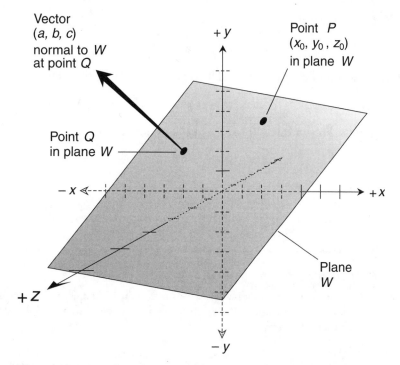

Figure 17-1 A plane *W* can be uniquely defined on the basis of a point *P* in the plane and a vector (*a,b,c*) normal to the plane.

Plotting a plane

When we want to construct a plane in Cartesian *xyz* space based on its equation, we can do it by figuring out the coordinates of points where the plane crosses each of the three coordinate axes. These points are the *x*-intercept, the *y*-intercept, and the *z*-intercept. When we plot these intercept points on the axes, we can envision the position and orientation of the plane.

There's a potential "hangup" with this scheme for plane-graphing. Not all planes cross all three axes in Cartesian *xyz*-space. If a plane is parallel to one of the axes, then it does not cross that axis, although must cross at least one of the other two. If a plane is parallel to the plane formed by two coordinate axes, then that plane crosses only the axis with respect to which it is not parallel.

An example

Suppose that a plane contains the point (3,−6,2), and the standard form of a vector normal to the plane is $4\mathbf{i} + 3\mathbf{j} + 2\mathbf{k}$. Let's find the plane's equation in the standard form given above. To begin, we know that the vector

$$4\mathbf{i} + 3\mathbf{j} + 2\mathbf{k}$$

is equivalent to the ordered triple

$$(a,b,c) = (4,3,2)$$

We've been told that

$$(x_0, y_0, z_0) = (3, -6, 2)$$

and that this point lies in the plane. The general formula for the plane is

$$a(x - x_0) + b(y - y_0) + c(z - z_0) = 0$$

Plugging in the known values for a, b, c, x_0, y_0, and z_0, we get

$$4(x - 3) + 3[y - (-6)] + 2(z - 2) = 0$$

which simplifies to

$$4x + 3y + 2z + 2 = 0$$

Are you confused?

The standard-form equation of a plane in xyz space looks like an extrapolation of the standard-form equation of a straight line in the xy plane. This can confuse some people. Don't let it baffle you! An equation of the form

$$ax + by + cz + d = 0$$

where a, b, c, and d are constants represents a plane, not a line. In Chap. 18, you'll learn how to describe straight lines in Cartesian xyz space.

Here's a challenge!

Draw a graph of the plane represented by the following equation:

$$-2x - 4y + 3z - 12 = 0$$

Solution

Let's work out the graph by finding the coordinate-axis intercepts. The x-intercept, or the point where the plane intersects the x axis, can be found by setting $y = 0$ and $z = 0$, and then solving the resultant equation for x. Let's call this point P. We have

$$-2x - 4 \times 0 + 3 \times 0 - 12 = 0$$

Solving step-by-step, we get

$$-2x - 12 = 0$$
$$-2x = 12$$
$$x = 12/(-2) = -6$$

Therefore

$$P = (-6, 0, 0)$$

The *y*-intercept, or the point where the plane intersects the *y* axis, can be found by setting $x = 0$ and $z = 0$, and then solving the resultant equation for *y*. Let's call this point *Q*. We have

$$-2 \times 0 - 4y + 3 \times 0 - 12 = 0$$

Solving, we get

$$-4y - 12 = 0$$
$$-4y = 12$$
$$y = 12/(-4) = -3$$

Therefore

$$Q = (0, -3, 0)$$

The *z*-intercept, or the point where the plane intersects the *z* axis, can be found by setting $x = 0$ and $y = 0$, and then solving the resultant equation for *z*. Let's call this point *R*. We have

$$-2 \times 0 - 4 \times 0 + 3z - 12 = 0$$

Solving, we get

$$3z - 12 = 0$$
$$3z = 12$$
$$z = 12/3 = 4$$

Therefore

$$R = (0, 0, 4)$$

These three points are shown in Fig. 17-2. We can now envision the plane because, as we recall from our courses in spatial geometry, a plane in three dimensions can be uniquely defined on the basis of three points.

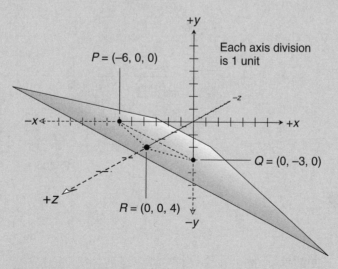

Figure 17-2   Here's the graph of a plane, based on the locations of the three axis intercept points *P*, *Q*, and *R*.

Spheres

A spherical surface is defined as the set of all points that lie at a fixed distance from a known central point in three dimensions. When we recall the formula for the distance between a point and the origin, it's easy to work out equations for spheres in Cartesian xyz space.

Center at the origin

Imagine a sphere whose center lies at the origin $(0,0,0)$, as shown in Fig. 17-3. Any point on the sphere's surface is at the same distance from the origin as any other point on the sphere's surface. Suppose that P is one such point whose coordinates are given by

$$P = (x_p, y_p, z_p)$$

In Chap. 7, we learned that the distance r of the point P from the origin in Cartesian xyz space is

$$r = (x_p^2 + y_p^2 + z_p^2)^{1/2}$$

We can square both sides of the above equation to get

$$r^2 = x_p^2 + y_p^2 + z_p^2$$

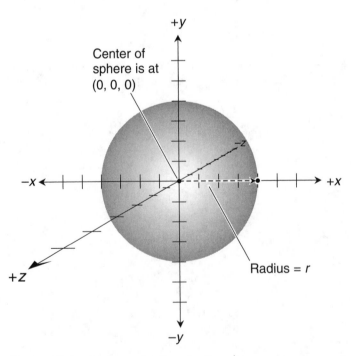

Figure 17-3   A sphere of radius r in Cartesian xyz space, centered at the origin. All points on the sphere's surface are at distance r from the center point $(0,0,0)$.

Transposing the left- and right-hand sides, we have

$$x_p^2 + y_p^2 + z_p^2 = r^2$$

Every point on the sphere's surface is the same distance from the origin as P, so we can generalize the above equation to get

$$x^2 + y^2 + z^2 = r^2$$

which defines the set of all points in three dimensions that lie at a fixed distance r from the origin. That's all there is to it! We've found the standard-form equation for a sphere of radius r, centered at the origin in Cartesian xyz space.

Center away from the origin

Consider a sphere whose center is somewhere other than the origin in Cartesian xyz space. Suppose that the coordinates of the center point are (x_0, y_0, z_0), as shown in Fig. 17-4. Whatever point P that we choose on the sphere's surface, the distance between P and the center is equal to the sphere's radius r. Adapting the distance-between-points formula for Cartesian xyz space from Chap. 7, we get

$$r = [(x_p - x_0)^2 + (y_p - y_0)^2 + (z_p - z_0)^2]^{1/2}$$

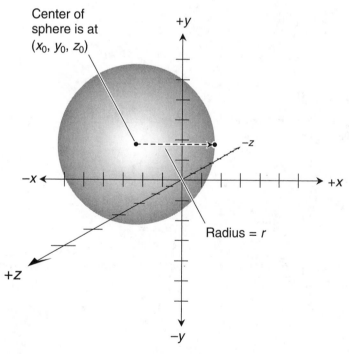

Figure 17-4 A sphere of radius r in Cartesian xyz space, centered away from the origin. All points on the sphere's surface are at distance r from the center point (x_0, y_0, z_0).

Squaring both sides of this equation and then transposing the left- and right-hand sides, we obtain

$$(x_p - x_0)^2 + (y_p - y_0)^2 + (z_p - z_0)^2 = r^2$$

Every point on the sphere's surface is the same distance from P as (x_0, y_0, z_0), so we can generalize to get

$$(x - x_0)^2 + (y - y_0)^2 + (z - z_0)^2 = r^2$$

This is the standard-form equation for a sphere of radius r, centered at the point (x_0, y_0, z_0) in Cartesian xyz space.

An example

Suppose we have a sphere whose center is at the origin, and whose radius is 7 units. If we let $r = 7$ in the general equation for a sphere centered at the origin, then we have

$$x^2 + y^2 + z^2 = 7^2$$

which can be simplified to

$$x^2 + y^2 + z^2 = 49$$

Another example

Consider a sphere centered at the point $(-2, 4, -1)$ with a radius of 5 units in Cartesian xyz space. We can let

$$x_0 = -2$$
$$y_0 = 4$$
$$z_0 = -1$$
$$r = 5$$

in the general equation for a sphere centered at a point other than the origin. When we plug in the numbers, we obtain

$$[x - (-2)]^2 + (y - 4)^2 + [z - (-1)]^2 = 5^2$$

which simplifies to

$$(x + 2)^2 + (y - 4)^2 + (z + 1)^2 = 25$$

Are you confused?

The radius of a sphere is usually defined as a positive real number. If we define the radius of a particular sphere as a negative real number, we get the same equation as we would if we defined the radius as the absolute value of that number. That's because we square the radius when we work out the formula. For example, if we have a sphere centered at the origin with a radius of 4 units, then its equation is

$$x^2 + y^2 + z^2 = 4^2$$

which simplifies to

$$x^2 + y^2 + z^2 = 16$$

If we find a companion "antisphere" centered at the origin with radius −4 units, then its equation is

$$x^2 + y^2 + z^2 = (-4)^2$$

which also simplifies to

$$x^2 + y^2 + z^2 = 16$$

In physics and engineering, it's possible to come up with spheres having negative radii, as well as negative dimensions for other physical objects. These results are usually mere artifacts of the calculation process, and don't have any significance in the real world. However, if you ever encounter a sphere whose radius is represented by an *imaginary* number such as $j4$, then you have good reason to be confused until you know what sort of object or phenomenon the equation describes!

Here's a challenge!

Suppose we're told that the following equation represents a sphere in Cartesian *xyz* space:

$$x^2 - 6x + y^2 - 2y + z^2 + 4z = 86$$

We're also informed that the sphere has a radius of 10 units. What are the coordinates of the center of the sphere?

Solution

To solve this problem, we need some intuition. We know that the radius r of the sphere is 10 units. Therefore, $r^2 = 100$. If we add 14 to both sides of the original equation, we get r^2 on the right-hand side:

$$x^2 - 6x + y^2 - 2y + z^2 + 4z + 14 = 100$$

We can split the stand-alone constant 14 into the sum of 9, 1, and 4, getting

$$x^2 - 6x + y^2 - 2y + z^2 + 4z + 9 + 1 + 4 = 100$$

Rearranging the addends on the left-hand side produces the following equation:

$$x^2 - 6x + 9 + y^2 - 2y + 1 + z^2 + 4z + 4 = 100$$

When we group the terms on the left-hand side by threes, we obtain

$$(x^2 - 6x + 9) + (y^2 - 2y + 1) + (z^2 + 4z + 4) = 100$$

This is a sum of three perfect squares! Factoring them individually gives us

$$(x - 3)^2 + (y - 1)^2 + (z + 2)^2 = 100$$

The coordinates of the center point are therefore

$$x_0 = 3$$
$$y_0 = 1$$
$$z_0 = -2$$

Expressed as an ordered triple, it's (3,1,−2).

Distorted Spheres

Spheres can be made "out of the round" by increasing or decreasing the *axial radii* in the x, y, and z directions individually.

Alternative equation for a sphere centered at the origin

Once again, consider the general equation of a perfect sphere centered at the origin. That equation, in standard form, is

$$x^2 + y^2 + z^2 = r^2$$

where r is the radius. If we divide through by r^2, we get

$$x^2/r^2 + y^2/r^2 + z^2/r^2 = 1$$

This equation tells us that the radius is always the same, whether we measure it in the direction of the x axis, y axis, or z axis. To emphasize the fact that we can, if desired, change any of all of these axial radii, let's rewrite the above equation as

$$x^2/a^2 + y^2/b^2 + z^2/c^2 = 1$$

where a, b, and c are positive real numbers representing the radii along the x, y, and z axes, respectively. In the case of a perfect sphere, we have

$$a = b = c$$

If these three positive real-number constants a, b, and c are not all the same, then we have a distorted sphere.

Oblate sphere centered at the origin

Suppose that we take a perfect sphere and then shorten one of the three axial radii. This process gives us an object called an *oblate sphere*. It's flattened, like a soft rubber ball when pressed between our hands. Figure 17-5 shows an example. This is what we get if we take the sphere from Fig. 17-3 and reduce the axial radius b (the one that goes along the y axis), while leaving the axial radii a and c unchanged. The center of the object is still at the origin, but we can no longer say that all the points on its surface are equidistant from the origin. The general equation for an oblate sphere centered at the origin is

$$x^2/a^2 + y^2/b^2 + z^2/c^2 = 1$$

where a is the x-axial radius, b is the y-axial radius, c is the z-axial radius, and exactly one of the following relationships holds true among them:

$$a < b = c$$
$$b < a = c$$
$$c < a = b$$

Alternative equation for a sphere centered away from the origin

Earlier in this chapter, we learned that the general equation of a sphere centered at some point other than the origin in Cartesian xyz space is

$$(x - x_0)^2 + (y - y_0)^2 + (z - z_0)^2 = r^2$$

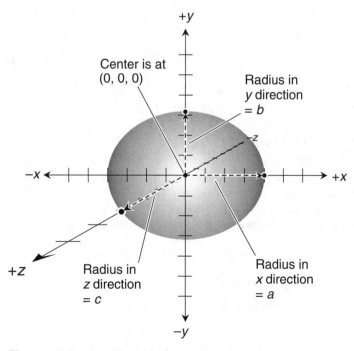

Figure 17-5 An oblate sphere in Cartesian xyz space, centered at the origin.

where r is the radius, and (x_0, y_0, z_0) are the coordinates of the center. Dividing through by r^2, we obtain

$$(x - x_0)^2/r^2 + (y - y_0)^2/r^2 + (z - z_0)^2/r^2 = 1$$

As we did with the sphere centered at the origin, we can rewrite this equation, getting

$$(x - x_0)^2/a^2 + (y - y_0)^2/b^2 + (z - z_0)^2/c^2 = 1$$

where a, b, and c are the radii parallel to the x, y, and z axes, respectively. As before, with a perfect sphere, we have

$$a = b = c$$

If a, b, and c are not all the same, then the sphere is distorted.

Oblate sphere centered away from the origin

If we take a sphere that's centered at (x_0, y_0, z_0) and shorten one of the axial radii, we get an oblate sphere defined by the general equation

$$(x - x_0)^2/a^2 + (y - y_0)^2/b^2 + (z - z_0)^2/c^2 = 1$$

where exactly one of the following is true:

$$a < b = c$$
$$b < a = c$$
$$c < a = b$$

Figure 17-6 should give you a general idea of what happens in a case like this. Imagine a sphere centered at (x_0, y_0, z_0) that has been squashed in the direction defined by a line parallel to the y axis.

Ellipsoid centered at the origin

Again, imagine that we have a perfect sphere centered at the origin in Cartesian *xyz* space. Let's lengthen one of the axial radii while leaving the other two unchanged. This stretching process produces an *ellipsoid*. It's elongated, like a football with blunted ends. Figure 17-7 shows an example. Imagine that we take the sphere from Fig. 17-3 and then stretch it in the z direction. The general equation for an ellipsoid centered at the origin is

$$x^2/a^2 + y^2/b^2 + z^2/c^2 = 1$$

where a is the x-axial radius, b is the y-axial radius, c is the z-axial radius, and exactly one of the following relationships is true:

$$a > b = c$$
$$b > a = c$$
$$c > a = b$$

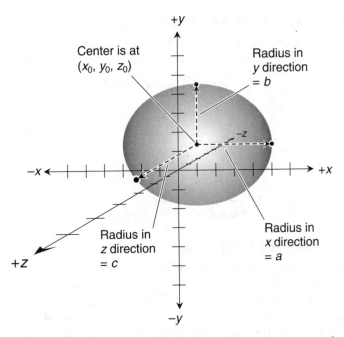

Figure 17-6   An oblate sphere in Cartesian *xyz* space, centered at (x_0, y_0, z_0).

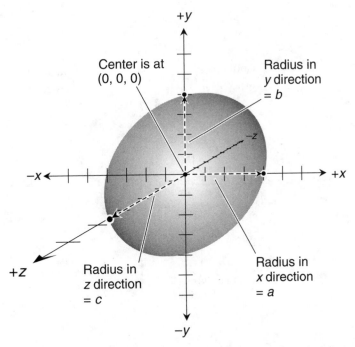

Figure 17-7   An ellipsoid in Cartesian *xyz* space, centered at the origin.

Ellipsoid centered away from the origin

Consider a sphere centered at (x_0, y_0, z_0). If we make one of the axial radii longer while leaving the other two unchanged, we get an ellipsoid defined by the general equation

$$(x - x_0)^2/a^2 + (y - y_0)^2/b^2 + (z - z_0)^2/c^2 = 1$$

where exactly one of the following is true:

$$a > b = c$$
$$b > a = c$$
$$c > a = b$$

Figure 17-8 portrays a situation in which a sphere centered at (x_0, y_0, z_0) has been stretched along a line parallel to the z axis to obtain an ellipsoid.

Oblate ellipsoid centered at the origin

One more time, imagine a sphere centered at the origin. We start out with all three axial radii equal in measure. Then we lengthen one of them, shorten another, and leave the third one unchanged. This process gives us an *oblate ellipsoid*. Figure 17-9 shows an example where we take the sphere from Fig. 17-3, squash the radius in the y direction, stretch the radius in the z direction, and leave the radius unchanged in the x direction. The general equation for an oblate ellipsoid centered at the origin is

$$x^2/a^2 + y^2/b^2 + z^2/c^2 = 1$$

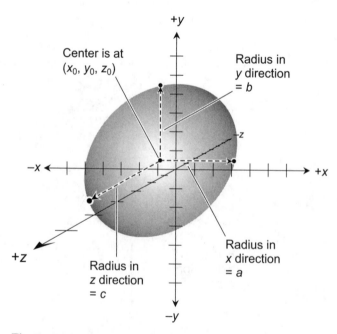

Figure 17-8   An ellipsoid in Cartesian *xyz* space, centered at (x_0, y_0, z_0).

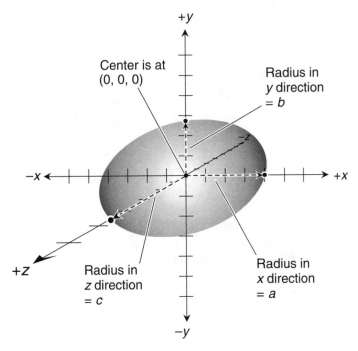

Figure 17-9 An oblate ellipsoid in Cartesian *xyz* space, centered at the origin.

where a is the *x*-axial radius, b is the *y*-axial radius, c is the *z*-axial radius, and all of the following are true:

$$a \neq b$$
$$b \neq c$$
$$a \neq c$$

Oblate ellipsoid centered away from the origin

Finally, imagine a sphere that's centered at (x_0, y_0, z_0). If we lengthen one of the axial radii, shorten another, and leave the third one unchanged, we get an oblate ellipsoid defined by

$$(x - x_0)^2/a^2 + (y - y_0)^2/b^2 + (z - z_0)^2/c^2 = 1$$

where all of the following are true:

$$a \neq b$$
$$b \neq c$$
$$a \neq c$$

Figure 17-10 shows an example of what happens when we move the center of the oblate ellipsoid from Fig. 17-9 away from the origin.

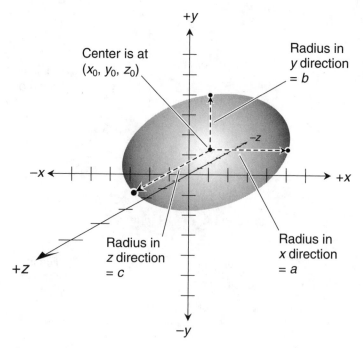

Figure 17-10   An oblate ellipsoid in Cartesian *xyz* space, centered at (x_0, y_0, z_0).

An example

Suppose that the coordinates of the center of a certain oblate sphere in Cartesian *xyz* space are (1,2,3). The axial radius in the *x* direction is 4, the axial radius in the *y* direction is 4, and the axial radius in the *z* direction is 2. The general equation is

$$(x - x_0)^2/a^2 + (y - y_0)^2/b^2 + (z - z_0)^2/c^2 = 1$$

where (x_0, y_0, z_0) are the coordinates of the center, *a* is the is the axial radius in the *x* direction, *b* is the axial radius in the *y* direction, and *c* is the axial radius in the *z* direction. We know that

$$(x_0, y_0, z_0) = (1,2,3)$$
$$a = 4$$
$$b = 4$$
$$c = 2$$

Plugging these values into the general equation, we conclude that our oblate sphere can be represented by the following equation:

$$(x - 1)^2/4^2 + (y - 2)^2/4^2 + (z - 3)^2/2^2 = 1$$

which simplifies to

$$(x - 1)^2/16 + (y - 2)^2/16 + (z - 3)^2/4 = 1$$

Another example

The coordinates of the center of an ellipsoid are $(-3,-2,-6)$. The axial radius in the x direction is 3, the axial radius in the y direction is 7, and the axial radius in the z direction is 3. The general equation is

$$(x - x_0)^2/a^2 + (y - y_0)^2/b^2 + (z - z_0)^2/c^2 = 1$$

This time, we have

$$(x_0, y_0, z_0) = (-3, -2, -6)$$
$$a = 3$$
$$b = 7$$
$$c = 3$$

Plugging these values into the general equation, we obtain

$$[x - (-3)]^2/3^2 + [y - (-2)]^2/7^2 + [z - (-6)]^2/3^2 = 1$$

which simplifies to

$$(x + 3)^2/9 + (y + 2)^2/49 + (z + 6)^2/9 = 1$$

Still another example

The coordinates of the center of an oblate ellipsoid are $(0,-3,11)$. The axial radius in the x direction is 5, the axial radius in the y direction is 8, and the axial radius in the z direction is 1. The general equation is

$$(x - x_0)^2/a^2 + (y - y_0)^2/b^2 + (z - z_0)^2/c^2 = 1$$

In this case, we have

$$(x_0, y_0, z_0) = (0, -3, 11)$$
$$a = 5$$
$$b = 8$$
$$c = 1$$

Plugging these values into the general equation gives us

$$[x - 0]^2/5^2 + [y - (-3)]^2/8^2 + (z - 11)^2/1^2 = 1$$

which simplifies to

$$x^2/25 + (y + 3)^2/64 + (z - 11)^2 = 1$$

Are you astute?

So far, we've described various surfaces by adding squared binomials to each other. You have every right to ask, "What will happen if we *subtract* any of the squared binomials in equations like these?" We'll do that shortly, and you'll see a few examples of what can take place. When we add squared binomials, the graphs always turn out to be spheres, oblate spheres, ellipsoids, or oblate ellipsoids in Cartesian *xyz* space. These are *closed surfaces*. They're "air-tight." If we subtract one or more of the squared binomials, we get *open surfaces* that "can't hold air." Such surfaces can take diverse, interesting forms.

Here's a challenge!

Consider a distorted sphere represented by the following equation:

$$12x^2 + 72x + 20y^2 - 80y + 15z^2 - 30z = -143$$

What are the coordinates of the center? What are the axial radii? Is the object an oblate sphere, an ellipsoid, or an oblate ellipsoid?

Solution

This problem requires a lot of insight to solve! Let's begin by adding 203 to each side of the equation to obtain

$$12x^2 + 72x + 20y^2 - 80y + 15z^2 - 30z + 203 = 60$$

The number we've added, 203, happens to be the sum of 108, 80, and 15. Let's add these three numbers into the above equation just after the terms $72x$, $80y$, and $-30z$, respectively. The equation then becomes

$$12x^2 + 72x + 108 + 20y^2 - 80y + 80 + 15z^2 - 30z + 15 = 60$$

Grouping the addends on the left-hand side by threes gives us

$$(12x^2 + 72x + 108) + (20y^2 - 80y + 80) + (15z^2 - 30z + 15) = 60$$

which is equivalent to

$$12(x^2 + 6x + 9) + 20(y^2 - 4y + 4) + 15(z^2 - 2z + 1) = 60$$

The three trinomials factor into perfect squares, so we can further morph the equation to obtain

$$12(x + 3)^2 + 20(y - 2)^2 + 15(z - 1)^2 = 60$$

Dividing through by 60, we get

$$(x + 3)^2/5 + (y - 2)^2/3 + (z - 1)^2/4 = 1$$

We recall that general formula for a distorted sphere in Cartesian *xyz* space is

$$(x - x_0)^2/a^2 + (y - y_0)^2/b^2 + (z - z_0)^2/c^2 = 1$$

where (x_0, y_0, z_0) are the coordinates of the center, *a* is the axial radius in the *x* direction, *b* is the axial radius in the *y* direction, and *c* is the axial radius in the *z* direction. In this situation, we have

$$(x_0, y_0, z_0) = (-3, 2, 1)$$
$$a = 5^{1/2}$$
$$b = 3^{1/2}$$
$$c = 4^{1/2} = 2$$

Our object is an oblate ellipsoid centered at $(-3, 2, 1)$. The radius in the *x* direction is $5^{1/2}$. The radius in the *y* direction is $3^{1/2}$. The radius in the *z* direction is 2.

Other Surfaces

Let's look at three general objects that arise in Cartesian *xyz* space from equations with sums and differences of terms containing x^2, y^2, and z^2.

Hyperboloid of one sheet

Figure 17-11 shows a generic example of a *hyperboloid of one sheet*. In this context, the term *sheet* refers to an unbroken surface. We get this type of object when we graph an equation of the form

$$x^2/a^2 + y^2/b^2 - z^2/c^2 = 1$$

where *a*, *b*, and *c* are positive real-number constants. This equation is like the one for a distorted sphere, except that one of the plus signs has been replaced by a minus sign. That sign change makes a huge difference! Instead of a closed surface centered at the origin, we get an infinitely tall, pinched cylinder whose axis lies along the coordinate *z* axis, and whose center coincides with the origin. The dimensions and shape of the hyperboloid depend on the values of *a*, *b*, and *c*. The perpendicular cross sections are always circles or ellipses.

If we move the minus sign so that it's in front of the term containing y^2 instead of the term containing z^2, we get the general equation

$$x^2/a^2 - y^2/b^2 + z^2/c^2 = 1$$

Again, we get a hyperboloid of one sheet, but its axis is along the coordinate *y* axis, and its center is at the origin. If we move the minus sign one more place to the left, putting it in front of the term containing x^2, the general equation becomes

$$-x^2/a^2 + y^2/b^2 + z^2/c^2 = 1$$

This is the general form of the equation for a hyperboloid of one sheet whose axis coincides with the coordinate *x* axis, and whose center is at the origin.

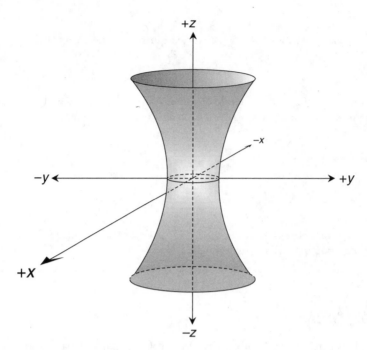

Figure 17-11   A hyperboloid of one sheet in Cartesian *xyz* space, centered at the origin.

Are you astute?

Figure 17-11 shows a perspective on Cartesian *xyz* space that we haven't seen before. We're looking "down" on the *yz* plane from somewhere near the positive *x* axis. Nevertheless, the axes are correctly oriented with respect to each other, as you can verify by referring back to Chap. 7. Let's stay with this axis orientation as we look at the next couple of objects.

Hyperboloid of two sheets

Figure 17-12 shows a *hyperboloid of two sheets*, which is the graph in Cartesian *xyz* space of an equation having the form

$$-x^2/a^2 + y^2/b^2 - z^2/c^2 = 1$$

where *a*, *b*, and *c* are positive real-number constants. Here, we have two surfaces that resemble bowls facing in opposite directions. In theory, the bowls extend infinitely toward the left and the right in this illustration. Both surfaces share a common straight-line axis that coincides with the coordinate *y* axis, and the two sheets are exact mirror images of each other. The center of the entire hyperboloid is at the origin. The contours of the surfaces depend on the values of *a*, *b*, and *c*.

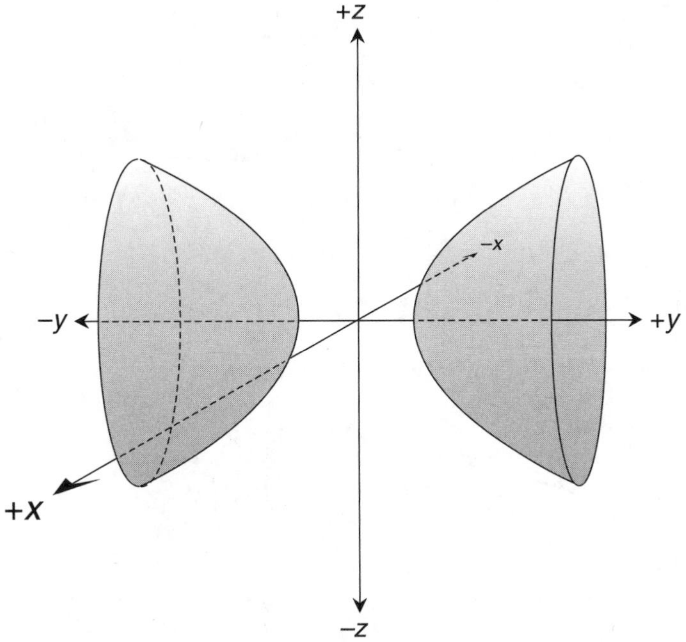

Figure 17-12   A hyperboloid of two sheets in Cartesian *xyz* space, centered at the origin.

If we make the term containing x^2 positive instead of the term containing y^2, we get the general equation

$$x^2/a^2 - y^2/b^2 - z^2/c^2 = 1$$

which produces a hyperboloid of two sheets whose axis lies along the coordinate *x* axis, and whose center is at the origin. If we move the plus sign so it's in front of the term containing z^2, the general equation becomes

$$-x^2/a^2 - y^2/b^2 + z^2/c^2 = 1$$

This maneuver gives us a hyperboloid of two sheets whose axis lies along the coordinate *z* axis, and whose center is at the origin.

Elliptic cone

Figure 17-13 shows an *elliptic cone*. It's what we get when we graph an equation of the form

$$x^2/a^2 + y^2/b^2 - z^2/c^2 = 0$$

where *a*, *b*, and *c* are positive real-number constants. The perpendicular cross sections of the cone are always circles or ellipses. The cone's axis coincides with the coordinate *z* axis, and the cone's vertex coincides with the origin. The flare angles, as well as the eccentricity of the cross-sectional ellipses, depend on the values of *a*, *b*, and *c*.

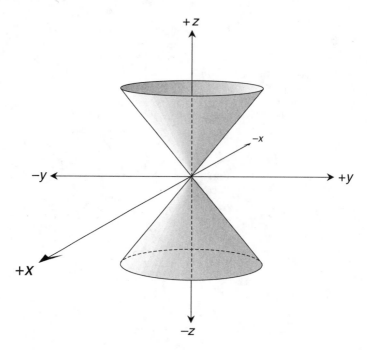

Figure 17-13 An elliptic cone in Cartesian *xyz* space, centered at the origin.

If we move the minus sign so it's in front of the term containing y^2, we get the general equation

$$x^2/a^2 - y^2/b^2 + z^2/c^2 = 0$$

whose graph is an elliptic cone with the axis along the coordinate *y* axis, and whose center is at the origin. If we move the minus sign so that it's in front of the term containing x^2, the general equation becomes

$$-x^2/a^2 + y^2/b^2 + z^2/c^2 = 0$$

and the graph becomes an elliptic cone whose axis lies along the coordinate *x* axis, and whose center is at the origin.

An example

Consider the object in Cartesian *xyz* space represented by

$$36x^2 - 16y^2 + 36z^2 = 0$$

We can divide through by 144 to obtain

$$x^2/4 - y^2/9 + z^2/4 = 0$$

This is the equation for an elliptic cone whose vertex is at the origin, and whose axis coincides with the coordinate *y* axis.

Another example

Consider the object in Cartesian *xyz* space represented by

$$-x^2 + y^2 + z^2 = -7$$

When we divide through by -7, we get

$$-x^2/(-7) + y^2/(-7) + z^2/(-7) = 1$$

which simplifies to

$$x^2/7 - y^2/7 - z^2/7 = 1$$

This equation describes a hyperboloid of two sheets whose center is at the origin, and whose axis lies along the coordinate *x* axis.

Still another example

Consider the object in Cartesian *xyz* space represented by

$$15x^2 + 10y^2 = 6z^2 + 30$$

We can subtract $6z^2$ from each side, getting

$$15x^2 + 10y^2 - 6z^2 = 30$$

Dividing through by 30 gives us

$$x^2/2 + y^2/3 - z^2/5 = 1$$

This is the equation for a hyperboloid of one sheet whose center is at the origin, and whose axis lies along the coordinate *z* axis.

Are you confused?

It's reasonable to ask, "What if the center of a hyperboloid, or the vertex of an elliptic cone, lies somewhere other than the origin, say at (x_0, y_0, z_0)? What happens to the equation in that case?" If you're willing to exercise your mathematical intuition, you can probably guess the answer. Make the following substitutions in the equation:

- Replace every occurrence of *x* with $x - x_0$
- Replace every occurrence of *y* with $y - y_0$
- Replace every occurrence of *z* with $z - z_0$

Consider a hyperboloid of two sheets such as the one in Fig. 17-12. The straight-line axis of the "bowls" lies along the coordinate *y* axis, and the center of the entire object is at the origin. If $a = 2$, $b = 3$, and $c = 4$, the equation is

$$-x^2/2^2 + y^2/3^2 - z^2/4^2 = 1$$

which simplifies to

$$-x^2/4 + y^2/9 - z^2/16 = 1$$

Now suppose that you change the equation to

$$-(x-7)^2/4 + (y+1)^2/9 - (z-5)^2/16 = 1$$

You've moved the entire hyperboloid, without altering its overall shape or orientation. It has a new center whose coordinates are (7,−1,5) instead of (0,0,0). The straight-line axis of the two bowls is parallel to, but no longer coincides with, the coordinate y axis. If you want to disguise this equation, you can multiply it through by the product of the denominators on the left-hand side, getting

$$-144(x-7)^2 + 64(y+1)^2 - 36(z-5)^2 = 576$$

Here's a challenge!

Consider the object in Cartesian xyz space represented by

$$3x^2 + 6x - 4y^2 - 16y + 2z^2 - 4z = 35$$

What do we get when we graph this equation in Cartesian xyz space? Where is the center of the object? How is its axis oriented?

Solution

As with some of the examples we've seen, we need lot of intuition to solve this problem. Let's subtract 11 from both sides of the equation. That gives us

$$3x^2 + 6x - 4y^2 - 16y + 2z^2 - 4z - 11 = 24$$

When we subtract 11, we in effect add −11, which happens to be the sum of 3, −16, and 2. That means we can rewrite the above equation as

$$3x^2 + 6x - 4y^2 - 16y + 2z^2 - 4z + 3 - 16 + 2 = 24$$

which can be rearranged to get

$$3x^2 + 6x + 3 - 4y^2 - 16y - 16 + 2z^2 - 4z + 2 = 24$$

Grouping the terms on the left-hand side into trinomials, and paying special attention to the signs associated with the variable y as we group the second three terms, we get

$$(3x^2 + 6x + 3) - (4y^2 + 16y + 16) + (2z^2 - 4z + 2) = 24$$

which morphs to

$$3(x^2 + 2x + 1) - 4(y^2 + 4y + 4) + 2(z^2 - 2z + 1) = 24$$

and further to

$$3(x + 1)^2 - 4(y + 2)^2 + 2(z - 1)^2 = 24$$

Dividing through by 24, we get

$$(x + 1)^2/8 - (y + 2)^2/6 + (z - 1)^2/12 = 1$$

This is the equation of a hyperboloid of one sheet whose center is at $(-1,-2,1)$, and whose axis is oriented along a line parallel to the coordinate y axis.

Practice Exercises

This is an open-book quiz. You may (and should) refer to the text as you solve these problems. Don't hurry! You'll find worked-out answers in App. B. The solutions in the appendix may not represent the only way a problem can be figured out. If you think you can solve a particular problem in a quicker or better way than you see there, by all means try it!

1. Suppose that a plane contains the point $(0,0,0)$, and the standard form of a vector normal to the plane is $-4\mathbf{i} + 4\mathbf{j} - 4\mathbf{k}$. Find the plane's equation in standard form.

2. Suppose that a plane contains the point $(4,5,6)$, and the standard form of a vector normal to the plane is $-2\mathbf{i} + 0\mathbf{j} + 0\mathbf{k}$. Find the plane's equation in standard form.

3. Consider a sphere whose equation is

$$x^2 + 2x + 1 + y^2 - 2y + 1 + z^2 + 8z + 16 = 64$$

What are the coordinates of the center of this sphere? What's its radius?

4. What's the equation of a sphere centered at the point $(5,7,-3)$ and whose radius is equal to the positive square root of 23?

5. Consider the equation

$$8(x - 1)^2 + 8(y + 2)^2 + 6(z + 7)^2 = 24$$

What sort of object does this equation describe? Does the object have a center? If so, what are the coordinates of the center point? Does the object have axial radii? If so, what are they?

6. Consider the equation

$$400(x + 2)^2 + 225(y - 4)^2 + 144z^2 - 3\backslash600 = 0$$

What sort of object does this equation describe? Does the object have a center? If so, what are the coordinates of the center point? Does the object have axial radii? If so, what are they?

7. Consider a surface whose equation is

$$x^2 + 2x + 1 + y^2 - 2y + 1 - z^2 + 6z - 9 = 36$$

What sort of object is this? What are the coordinates of the center? How is the axis oriented?

8. Write down a generalized equation for an elliptic cone whose axis is parallel to the coordinate y axis, and whose vertex is at $(-2,3,4)$.

9. Suppose we slice the elliptic cone described in Problem 8 straight through with the coordinate xz plane. The cone's surface intersects the xz plane in a curve. Derive a generalized equation of that curve in the variables x and z. What sort of curve is it? Here's a hint: At every point in the xz plane, $y = 0$.

10. Suppose we slice the elliptic cone described in Problem 8 straight through with the coordinate xy plane. The cone's surface intersects the xy plane in a curve. Derive a generalized equation of that curve in the variables x and y. What sort of curve is it? Here's a hint: At every point in the xy plane, $z = 0$.

Lines and Curves in Three-Space

In Chap. 16, we learned how parametric equations can define curves that are difficult to portray as conventional relations. "Parametric power" becomes more apparent when we graduate to three dimensions.

Straight Lines

Finding an equation for a straight line in Cartesian three-space is harder than it is in the Cartesian plane. The extra dimension makes expressing the line's location and orientation more complicated. There are at least two ways we can do it: the *symmetric method* and the *parametric method*.

Symmetric method

A straight line in Cartesian *xyz* space can be represented by a three-part *symmetric-form equation*. Suppose that (x_0, y_0, z_0) are the coordinates of a known point on the line, and a, b, and c are nonzero real-number constants. Given this information, we can represent the line as

$$(x - x_0)/a = (y - y_0)/b = (z - z_0)/c$$

If $a = 0$ or $b = 0$ or $c = 0$, then we get a zero denominator somewhere, and the system becomes meaningless.

Direction numbers

In the symmetric-form equation of a straight line, the constants a, b, and c are known as the *direction numbers*. Imagine a vector **m** whose originating point is at the origin $(0,0,0)$ and whose terminating point has coordinates (a,b,c). Under these circumstances, the vector **m** either lies right along, or is parallel to, the line denoted by the symmetric-form equation.

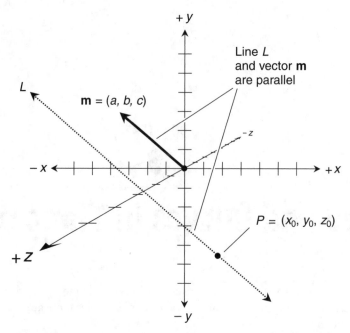

Figure 18-1   We can uniquely define a line L in Cartesian xyz
space on the basis of a point P on L and a vector
$\mathbf{m} = (a,b,c)$ parallel to L.

(In three-space, a vector $\mathbf{m}$ and a straight line L are parallel if and only if the line containing
$\mathbf{m}$ occupies the same plane as L but does not intersect L.) We have

$$\mathbf{m} = a\mathbf{i} + b\mathbf{j} + c\mathbf{k}$$

where $\mathbf{m}$ is the three-dimensional equivalent of the slope of a line in the Cartesian plane.
Figure 18-1 shows a generic example.

Parametric method

Given any particular line L in Cartesian xyz space, we can find infinitely many vectors to play
the role of the direction-defining vector $\mathbf{m}$. If t is a nonzero real number, then any vector

$$t\mathbf{m} = (ta,tb,tc) = ta\mathbf{i} + tb\mathbf{j} + tc\mathbf{k}$$

works just as well as

$$\mathbf{m} = a\mathbf{i} + b\mathbf{j} + c\mathbf{k}$$

for the purpose of defining the direction of L, so we have an alternative way to describe a
straight line using the following equations:

$$x = x_0 + at$$
$$y = y_0 + bt$$
$$z = z_0 + ct$$

The variable t behaves as a "master controller" for the variables x, y, and z, so the above system is a set of parametric equations for a straight line in Cartesian xyz space. To completely define a straight, infinitely long line this way, we must let t vary throughout the entire set of real numbers, including $t = 0$ to "fill the hole" at the point (x_0, y_0, z_0).

An example

Let's find the symmetric-form equation for the line L shown in Fig. 18-2. As indicated in the drawing, L passes through the point

$$P = (-5, -4, 3)$$

and is parallel to the vector

$$\mathbf{m} = 3\mathbf{i} + 5\mathbf{j} - 2\mathbf{k}$$

The direction numbers of L are the coefficients of the vector $\mathbf{m}$, so we have

$$a = 3$$
$$b = 5$$
$$c = -2$$

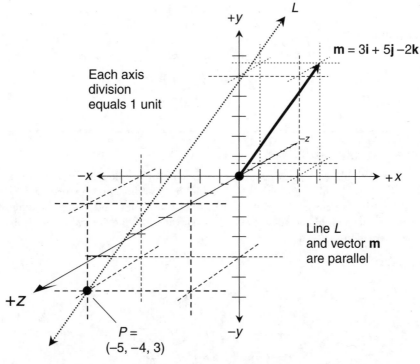

Figure 18-2   What are the symmetric and parametric equations for line L?

We are given a point P on the line L with the coordinates

$$x_0 = -5$$
$$y_0 = -4$$
$$z_0 = 3$$

The general symmetric-form equation for a line in Cartesian xyz space is

$$(x - x_0)/a = (y - y_0)/b = (z - z_0)/c$$

When we plug in the known values, we get the three-part equation

$$[x - (-5)]/3 = [y - (-4)]/5 = (z - 3)/(-2)$$

which simplifies to

$$(x + 5)/3 = (y + 4)/5 = (z - 3)/(-2)$$

Another example

Let's find a set of parametric equations for the line L shown in Fig. 18-2. In this case, our work is easy. We can take the values of $x_0, y_0, z_0, a, b,$ and c that we already know, and plug them into the generalized set of parametric equations

$$x = x_0 + at$$
$$y = y_0 + bt$$
$$z = z_0 + ct$$

The results are

$$x = -5 + 3t$$
$$y = -4 + 5t$$
$$z = 3 - 2t$$

Are you confused?

For any particular line in Cartesian xyz space, there are infinitely many valid ordered triples that can represent the direction numbers. If a line has the direction numbers $(2,3,4)$, then we can multiply all three entries by a real number other than 0 or 1, and we'll get another valid ordered triple of direction numbers. For example, all of the following ordered triples represent the same line orientation as $(2,3,4)$:

$$(4,6,8)$$
$$(-2,-3,-4)$$
$$(20,30,40)$$
$$(-20,-30,-40)$$
$$(2\pi,3\pi,4\pi)$$
$$(-2\pi,-3\pi,-4\pi)$$

"That's interesting," you say, "but which direction numbers are the best?" In theory, it doesn't matter; any of the above ordered triples is as "good" as any other. Nevertheless, from an esthetic point of view, it's a good idea to reduce an ordered triple of direction numbers so that the only common divisor is 1, and so that there is *at most* one negative element. According to that standard, (2,3,4) are the preferred direction numbers.

Here's a challenge!

Consider the following three-way equation that represents a straight line in Cartesian *xyz* space:

$$3x - 6 = 4y - 12 = 6z - 24$$

Find a point on the line. Determine the preferred direction numbers. Based on that information, write down the direction vector as a sum of multiples of **i**, **j**, and **k**.

Solution

Before we think about the direction numbers or any specific point on the line, let's try to get the equation into the standard symmetric form. We can multiply the left-hand part of the equation by 4/4, the middle part by 3/3, and the right-hand part by 2/2. That gives us

$$4(3x - 6)/4 = 3(4y - 12)/3 = 2(6z - 24)/2$$

Multiplying out the numerators, we get

$$(12x - 24)/4 = (12y - 36)/3 = (12z - 48)/2$$

We can factor out 12 from each of the numerators to obtain

$$12(x - 2)/4 = 12(y - 3)/3 = 12(z - 4)/2$$

Dividing the entire equation through by 12 gives us the standard symmetric form

$$(x - 2)/4 = (y - 3)/3 = (z - 4)/2$$

We remember that the generalized symmetric equation for a straight line in Cartesian *xyz* space is

$$(x - x_0)/a = (y - y_0)/b = (z - z_0)/c$$

where (x_0, y_0, z_0) are the coordinates of a specific point on the line, and *a*, *b*, and *c* are the direction numbers. Comparing the symmetric-form equation we derived with the generalized form, we can see that

$$x_0 = 2$$
$$y_0 = 3$$
$$z_0 = 4$$

This tells us that (2,3,4) is a point on the line. We can also see that

$$a = 4$$
$$b = 3$$
$$c = 2$$

so the line's direction numbers are (4,3,2). We can write down a standard-form direction vector **m** from these numbers as

$$\mathbf{m} = 4\mathbf{i} + 3\mathbf{j} + 2\mathbf{k}$$

Parabolas

From algebra, we remember that a *quadratic equation* in a variable x can always be written in the form

$$a_1 x^2 + a_2 x + a_3 = 0$$

where a_1, a_2, and a_3 are real-number constants called the *coefficients*, and $a_1 \neq 0$. If we replace the 0 on the right-hand side of this equation by another variable and then transpose the sides, we get an expression for a *quadratic function*. For example,

$$4x^2 + 2x + 1 = 0$$

is a quadratic equation in x, but

$$y = 4x^2 + 2x + 1$$

is a quadratic function in which the independent variable is x and the dependent variable is y. If we give our function a name (f, for example), then we can denote it as

$$f(x) = 4x^2 + 2x + 1$$

When we graph a quadratic function in Cartesian two-space, we always get a parabola that's fairly easy to graph, because there's only one plane to worry about (the xy plane, if our independent variable is x and our dependent variable is y). In xyz space, the situation is more complicated, because we have an extra variable. There are infinitely many different planes in which a parabola can lie, as well as infinitely many different shapes and orientations for a parabola in any particular plane. Let's look at a few simple cases.

Hold x constant

Imagine a parameter t that's allowed to wander all over the set of real numbers. Also imagine a generalized quadratic function f of this parameter, such that

$$f(t) = a_1 t^2 + a_2 t + a_3$$

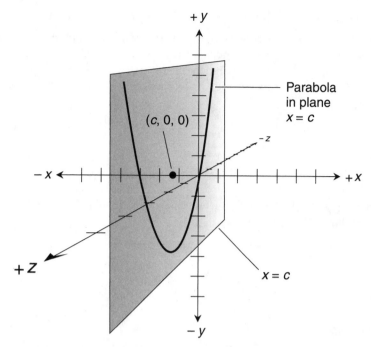

Figure 18-3 Parabola in a plane where x is held to a constant value c. The plane is perpendicular to the x axis, and intersects that axis at the point $(c,0,0)$.

where a_1, a_2, and a_3 are real-number coefficients. Let's go into Cartesian xyz space and restrict ourselves to a single plane in which the value of x is some real-number constant c. This plane is parallel to the yz plane, and it intersects the x axis at the point $(c,0,0)$. Consider a parabola in the plane $x = c$ whose axis is parallel to the y axis, as shown in Fig. 18-3. (The *axis of a parabola* is a straight line in the same plane as the parabola, and on either side of which the parabola is symmetrical.) In this situation, the value of z tracks right along with the value of t, while the variable y follows $f(t)$. Therefore

$$x = c$$
$$y = f(t) = a_1 t^2 + a_2 t + a_3$$
$$z = t$$

The above set of equations is a parametric description of our parabola. If we want to describe a parabola in the plane $x = c$ whose axis is parallel the z axis instead of the y axis, then y follows t while z follows $f(t)$, and we have

$$x = c$$
$$y = t$$
$$z = f(t) = a_1 t^2 + a_2 t + a_3$$

Hold *y* constant

Now suppose that we restrict our movements to a plane in which the value of y is always equal to a constant c. The equation of the plane is $y = c$. It's parallel to the xz plane, and it intersects the y axis at $(0,c,0)$. Imagine a parabola in this plane whose axis is parallel to the z axis, as shown in Fig. 18-4. In this situation, x follows t while z follows $f(t)$, and the curve can be described as

$$x = t$$
$$y = c$$
$$z = f(t) = a_1 t^2 + a_2 t + a_3$$

To describe a parabola in the plane $y = c$ whose axis is parallel the x axis, we can let z follow t and let x follow $f(t)$, getting the system

$$x = f(t) = a_1 t^2 + a_2 t + a_3$$
$$y = c$$
$$z = t$$

Hold *z* constant

Finally, let's confine our attention to a single plane in which the value of z is some real-number constant c. The plane $z = c$ is parallel to the xy plane, and it intersects the z axis at $(0,0,c)$.

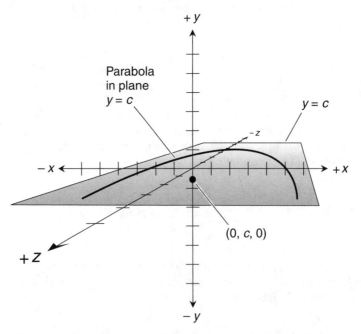

Figure 18-4   Parabola in a plane where y is held to a constant value c. The plane is perpendicular to the y axis, and intersects that axis at the point $(0,c,0)$.

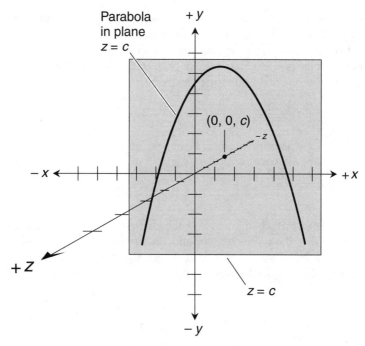

Figure 18-5   Parabola in a plane where z is held to a constant value c. The plane is perpendicular to the z axis, and intersects that axis at the point $(0,0,c)$.

Imagine a parabola in the plane $z = c$ whose axis is parallel to the y axis as shown in Fig. 18-5. Here, x follows t while y follows $f(t)$. We therefore have the parametric system

$$x = t$$
$$y = f(t) = a_1 t^2 + a_2 t + a_3$$
$$z = c$$

For a parabola in the plane $z = c$ whose axis is parallel to the x axis instead of the y axis, the value of y follows t while the value of x follows $f(t)$, so we have

$$x = f(t) = a_1 t^2 + a_2 t + a_3$$
$$y = t$$
$$z = c$$

An example

Consider a quadratic function in the plane $x = 2$. Suppose that the parametric equations are

$$x = 2$$
$$y = t$$
$$z = t^2 - 3t + 2$$

Using the knowledge we've gained so far in this chapter, along with our existing knowledge of algebra (such as we got from *Algebra Know-It-All* or a comparable algebra book), let's draw a graph of this function. Imagine that we're gazing broadside at the plane $x = 2$ from some distant point on the positive x axis. We've been told that $y = t$. If we stay in the plane $x = 2$, we can therefore write the quadratic function by direct substitution as

$$z = y^2 - 3y + 2$$

The coefficient of y^2 is positive, so the parabola opens in the positive z direction. The above polynomial equation factors into

$$z = (y - 1)(y - 2)$$

so we can see that $z = 0$ when $y = 1$, and also that $z = 0$ when $y = 2$. Because x is always equal to 2, we know that the points (2,1,0) and (2,2,0) are on the parabola. The curve opens in the positive z direction, so we know that the parabola must have an absolute minimum. The y-value at the point, y_{min}, is the average of the y-values of the points where $z = 0$. Therefore

$$y_{min} = (1 + 2)/2$$
$$= 3/2$$

To find the z-value at this point, we plug 3/2 into the quadratic function and get

$$z_{min} = (3/2)^2 - 3 \times 3/2 + 2 = 9/4 - 9/2 + 2$$
$$= 9/4 - 18/4 + 8/4 = (9 - 18 + 8)/4 = -1/4$$

We've determined that the coordinates of the absolute minimum are (2,3/2,−1/4). We also know that the points (2,1,0) and (2,2,0) lie on the parabola. Figure 18-6 shows these points. They're close together, so it's difficult to get a clear picture of the parabola based on their locations. But we can find another point to help us draw the curve. When we plug in 0 for y, we get

$$z = y^2 - 3y + 2 = 0^2 - 3 \times 0 + 2$$
$$= 0 - 0 + 2 = 2$$

This tells us that the point (2,0,2) is on the curve. It's also shown in Fig. 18-6.

Another example

Now let's look at a quadratic function in the plane where $y = 5$. Suppose that the parametric equations are

$$x = t$$
$$y = 5$$
$$z = 2t^2 + 4t + 3$$

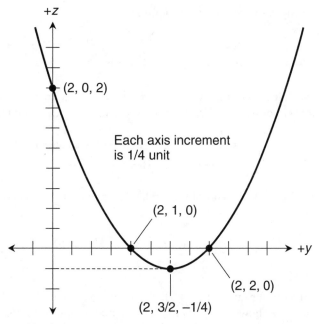

Figure 18-6 Graph of a parabola in a plane parallel to the *yz* plane, such that *x* has a constant value of 2. On both axes, each increment represents 1/4 unit.

Imagine that we're gazing broadside at the plane $y = 5$ from somewhere on the negative y axis. We have been told that $x = t$, so we can write the quadratic function as

$$z = 2x^2 + 4x + 3$$

This parabola opens in the positive z direction, because the coefficient of x^2 is positive. That means this parabola attains an absolute minimum for some value of x. Let's call it x_{min}. When x is the independent variable and z is the dependent variable, the general polynomial form for a quadratic function is

$$z = a_1 x^2 + a_2 x + a_3$$

where a_1, a_2, and a_3 are constants. From our algebra courses, we know that

$$x_{min} = -a_2 / (2a_1)$$

In this situation, we have

$$x_{min} = -4 / (2 \times 2) = (-4)/4 = -1$$

The z-value at the absolute minimum point is

$$z_{min} = 2x_{min}^2 + 4x_{min} + 3 = 2 \times (-1)^2 + 4 \times (-1) + 3$$
$$= 2 - 4 + 3 = -2 + 3 = 1$$

Now we know that the coordinates of the parabola's vertex are $(-1,5,1)$. As the basis for our next point, let's choose $x = -3$. We can plug it directly into the function to get

$$z = 2x^2 + 4x + 3 = 2 \times (-3)^2 + 4 \times (-3) + 3$$
$$= 18 - 12 + 3 = 6 + 3 = 9$$

This gives us $(-3,5,9)$ as the coordinates of a second point on the curve. Finally, let's set $x = 1$. Plugging it in, we obtain

$$z = 2x^2 + 4x + 3 = 2 \times 1^2 + 4 \times 1 + 3$$
$$= 2 + 4 + 3 = 9$$

The third point on our curve is $(1,5,9)$. We now have three points: $(-3,5,9)$, $(-1,5,1)$, and $(1,5,9)$. Figure 18-7 shows these points, along with a graph of the parabola passing through them, as seen in the plane where y maintains a constant value of 5.

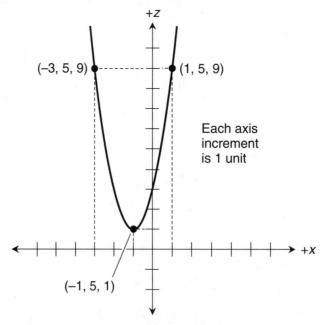

Figure 18-7 Graph of a parabola in a plane parallel to the *xz* plane, such that *y* has a constant value of 5. On both axes, each increment represents 1 unit.

Are you curious and ambitious?

Think about the graphs of higher-degree polynomial functions confined to specific planes in Cartesian *xyz* space. For example, consider the *cubic function* in the plane where $x = 2$, such that

$$x = 2$$
$$y = t$$
$$z = t^3$$

or the *quartic function* in the plane where $z = -7$, such that

$$x = 3t^4 + 6$$
$$y = t$$
$$z = -7$$

Can you draw graphs of these curves?

Circles

In Chap. 13, we learned that the equation of a circle centered at the origin in the Cartesian *xy* plane can be written in the form

$$x^2 + y^2 = r^2$$

where *r* is the radius. In Chap. 16, we learned that the parametric equations for such a circle are

$$x = r \cos t$$

and

$$y = r \sin t$$

where *t* is the parameter. Let's expand these notions to deal with any circle in *xyz* space that's centered on, and exists entirely in a plane perpendicular to, one of the three coordinate axes.

Hold *x* constant

Consider a plane $x = c$ in Cartesian *xyz* space, where *c* is a constant. This plane is parallel to the *yz* plane, and it intersects the *x* axis at $(c,0,0)$. Imagine a circle of radius *r* in the plane $x = c$ that's centered on the *x* axis as shown in Fig. 18-8. The variable *y* follows along with

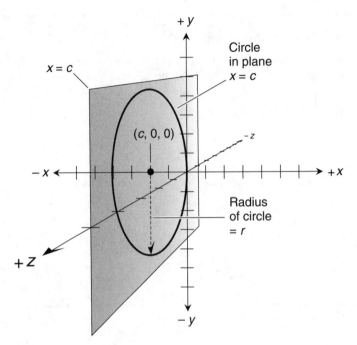

Figure 18-8   Circle in a plane where *x* is held to a constant value *c*. The plane is perpendicular to the *x* axis, and intersects that axis at the point (*c*,0,0). The circle has radius *r* and is centered at (*c*,0,0).

$r \cos t$, while the variable z follows along with $r \sin t$. Therefore, we can define our circle with the system of parametric equations

$$x = c$$
$$y = r \cos t$$
$$z = r \sin t$$

For the circle to be fully circumscribed, the parameter t must range continuously over a span of values sufficient to ensure that a moving point makes at least one full revolution around the x axis. The smallest such span is any half-open interval that's at least 2π units wide.

Hold y constant

Now suppose that we restrict ourselves to a plane such that $y = c$, where c is a constant. This plane is parallel to the xz plane, and it intersects the y axis at (0,*c*,0). Imagine a circle in the plane $y = c$ that's centered on the y axis, as shown in Fig. 18-9. In this case, the circle is described by the system

$$x = r \cos t$$
$$y = c$$
$$z = r \sin t$$

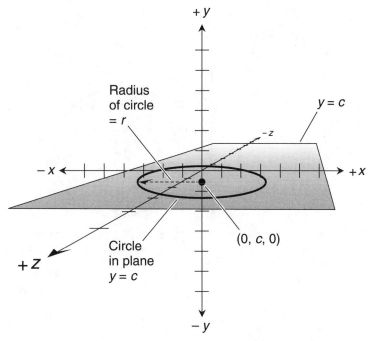

Figure 18-9   Circle in a plane where *y* is held to a constant value *c*.
The plane is perpendicular to the *y* axis, and intersects
that axis at the point (0,*c*,0). The circle has radius *r* and
is centered at (0,*c*,0).

For a complete circle to be described, the parameter *t* must range continuously over a
span of values sufficient to ensure that a moving point makes at least one full revolution around the *y* axis. The smallest such span is any half-open interval that's at least
2π units wide.

Hold *z* constant

Finally, consider a plane in which *z* = *c*. It's parallel to the *xy* plane, and it intersects the *z* axis
at (0,0,*c*). Imagine a circle in the plane *z* = *c* that's centered on the *z* axis as shown in Fig. 18-10.
Here, we have

$$x = r \cos t$$
$$y = r \sin t$$
$$z = c$$

For a complete circle to be described, the parameter *t* must range continuously over a
span of values sufficient to ensure that a moving point makes at least one full revolution around the *z* axis. The smallest such span is any half-open interval that's at least
2π units wide.

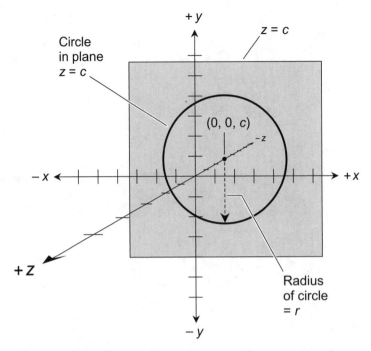

Figure 18-10   Circle in a plane where *z* is held to a constant value *c*. The plane is perpendicular to the *z* axis, and intersects that axis at the point (0,0,*c*). The circle has radius *r* and is centered at (0,0,*c*).

An example

Imagine a circle in the plane $x = 3$. Suppose that the circle is centered on the *x* axis, and its parametric equations are

$$x = 3$$
$$y = 3 \cos t$$
$$z = 3 \sin t$$

In the plane $x = 3$, our circle can be described by the two parametric equations

$$y = 3 \cos t$$

and

$$z = 3 \sin t$$

When we express this system as a relation between *y* and *z*, we have

$$y^2 + z^2 = 9$$

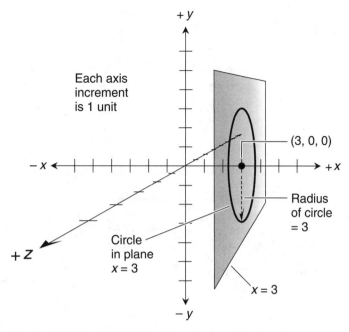

Figure 18-11 Graph of a circle of radius 3 in the plane $x = 3$, centered on (3,0,0). Each axis increment represents 1 unit.

Figure 18-11 is a perspective rendition of this circle's graph in Cartesian *xyz* space. The radius is 3, and the center is at (3,0,0).

Another example

Now let's look at a circle having a radius of 3 units, and contained in the plane where y maintains a constant value of -2. The parametric equations are

$$x = 3 \cos t$$
$$y = -2$$
$$z = 3 \sin t$$

In the plane $y = -2$, our circle can be described by the parametric system

$$x = 3 \cos t$$

and

$$z = 3 \sin t$$

As a relation between x and z, this system can be represented by

$$x^2 + z^2 = 9$$

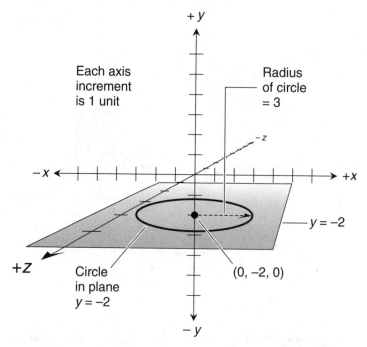

Figure 18-12   Graph of a circle of radius 3 in the plane $y = -2$, centered on $(0,-2,0)$. Each axis increment represents 1 unit.

Figure 18-12 is a perspective graph of this circle in Cartesian xyz space.

Are you confused?

All of the parabolas and circles described in this chapter are confined to planes parallel to the xy plane, the xz plane, or the yz plane. Finding equations for curves in other planes is sometimes easy, but more often it's difficult. The process can be streamlined by adding a function called a *coordinate transformation* to the relation describing the curve. That way, any curve that lies in a single plane (no matter how the plane is oriented in space, and no matter where the curve is positioned within the plane) can be described in terms of a curve in the xy plane, the xz plane, or the yz plane. You'll learn how to do coordinate transformations in advanced calculus or analysis courses.

Here's a challenge!

Consider a curve whose parametric equations are

$$x = 2 \cos t$$
$$y = 3 \sin t$$
$$z = -3$$

What sort of curve is this? Sketch its graph in Cartesian xyz space.

Solution

For a moment, suppose that the coefficient in the first equation was 3 rather than 2. In that case, the set of parametric equations would be

$$x = 3 \cos t$$
$$y = 3 \sin t$$
$$z = -3$$

and we'd have a circle in the plane $z = -3$. Figure 18-10, on page 360 is an approximate graph of this circle if we imagine each coordinate axis division to represent 1 unit. However, the coefficient in the first equation is 2, not 3. Therefore, the curve is squashed in the x direction; it's only 2/3 as wide as the above described circle. This squashed circle is an ellipse centered on the point $(0,0,-3)$. Figure 18-13 shows how its graph looks in Cartesian xyz space, from a vantage point far from the origin but close to the positive z axis.

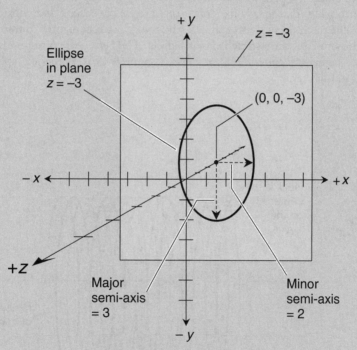

Figure 18-13   Graph of an ellipse in the plane $z = -3$, centered on $(0,0,-3)$. Each axis increment is 1 unit.

Circular Helixes

When we created the generalized circles and graphed them as shown in Figs. 18-8 through 18-10, we held one variable constant and forced the other two variables to follow the parametric equations for a circle in a plane. Now imagine that, instead of holding one variable constant, we let it change according to a constant multiple of the parameter. When we do this, we get a three-dimensional object called a *circular helix*.

Center on x axis

Consider a moving plane $x = ct$ in Cartesian xyz-space, where c is a constant and t is the parameter. This plane is always perpendicular to the x axis, so it's always parallel to the yz plane. It intersects the x axis at a moving point $(ct,0,0)$. Imagine a moving a circle of radius r in the moving plane $x = ct$ that's centered on the x axis. On this circle, the value of y tracks along with $r \cos t$, while the value of z tracks along with $r \sin t$. The complete set of parametric equations is

$$x = ct$$
$$y = r \cos t$$
$$z = r \sin t$$

When we graph the path of a point on this moving circle as t varies, we get a circular helix of uniform *pitch* (that means its "coil turns" are evenly spaced, like those of a well-designed spring). The pitch depends on c. Small values of c produce tightly compressed helixes, while large values of c produce stretched-out helixes. The *helix axis* corresponds to the coordinate x axis, so the helix is *centered* on the x axis. Figure 18-14 is a generic graph of a circular helix oriented in this way.

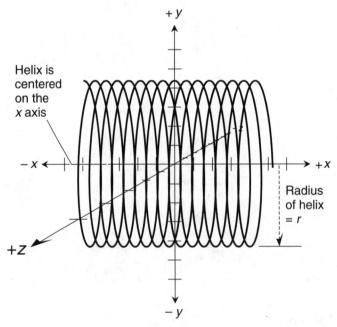

Figure 18-14   Circular helix of radius r, centered on the x axis. The pitch depends on the constant by which t is multiplied to obtain x.

Center on *y* axis

Now imagine a moving plane $y = ct$ that's perpendicular to the y axis, parallel to the xz plane, and intersects the y axis at a moving point $(0,ct,0)$. The value of x tracks along with $r \cos t$, while the value of z tracks along with $r \sin t$, so we have the system

$$x = r \cos t$$
$$y = ct$$
$$z = r \sin t$$

The graph of this set of parametric equations is a circular helix of uniform pitch, centered on the y axis as shown in Fig. 18-15.

Center on *z* axis

Finally, envision a moving plane $z = ct$ that's perpendicular to the z axis, parallel to the xy plane, and intersects the z axis at a moving point $(0,0,ct)$. The value of x follows $r \cos t$, while y follows $r \sin t$. Our parametric equations are therefore

$$x = r \cos t$$
$$y = r \sin t$$
$$z = ct$$

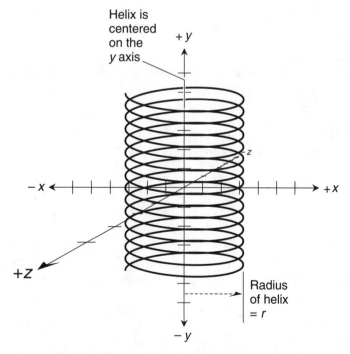

Figure 18-15　Circular helix of radius r, centered on the y axis. The pitch depends on the constant by which t is multiplied to obtain y.

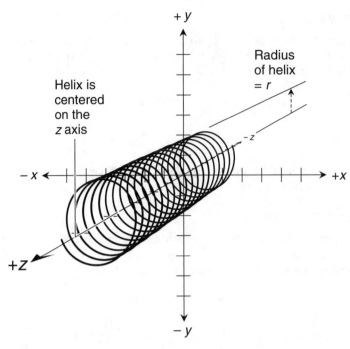

Figure 18-16 Circular helix of radius *r*, centered on the *z* axis. The pitch depends on the constant by which *t* is multiplied to obtain *z*.

In Cartesian *xyz* space, these equations produce a circular helix of uniform pitch, centered on the *z* axis. Figure 18-16 is a generic graph.

An example

Consider a circular helix centered on the *x* axis, described by the parametric equations

$$x = t/(2\pi)$$
$$y = \cos t$$
$$z = \sin t$$

Here are some values of *x*, *y*, and *z* that we can calculate as *t* varies, causing a point on the helix to complete a single revolution in a plane perpendicular to the *x* axis:

- When $t = 0$, we have $x = 0$, $y = 1$, and $z = 0$.
- When $t = \pi/2$, we have $x = 1/4$, $y = 0$, and $z = 1$.
- When $t = \pi$, we have $x = 1/2$, $y = -1$, and $z = 0$.
- When $t = 3\pi/2$, we have $x = 3/4$, $y = 0$, and $z = -1$.
- When $t = 2\pi$, we have $x = 1$, $y = 1$, and $z = 0$.

Every time t increases by 2π, our point makes one complete revolution in a moving plane that's always perpendicular to the x axis. Also, every time t increases by 2π, our point gets 1 unit farther away from the yz plane. The pitch of the helix is therefore equal to 1 linear unit per revolution.

Another example

Consider a circular helix centered on the y axis, described by the parametric equations

$$x = 2 \cos t$$
$$y = t$$
$$z = 2 \sin t$$

Here are some values of x, y, and z that we can calculate as t varies, causing a point on the helix to complete a single revolution in a plane perpendicular to the y axis:

- When $t = 0$, we have $x = 2$, $y = 0$, and $z = 0$.
- When $t = \pi/2$, we have $x = 0$, $y = \pi/2$, and $z = 2$.
- When $t = \pi$, we have $x = -2$, $y = \pi$, and $z = 0$.
- When $t = 3\pi/2$, we have $x = 0$, $y = 3\pi/2$, and $z = -2$.
- When $t = 2\pi$, we have $x = 2$, $y = 2\pi$, and $z = 0$.

Every time t increases by 2π, our point makes a complete revolution in a moving plane that's always perpendicular to the y axis. Also, every time t increases by 2π, our point moves 2π units farther away from the xz plane. The pitch of the helix is therefore equal to 2π linear units per revolution.

Are you confused?

You might ask, "When describing a helix with parametric equations, does it make any difference if we multiply t by a positive constant or a negative constant?" That's an excellent question. The answer is yes; it matters a lot!

The polarity of the constant affects the *sense* in which the helix rotates as we move in the positive direction. For example, suppose we have a helix described by the parametric equations

$$x = 3t$$
$$y = 3 \cos t$$
$$z = 3 \sin t$$

In this case, the helix turns counterclockwise as we move in the positive x direction. If we observe the situation from somewhere on the positive x axis while the value of t increases, a point on the helix will appear to approach us and rotate counterclockwise. If the value of t decreases, a point on the helix will appear to retreat from us and rotate clockwise.

Now suppose that we reverse the sign of the constant in the first equation, so our system becomes

$$x = -3t$$
$$y = 3 \cos t$$
$$z = 3 \sin t$$

If we watch this scene from somewhere on the positive x axis while the value of t increases, a point on the helix will appear to retreat from us and rotate counterclockwise. If the value of t decreases, a point on the helix will appear to approach us and rotate clockwise.

Are you astute?

Imagine yourself at some point far from the origin on the $+x$ axis in Fig. 18-14, or far from the origin on the $+y$ axis in Fig. 18-15, or far from the origin on the $+z$ axis in Fig. 18-16. If you have excellent spatial perception, you'll be able to figure out that in all three of these situations, the constant c is negative! In each case, a retreating point on the helix will appear to revolve counterclockwise, and an approaching point on the helix will appear to revolve clockwise.

Here's a challenge!

Suppose that we encounter an object in Cartesian *xyz* space whose parametric equations are

$$x = 2 \cos t$$
$$y = 3 \sin t$$
$$z = -3t$$

What sort of object is this?

Solution

Let's divide the first two equations through by their respective constants. That gives us

$$x/2 = \cos t$$

and

$$y/3 = \sin t$$

Squaring both sides of both equations, we obtain

$$(x/2)^2 = \cos^2 t$$

and

$$(y/3)^2 = \sin^2 t$$

When we add these two equations, left-to-left and right-to-right, we get

$$(x/2)^2 + (y/3)^2 = \cos^2 t + \sin^2 t$$

The rules of trigonometry tell us that

$$\cos^2 t + \sin^2 t = 1$$

so the preceding equation can be rewritten as

$$(x/2)^2 + (y/3)^2 = 1$$

and further morphed into

$$x^2/4 + y^2/9 = 1$$

This equation describes an ellipse in the *xy* plane whose horizontal (*x*-coordinate) semi-axis measures 2 units, and whose vertical (*y*-coordinate) semi-axis measures 3 units. Now let's consider the *z* coordinate. The equation for *z* in terms of *t* is

$$z = -3t$$

This equation tells us that a point on our object travels in the negative *z* direction as the value of the parameter *t* increases. The complete set of three parametric equations therefore describes an *elliptical helix* centered on the *z* axis. As we move in the positive *z* direction, the helix rotates clockwise, because the coefficient of *t* is negative. It looks something like the helix in Fig. 18-16, except that it's stretched by approximately 50 percent in the positive and negative *y* directions (vertically in this particular illustration).

Here's an extra-credit challenge!

Sketch three-dimensional perspective graphs of the helixes described in the foregoing two examples and challenge.

Solution

You're on your own. That's why you get extra credit!

Practice Exercises

This is an open-book quiz. You may (and should) refer to the text as you solve these problems. Don't hurry! You'll find worked-out answers in App. B. The solutions in the appendix may not represent the only way a problem can be figured out. If you think you can solve a particular problem in a quicker or better way than you see there, by all means try it!

1. Consider the following three-way equation for a straight line in Cartesian *xyz* space:

$$x - 1 = y - 2 = z - 4$$

   Find a point on the line, find the preferred direction numbers, and determine the direction vector as a sum of multiples of **i**, **j**, and **k**.

2. Consider the following three-way equation for a straight line in Cartesian *xyz* space:

$$4x = 5y = 6z$$

   Find a point on the line, find the preferred direction numbers, and determine the direction vector as a sum of multiples of **i**, **j**, and **k**.

3. Consider the following three-way equation for a straight line in Cartesian *xyz* space:

$$(x - 2)/3 = (4y - 8)/4 = (z + 5)/(-2)$$

   Find a point on the line, find the preferred direction numbers, and determine the direction vector as a sum of multiples of **i**, **j**, and **k**.

4. Consider a relation in Cartesian *xyz* space described by the system of parametric equations

$$x = -4$$
$$y = t$$
$$z = -t^2 - 1$$

   Draw a two-dimensional graph of this relation as it appears when we look broadside at the plane containing it.

5. Consider a relation in Cartesian *xyz* space described by the system of parametric equations

$$x = t^2 + 2t$$
$$y = t$$
$$z = 0$$

Draw a two-dimensional graph of this relation as it appears when we look broadside at the plane containing it.

6. Consider a relation in Cartesian *xyz* space described by the system of parametric equations

$$x = t$$
$$y = -7$$
$$z = t^2/2 - 5$$

Draw a two-dimensional graph of this relation as it appears when we look broadside at the plane containing it.

7. Consider a relation in Cartesian *xyz* space described by the system of parametric equations

$$x = 4 \cos t$$
$$y = 4 \sin t$$
$$z = 1$$

Draw a two-dimensional graph of this relation as it appears when we look broadside at the plane containing it.

8. Consider a relation in Cartesian *xyz* space described by the system of parametric equations

$$x = 5 \cos t$$
$$y = 0$$
$$z = 5 \sin t$$

Draw a two-dimensional graph of this relation as it appears when we look broadside at the plane containing it.

9. Consider a relation in Cartesian *xyz* space described by the system of parametric equations

$$x = 5 \cos t$$
$$y = 3 \sin t$$
$$z = \pi$$

Draw a two-dimensional graph of this relation as it appears when we look broadside at the plane containing it.

10. Consider a relation in Cartesian *xyz* space described by the system of parametric equations

$$x = 2 \cos t$$
$$y = t/(2\pi)$$
$$z = 2 \sin t$$

Draw a perspective view of this relation's three-dimensional graph. Here's a hint: You can probably tell that the graph is a circular helix, but as you draw it, pay attention to the orientation, the pitch, and the sense of rotation.

19

Sequences, Series, and Limits

Have you ever tried to find the missing number in a list? Have you ever figured out how much money an interest-bearing bank account will hold after 10 years? Have you ever calculated the value that a function approaches but never reaches? If you can answer "Yes" to any of these questions, you've worked with *sequences* (also called *progressions*), *series*, or *limits*.

Repeated Addition

A sequence is a list of numbers. Some sequences are finite; others are infinite. The simplest sequences have values that repeatedly increase or decrease by a fixed amount. Here are some examples:

$$A = 1, 2, 3, 4, 5, 6$$
$$B = 0, -1, -2, -3, -4, -5$$
$$C = 2, 4, 6, 8$$
$$D = -5, -10, -15, -20$$
$$E = 4, 8, 12, 16, 20, 24, 28, ...$$
$$F = 2, 0, -2, -4, -6, -8, -10, ...$$

The first four sequences are finite. The last two are infinite, as indicated by an *ellipsis* (three dots) at the end.

Arithmetic sequence

In each of the sequences shown above, the values either increase steadily (in A, C, and E) or decrease steadily (in B, D, and F). In all six sequences, the spacing between numbers is constant throughout. Here's how each sequence changes as we move along from term to term:

- The values in A always increase by 1.
- The values in B always decrease by 1.
- The values in C always increase by 2.

- The values in D always decrease by 5.
- The values in E always increase by 4.
- The values in F always decrease by 2.

Each sequence has an initial value. After that, we can easily predict subsequent values by repeatedly adding a constant. If the constant is positive, the sequence increases. If the added constant is negative, the sequence decreases.

Suppose that s_0 is the first number in a sequence S. Let c be a real-number constant. If S can be written in the form

$$S = s_0, (s_0 + c), (s_0 + 2c), (s_0 + 3c), \ldots$$

then it's an *arithmetic sequence* or an *arithmetic progression*. In this context, the word "arithmetic" is pronounced "err-ith-MET-ick."

The numbers s_0 and c can be integers, but that's not a requirement. They can be fractions such as 2/3 or −7/5. They can be irrational numbers such as the square root of 2. As long as the separation between any two adjacent terms is the same wherever we look, we have an arithmetic sequence, even in the trivial case

$$S_0 = 0, 0, 0, 0, 0, 0, 0, \ldots$$

Arithmetic series

A series is the sum of all the terms in a sequence. For an arithmetic sequence, the corresponding *arithmetic series* can be defined only if the sequence has a finite number of terms. For the above sequences A through F, let the corresponding series be called A_+ through F_+. The total sums are as follows.

$$A_+ = 1 + 2 + 3 + 4 + 5 + 6 = 21$$
$$B_+ = 0 + (-1) + (-2) + (-3) + (-4) + (-5) = -15$$
$$C_+ = 2 + 4 + 6 + 8 = 20$$
$$D_+ = (-5) + (-10) + (-15) + (-20) = -50$$
$$E_+ \text{ is not defined}$$
$$F_+ \text{ is not defined}$$

Now consider the infinite series

$$S_{0+} = 0 + 0 + 0 + 0 + 0 + 0 + 0 + \cdots$$

We might think of S_{0+} as "infinity times 0," because it's the sum of 0 added to itself infinitely many times. It's tempting to suppose that $S_{0+} = 0$, but we can't prove it. When we add up any finite number of "nothings", we get "nothing", of course. However, when we try to find the sum of infinitely many nothings, we encounter a mystery. The best we can do is say that S_{0+} is undefined.

Graphing an arithmetic sequence

When we plot the values of an arithmetic sequence as a function of the term number in rectangular coordinates, we get a set of discrete points. We can depict the term number along the horizontal axis going toward the right, so the term number plays the role of the independent variable. We can plot the term value along the vertical axis, so it plays the role of the dependent variable.

Figure 19-1 illustrates two arithmetic sequences as they appear when graphed in this way. (The dashed lines connect the dots, but they aren't actually parts of the sequences.) One sequence is increasing, and the dashed line connecting this set of points ramps upward as we go toward the right. Because this sequence is finite, the dashed line ends at (6,6). The other sequence is decreasing, and its dashed line ramps downward as we go toward the right. This sequence is infinite, as shown by the ellipsis at the end of the string of numbers, and also by the arrow at the right-hand end of the dashed line.

When any arithmetic sequence is graphed according to the scheme shown in Fig. 19-1, its points lie along a straight line. The slope m of the line depends on whether the sequence increases (positive slope) or decreases (negative slope). In fact, m is equal to the constant c in the general arithmetic series form:

$$S = s_0, (s_0 + c), (s_0 + 2c), (s_0 + 3c), \ldots$$

regardless of how many terms the sequence contains.

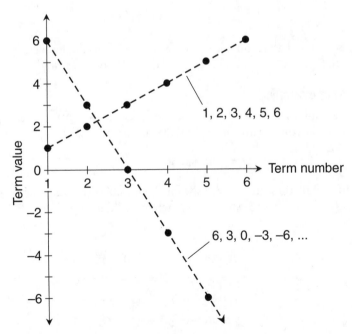

Figure 19-1   Rectangular-coordinate plots of two arithmetic sequences.

An example

Suppose that in an infinite sequence S, we have $s_0 = 5$ and $c = 3$. The first 10 terms are

$$s_0 = 5$$
$$s_1 = s_0 + 3 = 5 + 3 = 8$$
$$s_2 = s_1 + 3 = 8 + 3 = 11$$
$$s_3 = s_2 + 3 = 11 + 3 = 14$$
$$s_4 = s_3 + 3 = 14 + 3 = 17$$
$$s_5 = s_4 + 3 = 17 + 3 = 20$$
$$s_6 = s_5 + 3 = 20 + 3 = 23$$
$$s_7 = s_6 + 3 = 23 + 3 = 26$$
$$s_8 = s_7 + 3 = 26 + 3 = 29$$
$$s_9 = s_8 + 3 = 29 + 3 = 32$$

Therefore

$$S = 5, 8, 11, 14, 17, 20, 23, 26, 29, 32, \ldots$$

Another example

Consider the following sequence T. Someone asks, "Is this an arithmetic sequence? If so, what are the values t_0 (the starting value) and c_t (the constant of change)?"

$$T = 2, 4, 8, 16, 32, 64, 128, 256, 512, \ldots$$

In this case, T is not an arithmetic sequence. The numbers do not increase at a steady rate. There is a pattern, however. Each number in the sequence is twice as large as the number before it.

Still another example

Consider the following sequence U. Someone asks, "Is this an arithmetic sequence? If so, what are the values u_0 (the starting value) and c_u (the constant of change)?"

$$U = 100, 65, 30, -5, -40, -75, -110, \ldots$$

This is an arithmetic sequence, at least for the numbers shown (the first seven terms). In this case, $s_0 = 100$ and $c_u = -35$, so we can generate the following list:

$$s_0 = 100$$
$$u_1 = u_0 + (-35) = 100 - 35 = 65$$
$$u_2 = u_1 + (-35) = 65 - 35 = 30$$
$$u_3 = u_2 + (-35) = 30 - 35 = -5$$
$$u_4 = u_3 + (-35) = -5 - 35 = -40$$
$$u_5 = u_4 + (-35) = -40 - 35 = -75$$
$$u_6 = u_5 + (-35) = -75 - 35 = -110$$

Are you confused?

You ask, "What happens if we start a sequence with a fixed number and then alternately add and subtract a constant? Is the result an arithmetic sequence?" Here's an example:

$$V = -1/2,\ 1/2,\ -1/2,\ 1/2,\ -1/2,\ 1/2,\ -1/2,\ \ldots$$

In this case, the first term, v_0, is equal to $-1/2$. We might say that the constant, c_v, is equal to 1, but we alternately add and subtract it to generate the terms. This is a definable sequence, but it's not an arithmetic sequence. In order to generate a true arithmetic sequence, we must repeatedly add the constant, whether it's positive, negative, or 0. When the constant is positive, the terms steadily increase. When the constant is negative, the terms steadily decrease. Arithmetic sequences never alternate as V does.

Here's a challenge!

When we have a sequence and we start to add up its numbers, we get another sequence of numbers representing the sums. These sums are called *partial sums*. List the first five partial sums of the following sequences:

$$S = 5,\ 8,\ 11,\ 14,\ 17,\ 20,\ 23,\ 26,\ 29,\ 32,\ \ldots$$
$$T = 2,\ 4,\ 8,\ 16,\ 32,\ 64,\ 128,\ 256,\ 512,\ \ldots$$
$$U = 100,\ 65,\ 30,\ -5,\ -40,\ -75,\ -110,\ \ldots$$
$$V = -1/2,\ 1/2,\ -1/2,\ 1/2,\ -1/2,\ 1/2,\ -1/2,\ \ldots$$

Solution

We simply add increasing numbers of terms and list the sums. For the sequence S, the first five partial sums are

$$s_{0+} = 5$$
$$s_{1+} = 5 + 8 = 13$$
$$s_{2+} = 5 + 8 + 11 = 24$$
$$s_{3+} = 5 + 8 + 11 + 14 = 38$$
$$s_{4+} = 5 + 8 + 11 + 14 + 17 = 55$$

For the sequence T, the first five partial sums are

$$t_{0+} = 2$$
$$t_{1+} = 2 + 4 = 6$$
$$t_{2+} = 2 + 4 + 8 = 14$$
$$t_{3+} = 2 + 4 + 8 + 16 = 30$$
$$t_{4+} = 2 + 4 + 8 + 16 + 32 = 62$$

For the sequence U, the first five partial sums are

$$u_{0+} = 100$$
$$u_{1+} = 100 + 65 = 165$$
$$u_{2+} = 100 + 65 + 30 = 195$$
$$u_{3+} = 100 + 65 + 30 + (-5) = 190$$
$$u_{4+} = 100 + 65 + 30 + (-5) + (-40) = 150$$

For the sequence V, the first five partial sums are

$$v_{0+} = -1/2$$
$$v_{1+} = -1/2 + 1/2 = 0$$
$$v_{2+} = -1/2 + 1/2 + (-1/2) = -1/2$$
$$v_{3+} = -1/2 + 1/2 + (-1/2) + 1/2 = 0$$
$$v_{4+} = -1/2 + 1/2 + (-1/2) + 1/2 + (-1/2) = -1/2$$

Repeated Multiplication

Another common type of sequence has values that are repeatedly multiplied by some constant. Here are a few examples:

$$G = 1, 2, 4, 8, 16, 32$$
$$H = 1, -1, 1, -1, 1, -1, \ldots$$
$$I = 1, 10, 100, 1000$$
$$J = -5, -15, -45, -135, -405$$
$$K = 3, 9, 27, 81, 243, 729, 2187, \ldots$$
$$L = 1/2, 1/4, 1/8, 1/16, 1/32, \ldots$$

Sequences G, I, and J are finite. Sequences H, K, and L are infinite, as indicated by an ellipsis at the end of each list.

Geometric sequence

Upon casual observation, the above sequences appear to be much different from one another. But in all six sequences, each term is a constant multiple of the term before it:

- The values in G progress by a constant factor of 2.
- The values in H progress by a constant factor of -1.
- The values in I progress by a constant factor of 10.
- The values in J progress by a constant factor of 3.
- The values in K progress by a constant factor of 3.
- The values in L progress by a constant factor of 1/2.

If the constant is positive, the values either remain positive or remain negative. If the constant is negative, the values alternate between positive and negative.

Let t_0 be the first number in a sequence T, and let k be a constant. Imagine that T can be written in the general form:

$$T = t_0, \ t_0 k, \ t_0 k^2, \ t_0 k^3, \ t_0 k^4, \ ...$$

for as long as the sequence goes. Such a sequence is called a *geometric sequence* or a *geometric progression*.

If $k = 1$, the sequence consists of the same number over and over. (In that case, it's also an arithmetic sequence with a constant equal to 0!) If $k = -1$, the sequence alternates between t_0 and its negative. If t_0 is less than -1 or greater than 1, the values get farther from 0 as we move along in the series. If t_0 is between (but not including) -1 and 1, the values get closer to 0. If $t_0 = 1$ or $t_0 = -1$, the values stay the same distance from 0.

The numbers t_0 and k can be whole numbers, but this is not a requirement. As long as the multiplication factor between any two adjacent terms in a sequence is the same, the sequence is a geometric sequence. In the sequence L above, we have the constant $k = 1/2$. This is an especially interesting case, as we'll see in a moment.

Geometric series

In a geometric sequence, the corresponding *geometric series*, which is the sum of all the terms, can *always* be defined if the sequence is finite, and can *sometimes* be defined if the sequence is infinite.

For the above sequences G through L, let the corresponding series be called G_+ through L_+. Then we have

$$G_+ = 1 + 2 + 4 + 8 + 16 + 32 = 63$$
$$H_+ = 1 - 1 + 1 - 1 + 1 - 1 + \cdots = ?$$
$$I_+ = 1 + 10 + 100 + 1000 = 1111$$
$$J_+ = -5 - 15 - 45 - 135 - 405 = -605$$
$$K_+ \text{ is not defined}$$
$$L_+ = 1/2 + 1/4 + 1/8 + 1/16 + 1/32 + \cdots = ?$$

The finite series G_+, I_+, and J_+ are straightforward. There's no mystery there! The partial sums of H_+ alternate between 0 and 1, but can't settle on either of those values. It's tempting to say that H_+ has two values, just as certain equations have solution sets containing two roots. But we're looking for a single, identifiable number, not the solution set of an equation. On that basis, we're forced to conclude that H_+ is not definable. The infinite series K_+ goes "out of control." It's an example of a *divergent series*; its values keep getting farther from 0 without ever reaching a limit.

Convergence

For the above sequences H, K, and L, the sequences of partial sums, which we'll denote using asterisk subscripts, go as follows:

$$H_* = 1, 0, 1, 0, 1, 0, \ldots$$
$$K_* = 3, 12, 39, 120, 363, 1092, 3279, \ldots$$
$$L_* = 1/2, 3/4, 7/8, 15/16, 31/32, \ldots$$

The partial sums denoted by H_* and K_* don't settle down on anything. But the partial sums denoted by L_* seem to approach 1. They don't "run away" into uncharted territory, and they don't alternate between or among multiple numbers. The partial sums in L_* seem to have a clear destination that they could reach, if only they had an infinite amount of time to get there.

It turns out that the complete series L_+, representing the sum of the infinite string of numbers in the sequence L, is exactly equal to 1! We can get an intuitive view of this fact by observing that the partial sums approach 1. As the position in the sequence of partial sums, L_*, gets farther and farther along, the denominators keep doubling, and the numerator is always 1 less than the denominator. In fact, if we want to find the nth number L_{*n} in the sequence of partial sums L_*, we can calculate it by using the following formula:

$$L_{*n} = (2^n - 1)/2^n$$

As n becomes large, 2^n becomes large much faster, and the proportional difference between $2^n - 1$ and 2^n becomes smaller. When n reaches extremely large positive integer values, the quotient $(2^n - 1)/2^n$ is almost exactly equal to 1. We can make the quotient as close to 1 as we want by going out far enough in the series of partial sums, but we can never make it equal to or larger than 1. The sequence L_* is said to *converge* on the number 1. The sequence of partial sums L_* is an example of a *convergent sequence*. The series L_+ is an example of a *convergent series*.

Plotting a geometric sequence

A geometric sequence, like an arithmetic sequence, appears as a set of points when plotted on a Cartesian plane. Figure 19-2 shows examples of two geometric sequences as they appear

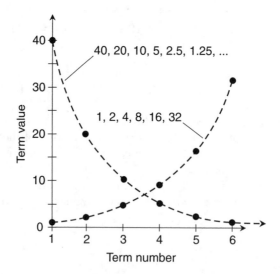

Figure 19-2 Rectangular-coordinate plots of two geometric sequences.

when graphed. Note that the dashed curves, which show the general trends of the sequences (but aren't actually parts of the sequences), aren't straight lines, but they are "smooth." They don't turn corners or make sudden leaps.

One of the sequences in Fig. 19-2 is increasing, and the dashed curve connecting this set of points goes upward as we move to the right. Because this sequence is finite, the dashed curve ends at the point (6,32), where the term number is 6 and the term value is 32. The other sequence is decreasing, and the dashed curve goes downward and approaches 0 as we move to the right. This sequence is infinite, as shown by the three dots at the end of the string of numbers, and also by the arrow at the right-hand end of the dashed curve.

If a geometric sequence has a negative factor, that is, if $k < 0$, the plot of the points alternates back and forth on either side of 0. The points fall along two different curves, one above the horizontal axis and the other below. If you want to see what happens in a case like this, try plotting an example. Set $t_0 = 64$ and $k = -1/2$, and plot the resulting points.

An example

Suppose you get a 5-year certificate of deposit (CD) at your local bank for $1000.00, and it earns interest at the annualized rate of exactly 5 percent per year. The CD will be worth $1276.28 after 6 years. To calculate this, multiply $1000 by 1.05, then multiply this result by 1.05, and repeat this process a total of 5 times. The resulting numbers form a geometric sequence:

- After 1 year: $1000.00 × 1.05 = $1050.00
- After 2 years: $1050.00 × 1.05 = $1102.50
- After 3 years: $1102.50 × 1.05 = $1157.63
- After 4 years: $1157.63 × 1.05 = $1215.51
- After 5 years: $1215.51 × 1.05 = $1276.28

Another example

Is the following sequence a geometric sequence? If so, what are the values t_0 (the starting value) and k (the factor of change)?

$$T = 3, -6, 12, -24, 48, -96, \ldots$$

This is a geometric sequence. The numbers change by a factor of -2. In this case, $t_0 = 3$ and $k = -2$.

Are you confused?

It's reasonable to ask, "Can we categorize all sequences as either arithmetic or geometric?" The answer is no! Consider

$$U = 10, 13, 17, 22, 28, 35, 43, \ldots$$

This sequence shows a pattern, but it's neither arithmetic nor geometric. The difference between the first and second terms is 3, the difference between the second and third terms is 4, the difference between the third and fourth terms is 5, and so on. The difference keeps increasing by 1 for each succeeding pair of terms. This is a fairly simple example of a nonarithmetic, nongeometric sequence with an identifiable pattern.

Here's a challenge!

Suppose a particular species of cell undergoes *mitosis* (splits in two) every half hour, precisely on the half hour. We take our first look at a cell culture at 12:59 p.m., and find three cells. At 1:00 p.m., mitosis occurs for all the cells at the same time, and then there are six cells in the culture. At 1:30 p.m., mitosis occurs again, and we have 12 cells. How many cells are there in the culture at 4:01 p.m.?

Solution

There are 3 hours and 2 minutes between 12:59 p.m. and 4:01 p.m. This means that mitosis takes place 7 times: at 1:00, 1:30, 2:00, 2:30, 3:00, 3:30, and 4:00. Table 19-1 illustrates the scenario. We look at the culture repeatedly at 1 minute past each half hour. There are 384 cells at 4:01 p.m., just after the mitosis event that occurs at 4:00 p.m.

Table 19-1   Cell division as a function of time, assuming mitosis occurs every half hour

Time	Number of cells
12:59	3
1:01	6
1:31	12
2:01	24
2:31	48
3:01	96
3:31	192
4:01	384

Limit of a Sequence

A *limit* is a specific, well-defined quantity that a sequence, series, relation, or function approaches. The value of the sequence, series, relation, or function can get arbitrarily close to the limit, but doesn't always reach it.

An example

Let's look at an infinite sequence A that starts with 1 and then keeps getting smaller. For any positive integer n, the nth term is $1/n$, so we have

$$A = 1, 1/2, 1/3, 1/4, 1/5, ..., 1/n, ...$$

This is a simple example of a special type of sequence called a *harmonic sequence*. In this particular case, the values of the terms approach 0. The hundredth term is 1/100; the thousandth term is 1/1000; the millionth term is 1/1,000,000. If we choose a tiny but positive real number, we can always find a term in the sequence that's closer to 0 than that number. But no matter how much time we spend generating terms, we'll never get 0. We say that "The limit of $1/n$, as n approaches infinity, is 0," and write it as

$$\lim_{n \to \infty} 1/n = 0$$

Another example

Consider the sequence B in which the numerators ascend one by one through the set of natural numbers, while every denominator is equal to the corresponding numerator plus 1. For any positive integer n, the nth term is $(n-1)/n$, so we have

$$B = 0/1, \ 1/2, \ 2/3, \ 3/4, \ 4/5, \ ..., \ (n-1)/n, \ ...$$

As n becomes extremely large, the numerator $(n-1)$ gets closer and closer to the denominator, when we consider the difference in proportion to the value of n. Therefore

$$\lim_{n \to \infty} (n-1)/n = n/n = 1$$

Still another example

Let's see what happens in a sequence C where every numerator is equal to the square of the term number, while every denominator is equal to twice the term number. For any positive integer n, the nth term is $n^2/(2n)$, so we have

$$C = 1/2, \ 4/4, \ 9/6, \ 16/8, \ 25/10, \ 36/12, \ 49/14, \ ..., \ n^2 \ / \ (2n), \ ...$$

Note that

$$n^2/(2n) = n/2$$

This tells us that

$$\lim_{n \to \infty} n^2/(2n) = \lim_{n \to \infty} n/2$$

As n grows larger without end, so does $n/2$. Therefore

$$\lim_{n \to \infty} n/2$$

is undefined, so we know that

$$\lim_{n \to \infty} n^2/(2n)$$

is also undefined. Alternatively, we can say that this limit doesn't exist, or that it's meaningless.

Are you confused?

By now, you should suspect that any given sequence must fall into one or the other of two categories: convergent (meaning that it has a limit) or divergent (meaning that it doesn't have a limit). But what if a sequence alternates between two numbers endlessly? Once again, look at the sequence

$$H = 1, -1, 1, -1, 1, -1, \ldots$$

We might be tempted to suggest that a sequence of this type "has two different limits," but it doesn't converge on any single number. However, that won't work because a limit must always be a single value that we can specify as a number. In cases like this, it's customary to say that the limit is not defined.

Here's a challenge!

Consider the sequence D in which the numerators alternate between -1 and 1, while the denominators start at 1 and increase by 1 with each succeeding term. For any positive integer n, the nth term is $(-1)^n/n$, so that

$$D = -1/1, 1/2, -1/3, 1/4, -1/5, \ldots, (-1)^n/n, \ldots$$

Does this sequence have a limit? If so, what is it? If not, why not?

Solution

As n becomes extremely large, the absolute value of the numerator is always 1, although the sign alternates. The denominator increases steadily, and without end. If we choose a tiny positive or negative real number, we can always find a term that's closer to 0 than that number, but we'll never actually reach 0 from either the positive side or the negative side. Therefore

$$\lim_{n \to \infty} (-1)^n/n = 0$$

Here's another challenge!

Consider the following sequence:

$$K = (-1 - 1/1), (1 + 1/2), (-1 - 1/3), (1 + 1/4), \ldots, [(-1)^n + (-1)^n/n], \ldots$$

The parentheses and brackets are not technically necessary here, but they visually isolate the terms from one another. Does K have a limit? If so, what is it? If not, why not?

Solution

Each term in K is expressed as a sum. The first addend alternates between -1 and 1, endlessly. The second addend is identical to the corresponding term in the sequence D that we evaluated in the previous challenge. We determined that D converges toward 0. The terms in K therefore approach two different values, −1 and 1, as we generate terms indefinitely. If we want to claim that a sequence has a limit, we must take that expression literally. "A limit" means "one and only one limit." We therefore conclude that

$$\underset{n \to \infty}{Lim}\ (-1)^n + (-1)^n/n$$

is not defined.

Summation "Shorthand"

Mathematicians have a "shorthand" way to denote long sums. This technique can save a lot of space and writing time. We can even write down an infinite sum in a compact statement. It's called *summation notation*.

Specify the series

Imagine a set of constants, all denoted by *a* with a subscript, such as

$$\{a_1, a_2, a_3, a_4, a_5, a_6, a_7, a_8\}$$

Suppose that we add up the elements of this set, and call the sum *b*. We can write this sum out term by term as

$$a_1 + a_2 + a_3 + a_4 + a_5 + a_6 + a_7 + a_8 = b$$

That's easy because we have only eight terms, but if the set contained 800 elements, writing down the entire sum would be exasperating. We could put an ellipsis in the middle of the sum, calling it *c* and then writing

$$a_1 + a_2 + a_3 + \cdots + a_{798} + a_{799} + a_{800} = c$$

If the series had infinitely many terms, we could use an ellipsis after the first few terms and leave the statement wide open after that, calling it *d* and then writing

$$a_1 + a_2 + a_3 + a_4 + a_5 + \cdots = d$$

Tag the terms

Let's invent a nonnegative-integer variable and call it *i*. Written as a subscript, *i* can serve as a counting tag in a series containing a large number of terms. Don't confuse this *i* with the symbol some texts use to represent the unit imaginary number, which is the positive square root of −1!

In the above-described series, we can call each term by the generic name a_i. In the first series, we add up eight a_i's to get the final sum *b*, and the counting tag *i* goes from 1 to 8. In the second series, we add up 800 a_i's to get the final sum c, and the counting tag i goes from 1 to 800. In the third series, we add up infinitely many a_i's to get the final sum d, and the counting tag i ascends through the entire set of positive integers. Suppose that we have a series with n terms, as follows:

$$a_1 + a_2 + a_3 + \cdots + a_{n-2} + a_{n-1} + a_n = k$$

In this case, we add up n a_i's to get the final sum k.

The big sigma

Let's go back to the series with eight terms. We can write it down in a cryptic but information-dense manner as

$$\sum_{i=1}^{8} a_i = b$$

We read this expression out loud as, "The summation of the terms a_i, from $i = 1$ to 8, is equal to b." The large symbol Σ is the uppercase Greek letter *sigma*, which stands for summation or sum. Now let's look at the series in which 800 terms are added:

$$\sum_{i=1}^{800} a_i = c$$

We can read this aloud as, "The summation of the terms a_i, from $i = 1$ to 800, is equal to c." In the third example containing infinitely many terms, we can write

$$\sum_{i=1}^{\infty} a_i = d$$

This statement can be read as, "The summation of the terms a_i, from $i = 1$ to infinity, is equal to d." Finally, in the general case, we can write

$$\sum_{i=1}^{n} a_i = k$$

and read it aloud as, "The summation of the terms a_i, from $i = 1$ to n, is equal to k."

A more sophisticated example

Suppose we want to determine the value of an infinite series starting with 1, then adding 1/2, then adding 1/4, then adding 1/8, and going on forever, each time cutting the value in half. As things work out, we get

$$1 + 1/2 + 1/4 + 1/8 + \cdots = 2$$

even though the series has infinitely many terms. We can also write

$$1/2^0 + 1/2^1 + 1/2^2 + 1/2^3 + \cdots = 2$$

In summation notation, we write

$$\sum_{i=0}^{\infty} 1/2^i = 2$$

Are you confused?

If you're baffled by the idea that we can add up infinitely many numbers and get a finite sum, you can use the "frog-and-wall" analogy. Imagine that a frog sits 8 meters (8 m) away from a wall. Then she jumps halfway to the wall, so she's 4 m away from it. Now imagine that she continues to make repeated jumps toward the wall, each time getting halfway there (Fig. 19-3). No finite number of jumps will allow the frog to reach the wall. To accomplish that goal, she would have to take infinitely many jumps. This scenario can be based on a sequence of partial sums of a series

$$S = 4 + 2 + 1 + 1/2 + 1/4 + 1/8 + \cdots$$

A real-world frog cannot reach the wall by jumping halfway to it, over and over. But in the imagination, she can. There are two ways this can happen. First, in the universe of mathematics, we have an infinite amount of time, so an infinite number of jumps can take place. Another way around the problem is to keep halving the length of time in between jumps, say from 4 seconds to 2 seconds, then to 1 second, then to 1/2 second, and so on. This will make it possible for our "cosmic superfrog" to hop an infinite number of times in a finite span of time. Either way, when she has finished her journey and her nose touches the wall, she'll have traveled exactly 8 m. Therefore, the sum total of the lengths of her jumps is

$$S = 4 + 2 + 1 + 1/2 + 1/4 + 1/8 + \cdots = 8$$

Here's a challenge!

Consider the series that we dealt with in "A more sophisticated example" a couple of paragraphs ago, but only up to the reciprocal of the nth power of 2. Let S_n be the partial sum of this series up to, and including, that term. Write S_n in summation notation.

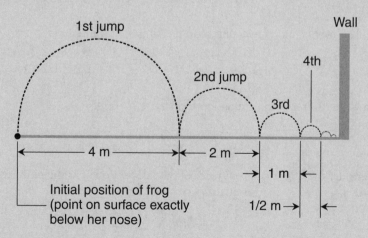

Figure 19-3   A frog jumps toward a wall, getting halfway there with each jump.

Solution

Let's use the letter i as the counting tag. We start at $i = 0$ and go up to $i = n$, with each term having the value $1/2^i$. Therefore, the summation notation is

$$\sum_{i=0}^{n} 1/2^i$$

Limit of a Series

If a series has a limit, we can sometimes figure it out by creating a sequence from the partial sums, and then finding the limit of that sequence.

An example

Think of the summation in the previous challenge, and imagine what happens as n increases endlessly—that is, as n approaches infinity. As n grows larger, the sequence of partial sums approaches 2. We can plug the summation into a limit template, and then state that

$$\lim_{n \to \infty} \sum_{i=0}^{n} 1/2^i = 2$$

Another example

Let's look once again at the infinite sequence V we saw a little while ago, where the numerators keep alternating between -1 and 1, as follows:

$$V = -1/2,\ 1/2,\ -1/2,\ 1/2,\ -1/2,\ 1/2,\ -1/2 \ldots$$

Let's replace every comma by a plus sign, creating the infinite series

$$V_+ = -1/2 + 1/2 - 1/2 + 1/2 - 1/2 + 1/2 - 1/2 + \cdots$$

We can write this series in summation form as

$$\sum_{i=1}^{\infty} (-1)^i/2$$

Now consider the limit of the sequence of partial sums of V_+ as the number of terms becomes arbitrarily large. We write this quantity symbolically as

$$\lim_{n \to \infty} \sum_{i=1}^{n} (-1)^i/2$$

This limit does not exist, because the sequence of partial sums alternates endlessly between two values, $-1/2$ and 0.

Are you confused?

Does the combination of limit and summation notation look intimidating? Besides getting used to the symbology, you have to keep track of two different indexes, i for the sum and n for the limit. It helps if you remember that the two indexes are independent of each other. You're finding the limit of a sum as you keep making that sum longer.

Here's a challenge!

Find the limit of the partial sums of the infinite series

$$1/100 + 1/100^2 + 1/100^3 + 1/100^4 + 1/100^5 + \cdots$$

as the number of terms in the partial sum increases without end. That is, find

$$\lim_{n \to \infty} \sum_{i=1}^{n} 1/100^i$$

Solution

In decimal form, $1/100 = 0.01$, $1/100^2 = 0.0001$, $1/100^3 = 0.000001$, and so on. Let's arrange these numbers in a column with each term underneath its predecessor, and all the decimal points along a vertical line, as the following:

$$
\begin{array}{l}
0.01 \\
0.0001 \\
0.000001 \\
0.00000001 \\
0.0000000001 \\
\downarrow \\
\hline
0.0101010101\ldots
\end{array}
$$

When we look at the series this way, we can see that it must ultimately add up to the nonterminating, repeating decimal $0.0101010101\ldots$. From our algebra or number theory courses, we recall that this endless decimal number is equal to $1/99$. That's the limit of the sequence of partial sums in the series:

$$\sum_{i=1}^{n} 1/100^i$$

as the positive integer n increases without end. It's also the value of the entire infinite series:

$$\sum_{i=1}^{\infty} 1/100^i$$

Limits of Functions

So far, we've looked at situations where we move from term to term in a sequence or series. Sometimes, such sequences and series have limits (they converge); in other cases they don't have limits (they diverge). Similar phenomena can occur when we have a variable that changes in a smooth, continuous manner, rather than jumping among discrete values.

Some functions have limits, and some don't

Certain functions increase or decrease without bound, while others reach specific values and stay there. Still others increase or decrease continuously without ever passing, or even reaching, a certain value. It's also possible for a function to "blow up" and have no limit at all.

The solid curve in Fig. 19-4 shows the reciprocal function in the first quadrant of the Cartesian plane, where the value of the independent variable is positive. The dashed curve shows the negative reciprocal function in the fourth quadrant, where, again, the value of the independent variable is positive. The functions are

$$f(x) = x^{-1}$$

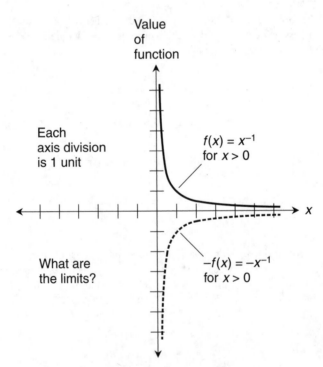

Figure 19-4   Graphs of the reciprocal function (solid curve) and its negative (dashed curve) in the first and fourth quadrants of the Cartesian plane, where $x > 0$. Each axis division represents 1 unit.

and

$$-f(x) = -x^{-1}$$

As the value of x increases without end, both of these functions approach, but never reach, 0. We can therefore write

$$\lim_{x \to \infty} f(x) = 0$$

and

$$\lim_{x \to \infty} -f(x) = 0$$

As the value of x becomes arbitrarily small but remains positive, both of these functions approach singularity. The reciprocal function "blows up positively" while the negative reciprocal function "blows up negatively." Therefore, we must conclude that neither

$$\lim_{x \to 0} f(x)$$

nor

$$\lim_{x \to 0} -f(x)$$

exists. It's tempting to claim that

$$\lim_{x \to 0} f(x) = +\infty$$

and

$$\lim_{x \to 0} -f(x) = -\infty$$

However, we haven't explicitly defined $+\infty$ ("positive infinity") or $-\infty$ ("negative infinity"), so such statements are informal at best.

Right-hand limit at a point

Consider again the reciprocal function

$$f(x) = x^{-1}$$

To specify that we approach 0 from the positive direction, we can refine the limit notation by placing a plus sign after the 0, as follows:

$$\lim_{x \to 0+} f(x)$$

This expression reads, "The limit of $f(x)$ as x approaches 0 from the positive direction." We can also say, "The limit of $f(x)$ as x approaches 0 from the right." (In most graphs where x is on the horizontal axis, the value of x becomes more positive as we move toward the right.)

This sort of limit is called a *right-hand limit*. Because f is singular where $x = 0$, this particular limit is not defined.

Left-hand limit at a point

Let's expand the domain of f to the entire set of reals except 0, for which f is not defined because 0^{-1} is meaningless. Suppose that we start out with negative real values of x and approach 0 from the left. As we do this, f decreases endlessly. Another way of saying this is that f increases negatively without limit, or that it "blows up negatively." Therefore,

$$\underset{x \to 0-}{Lim}\ f(x)$$

is not defined. We read the above symbolic expression as, "The limit of $f(x)$ as x approaches 0 from the negative direction." We can also say, "The limit of $f(x)$ as x approaches 0 from the left." This sort of limit is called a *left-hand limit*.

An example

Let's consider a function g that takes the reciprocal of twice the independent variable. If the independent variable is x, then we have

$$g(x) = (2x)^{-1}$$

Imagine that we allow x to be any positive real number. As x gets arbitrarily large positively, $g(x)$ gets arbitrarily small positively, approaching 0 but never quite getting there. We can say, "The limit of $g(x)$, as x approaches infinity, is 0," and write

$$\underset{x \to \infty}{Lim}\ g(x) = 0$$

This scenario is similar to what happens with the reciprocal function, except that this function g approaches 0 at a different rate than the reciprocal function as the independent variable becomes arbitrarily large.

Now let's see what happens when x gets smaller but stays positive, so that $g(x)$ gets larger. If we make x close enough to 0, we can make $g(x)$ as large as we want. This function, like the reciprocal function, "blows up" as x approaches 0 from the positive direction, but at a different rate. Therefore

$$\underset{x \to 0+}{Lim}\ g(x)$$

is not defined.

Another example

Suppose that x is a positive real-number variable, and we want to evaluate

$$\underset{x \to \infty}{Lim}\ 1/x^2$$

Let's start out with x at some positive real number for which the function is defined. As we increase the value of x, the value of $1/x^2$ decreases, but it always remains positive. If we choose some tiny positive real number r, no matter how close to 0 it might be, we can always find some large value of x for which $1/x^2$ is smaller than r. Therefore, as x grows without bound, $1/x^2$ approaches 0, telling us that

$$\underset{x \to \infty}{Lim}\ 1/x^2 = 0$$

Still another example

Again, let x be a positive real-number variable. This time, let's evaluate

$$\underset{x \to 0+}{Lim}\ 1/x^2$$

Suppose that we start out with x at some positive real number for which the function is defined and then decrease x, letting it get arbitrarily close to 0 but always remaining positive. As we decrease the value of x, the value of $1/x^2$ remains positive and increases. If we choose some large positive real number s, no matter how gigantic, we can always find some small, positive value of x for which $1/x^2$ is larger than s. As x becomes arbitrarily small positively, $1/x^2$ grows without bound, so

$$\underset{x \to 0+}{Lim}\ 1/x^2$$

is not defined.

Are you confused?

It's easy to get mixed up by the meanings of negative direction and positive direction, and how these relate to the notions of left hand and right hand. These terms are based on the assumption that we're talking about the horizontal axis in a graph, and that this axis represents the independent variable. In most graphs of this type, the value of the independent variable gets more negative as we move to the left, and it gets more positive as we move to the right.

As we travel along the horizontal axis, we might be in positive territory the whole time; we might be in negative territory the whole time; we might cross over from the negative side to the positive side or vice versa. Whenever we come toward a point from the left, we approach from the negative direction, even if that point corresponds to something like $x = 567$. Whenever we come toward a point from the right, we approach from the positive direction, even if the point is at $x = -53{,}535$. The location of the point doesn't matter. The important consideration is the *direction from which we approach the point*.

Here's a challenge!

Consider the base-10 logarithm function (symbolized $\log_{10}$). Sketch a graph of the function $f(x) = \log_{10} x$ for values of x from 0.1 to 10, and for values of f from -1 to 1. Then determine

$$\underset{x \to 5-}{Lim}\ \log_{10} x$$

Solution

Figure 19-5 is a graph of the function $f(x) = \log_{10} x$ for values of x from 0.1 to 10, and for values of f from -1 to 1. The function varies smoothly throughout this span. If we start at values of x a little smaller than 5 and work our way toward 5, the value of f approaches $\log_{10} 5$. Therefore,

$$\underset{x \to 5-}{Lim}\ \log_{10} x = \log_{10} 5$$

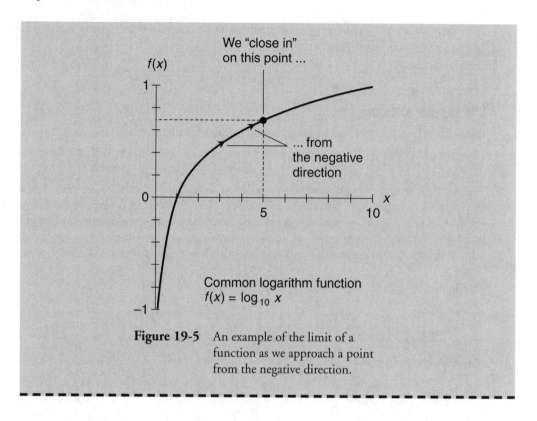

Figure 19-5 An example of the limit of a function as we approach a point from the negative direction.

Memorable Limits of Series

Certain limits of series are found often in calculus and analysis. If you plan to go on to *Calculus Know-It-All* after finishing this book, you're certain to see the three examples that follow!

An example

Imagine an infinite series where we take a positive integer i and then divide it by the square of another positive integer n. Symbolically, we write this as

$$\sum_{i=1}^{\infty} i/n^2$$

When we expand this series out, we write it as

$$1/n^2 + 2/n^2 + 3/n^2 + \cdots + n/n^2 + \cdots$$

which simplifies to

$$(1 + 2 + 3 + \cdots + n + \cdots)/n^2$$

Suppose that we let n grow endlessly larger, increasing the number of terms in the series. Let's consider

$$\operatorname*{Lim}_{n \to \infty} (1 + 2 + 3 + \cdots + n)/n^2$$

As things work out, this limit is equal to 1/2. Therefore

$$\operatorname*{Lim}_{n \to \infty} \sum_{i=1}^{n} i/n^2 = 1/2$$

Another example

Now imagine an infinite series where we square a positive integer i and then divide it by the cube of another positive integer n. Symbolically, we write this as

$$\sum_{i=1}^{\infty} i^2/n^3$$

We can expand it to

$$1^2/n^3 + 2^2/n^3 + 3^2/n^3 + \cdots + n^2/n^3 + \cdots$$

which simplifies to

$$(1^2 + 2^2 + 3^2 + \cdots + n^2 + \cdots)/n^3$$

As n grows endlessly larger, we have

$$\operatorname*{Lim}_{n \to \infty} (1^2 + 2^2 + 3^2 + \cdots + n^2)/n^3$$

This limit turns out to be 1/3. Therefore

$$\operatorname*{Lim}_{n \to \infty} \sum_{i=1}^{n} i^2/n^3 = 1/3$$

Still another example

Finally, let's look at an infinite series where we cube a positive integer i and then divide it by the fourth power of another positive integer n. Symbolically, we write this as

$$\sum_{i=1}^{\infty} i^3/n^4$$

When we write this series out, we obtain

$$1^3/n^4 + 2^3/n^4 + 3^3/n^4 + \cdots + n^3/n^4 + \cdots$$

which simplifies to

$$(1^3 + 2^3 + 3^3 + \cdots + n^3 + \cdots)/n^4$$

As n grows endlessly larger, we have

$$\underset{n \to \infty}{Lim} \, (1^3 + 2^3 + 3^3 + \cdots + n^3)/n^4$$

This limit turns out to be 1/4. Therefore

$$\underset{n \to \infty}{Lim} \sum_{i=1}^{n} i^3/n^4 = 1/4$$

Practice Exercises

This is an open-book quiz. You may (and should) refer to the text as you solve these problems. Don't hurry! You'll find worked-out answers in App. B. The solutions in the appendix may not represent the only way a problem can be figured out. If you think you can solve a particular problem in a quicker or better way than you see there, by all means try it!

1. Figure 19-6 is a graph of the first few elements of an infinite arithmetic sequence. If we call the sequence S, then

$$S = s_0, (s_0 + c), (s_0 + 2c), (s_0 + 3c), \ldots$$

where s_0 is the initial term value and c is a constant. Based on the information given in this graph, what is s_0? What is c? What is the value of the hundredth term in S?

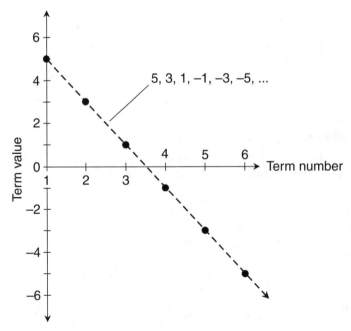

Figure 19-6   Illustration for Problem 1.

2. Does the infinite arithmetic sequence described in Problem 1 converge? If so, on what value does it converge? If not, why not?

3. The general form for an infinite geometric sequence T is

$$T = t_0, \, t_0 k, \, t_0 k^2, \, t_0 k^3, \, t_0 k^4, \, ...$$

where t_0 is the initial value and k is the constant of multiplication. Calculate, and write down, the first seven terms in an infinite geometric sequence T where $t_0 = 2$ and $k = -4$. Does this sequence converge? If so, on what value does it converge? If not, why not?

4. Suppose that in the scenario of Problem 3, we change k from -4 to $-1/4$. Calculate and list the first seven values of the resulting infinite sequence. Does it converge? If so, on what value does it converge? If not, why not?

5. Consider again the sequence we saw earlier in this chapter:

$$B = 0/1, \, 1/2, \, 2/3, \, 3/4, \, 4/5, \, ..., \, (n-1)/n, \, ...$$

We determined that the limit of B, as n grows without end, is

$$\underset{n \to \infty}{Lim} \, (n-1)/n = 1$$

so we know that B converges. Write down the series B_+ that we get when we add the elements of B. Then write down the first five terms of the sequence B_*, which is made up of the partial sums in B_+. Does the sequence B_* converge? If so, to what value does it converge? If not, why not?

6. Express the following series by writing out the first five terms followed by an ellipsis:

$$S_+ = \sum_{i=1}^{n} 1/10^i$$

First, express the terms as fractions. Then express them as powers of 10. Then express them as decimal quantities. Finally, write down the first five terms in the sequence S_* of partial sums.

7. Find the following limit if it exists. If no limit exists, explain:

$$\underset{n \to \infty}{Lim} \, \sum_{i=1}^{n} 1/10^i$$

8. Using a calculator, plug in $n = 2$, $n = 6$, $n = 10$, and $n = 20$ to informally illustrate that

$$\underset{n \to \infty}{Lim} \, (1 + 2 + 3 + \cdots + n)/n^2 = 1/2 = 0.5$$

and therefore that

$$\underset{n \to \infty}{Lim} \, \sum_{i=1}^{n} i/n^2 = 1/2 = 0.5$$

Work out the partial sums to obtain decimal quantities. Round off your results to five decimal places when you encounter repeating or lengthy decimals.

9. Using a calculator, plug in $n = 2$, $n = 6$, $n = 10$, and $n = 20$ to informally illustrate that

$$\underset{n \to \infty}{Lim} \, (1^2 + 2^2 + 3^2 + \cdots + n^2)/n^3 = 1/3 = 0.33333...$$

and therefore that

$$Lim_{n\to\infty} \sum_{i=1}^{n} i^2/n^3 = 1/3 = 0.33333...$$

Work out the partial sums to obtain decimal quantities. Round off your results to five decimal places when you encounter repeating or lengthy decimals.

10. Using a calculator, plug in $n = 2$, $n = 6$, $n = 10$, and $n = 20$ to informally illustrate that

$$Lim_{n\to\infty} (1^3 + 2^3 + 3^3 + \cdots + n^3)/n^4 = 1/4 = 0.25$$

and therefore that

$$Lim_{n\to\infty} \sum_{i=1}^{n} i^3/n^4 = 1/4 = 0.25$$

Work out the partial sums to obtain decimal quantities. Round off your results to five decimal places when you encounter repeating or lengthy decimals.

20

Review Questions and Answers

Part Two

This is not a test! It's a review of important general concepts you learned in the previous nine chapters. Read it though slowly and let it sink in. If you're confused about anything here, or about anything in the section you've just finished, go back and study that material some more.

Chapter 11

Question 11-1

What's a mathematical relation?

Answer 11-1

A relation is a clearly defined way of assigning, or mapping, some or all of the elements of a source set to some or all of the elements of a destination set. Suppose that X is the source set for a relation, and Y is the destination set for the same relation. In that case, the relation can be expressed as a collection of ordered pairs of the form (x,y), where x is an element of set X and y is an element of set Y.

Question 11-2

What's an injection, also known as an injective relation?

Answer 11-2

Imagine two sets X and Y. Suppose that a relation assigns each element of X to exactly one element of Y. Also suppose that, according to the same relation, an element of Y never has more than one mate in X. (Some elements of Y might have no mates in X.) In a situation like this, the relation is an injection.

Question 11-3

What's a surjection, also called an onto relation?

Answer 11-3

Again, imagine two sets X and Y. Suppose that according to a certain relation, every element of Y has at least one (and maybe more than one) mate in X, so that no element of Y is left out. A relation of this type is a surjection from X onto Y.

Question 11-4

What's a bijection, also called a one-to-one correspondence?

Answer 11-4

A bijection is a relation that's both an injection and a surjection. Given two sets X and Y, a bijection assigns every element of X to exactly one element of Y, and vice versa. This is why a bijection is sometimes called a one-to-one correspondence.

Question 11-5

What's a two-space function? Is every two-space function a relation? Is every two-space relation a function?

Answer 11-5

A two-space function is a relation between two sets that never maps any element of the source set to more than one element of the destination set. All two-space functions are relations. However, not all two-space relations are functions.

Question 11-6

What's the vertical-line test for the graph of a two-space function?

Answer 11-6

The vertical-line test is a quick way to determine, based on the graph of a two-space relation, whether or not the relation is a function. Imagine an infinitely long, movable line that's always parallel to the dependent-variable axis (usually the vertical axis). Suppose that we're free to move the line to the left or right, so it intersects the independent-variable axis (usually the horizontal axis) wherever we want. The graph is a function of the independent variable *if and only if* the movable vertical line never intersects the graph at more than one point.

Question 11-7

Based on the vertical-line test, which of the curves in Fig. 20-1 are functions of x within the span of values for which $-6 < x < 6$?

Answer 11-7

Only f is a function of x. If we construct a movable vertical line (always parallel to the y axis), it never intersects the curve for f at more than one point over the span of values for which $-6 < x < 6$. However, the movable vertical line intersects the curve for g at more than one point for some values of x where $-6 < x < 6$. The same is true of the curve for h.

Question 11-8

Suppose we're working in the polar coordinate plane, and we encounter the graph of a relation where the independent variable is represented by θ (the direction angle) and the dependent

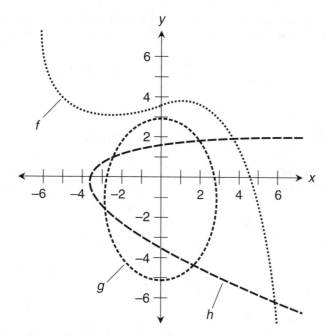

Figure 20-1 Illustration for Question and Answer 11-7.

variable is represented by r (the radial distance from the origin). How can we tell if the relation is a function of θ?

Answer 11-8

We can draw the graph of the relation in a Cartesian plane, plotting values of θ along the horizontal axis, and plotting values of r along the vertical axis. We can allow both θ and r to attain all possible real-number values. Then we can use the Cartesian vertical-line test to see if the relation is a function of θ.

Question 11-9

How do functions add, subtract, multiply, and divide?

Answer 11-9

To add one function to another, we add both sides of their equations. This can be done in either order, producing identical results. If f_1 and f_2 are functions of x, then

$$(f_1 + f_2)(x) = f_1(x) + f_2(x)$$

and

$$(f_2 + f_1)(x) = f_2(x) + f_1(x)$$

To subtract one function from another, we subtract both sides of their equations. This can be done in either order, usually producing different results. If f_1 and f_2 are functions of x, then

$$(f_1 - f_2)(x) = f_1(x) - f_2(x)$$

and

$$(f_2 - f_1)(x) = f_2(x) - f_1(x)$$

To multiply one function by another, we multiply both sides of their equations. This can be done in either order, producing identical results. If f_1 and f_2 are functions of x, then

$$(f_1 \times f_2)(x) = f_1(x) \times f_2(x)$$

and

$$(f_2 \times f_1)(x) = f_2(x) \times f_1(x)$$

To divide one function by another, we divide both sides of their equations. This can be done in either order, usually producing different results. If f_1 and f_2 are functions of x, then

$$(f_1/f_2)(x) = f_1(x)/f_2(x)$$

and

$$(f_2/f_1)(x) = f_2(x)/f_1(x)$$

Question 11-10

When we add, subtract, multiply, or divide functions, we must adhere to three important rules. What are they?

Answer 11-10

First, we must be sure that the functions both operate on the same thing. In other words, the independent variables must describe the same parameters or phenomena. Second, we must restrict the domain of the resultant function to only those values that are in the domains of both functions (the intersection of the domains). Third, if we divide a function by another function, we can't define the resultant function for any value of the independent variable where the denominator function becomes 0.

Chapter 12

Question 12-1

How can we informally define the inverse of a relation?

Answer 12-1

The inverse of a relation is another relation that undoes whatever the original relation does. Also, the original relation undoes whatever its inverse does.

Question 12-2

How can we rigorously define the inverse of a relation?

Answer 12-2

Let f be a relation where x is the independent variable and y is the dependent variable. The inverse relation for f is another relation f^{-1} such that

$$f^{-1}\left[f(x)\right] = x$$

for all values of x in the domain of f, and

$$f[f^{-1}(y)] = y$$

for all values of y in the range of f.

Question 12-3

Suppose we've drawn the graph of a relation f in the Cartesian xy plane. How can we create the graph of the inverse relation f^{-1}?

Answer 12-3

Imagine the line $y = x$ as a "point reflector." For any point on the graph of f, its counterpoint on the graph of f^{-1} lies on the opposite side of the line $y = x$ but the same distance away, as shown in Fig. 20-2. Mathematically, we can do this transformation by reversing the sequence of the ordered pair representing the point. When we want to obtain the graph of f^{-1} based on the graph of f, we can "flip the whole graph over in three dimensions" around the line $y = x$, as if that line were the hinge of a revolving door.

Question 12-4

Is it possible for a relation to be its own inverse?

Answer 12-4

Yes. The simplest example is the relation described in the Cartesian xy plane by the equation

$$y = x$$

which can also be written as

$$f(x) = x$$

Another, less obvious example, is

$$y = -x$$

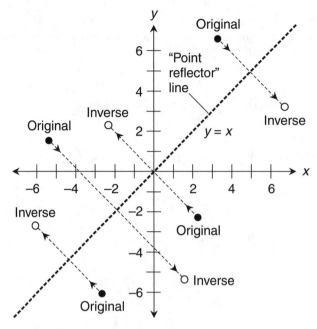

Figure 20-2 Illustration for Question and Answer 12-3.

which can also be written as

$$f(x) = -x$$

If a relation's graph is a circle centered at the origin, then that relation is its own inverse. Examples include all relations of the form

$$x^2 + y^2 = r^2$$

where r is the radius of the circle. We can also write such a relation in the form

$$f(x) = \pm(r^2 - x^2)^{1/2}$$

Question 12-5

How can we tell, simply by looking at the graph of a relation, whether or not that relation is its own inverse?

Answer 12-5

Suppose that when we "flip the graph over in three dimensions" along the line $y = x$ as if that line were the hinge of a revolving door, we end up with *exactly* the same graph as the one we started with. In any case like that, the relation is its own inverse. If we do the "revolving door" transformation and end up with a graph that's different *in any way* from the one we started

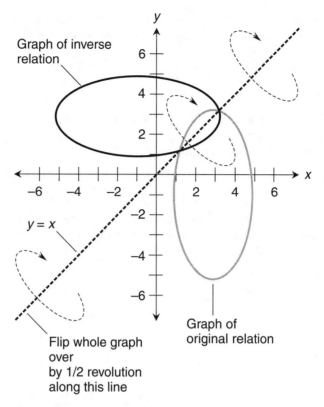

Graph of inverse relation

Graph of original relation

$y = x$

Flip whole graph over by 1/2 revolution along this line

Figure 20-3 Illustration for Question and Answer 12-5.

with, then the relation isn't its own inverse. Figure 20-3 shows an example of a graph of the second type, where the inverse obviously differs from the original relation.

Question 12-6

Suppose we have a relation that's not a function, because it maps some values of the independent variable x to more than one value of the dependent variable y. Is it possible to modify such a relation so that it becomes a function of x?

Answer 12-6

Yes, in most cases it's possible. If we can restrict the domain or the range to values such that the modified relation never maps any value of x to more than one value of y, then the modified relation is a function of x.

Question 12-7

Is the inverse of a function always a function?

Answer 12-7

No, not always. Suppose we have a function f that maps values of an independent variable x to values of a dependent variable y. Also imagine that, for any value of x in the domain, there's only one corresponding value of y in the range. On that basis, we know that f is a function of x. However, if some values of y are mapped from two or more values of x, then we don't have a function of y when we consider y as the independent variable and x as the dependent variable. Although the inverse f^{-1} is a relation, it's not a true function.

Question 12-8

Consider a function f that maps values of x to values of y. Suppose that f^{-1}, which maps values of y to values of x, is a relation but not a true function. Is it possible to modify the inverse relation f^{-1} so that it becomes a function of y?

Answer 12-8

In most cases, yes. If we can restrict the inverse relation's domain (the set of y values for which f^{-1} is defined) or the inverse relation's range (the set of x values for which f^{-1} is defined) so that the modified version of f^{-1} never maps any value of y to more than one value of x, then the modified inverse is a true function of y.

Question 12-9

Consider the two functions

$$f(x) = x$$

and

$$g(x) = -x$$

Both f and g are their own inverses, and the inverses are also true functions. Is it possible for any other true function to be its own inverse, with that inverse also constituting a true function?

Answer 12-9

Yes, this can happen. Consider the real-number function

$$h(x) = 1/x$$

where $x \neq 0$. This function is its own inverse, because

$$h^{-1}[h(x)] = h^{-1}(1/x) = 1/(1/x) = x$$

and

$$h[h^{-1}(x)] = h(1/x) = 1/(1/x) = x$$

Question 12-10

Imagine a function f such that

$$y = f(x)$$

and whose inverse f^{-1} is a true function, so that

$$f^{-1}(y) = x$$

for all values of x in the domain of f, and for all values of y in the range of f. Based on this information, what can we conclude about the nature of the mapping that f represents between the elements of its domain and the elements of its range?

Answer 12-10

Every element x in the domain maps to exactly one element y in the range, and every element y in the range is mapped from exactly one element x in the domain. Therefore, within the specified domain and range, the mapping that f represents is a one-to-one correspondence, technically known as a bijection.

Chapter 13

Question 13-1

What are the four basic types of conic sections? What do they look like in the Cartesian plane?

Answer 13-1

The conic sections are geometric curves representing the intersection of a plane with a double cone. There are four types: the circle, the ellipse, the parabola, and the hyperbola. Figure 20-4 shows generic graphs of each type of conic section in the Cartesian plane.

Question 13-2

How are the conic sections generated in 3D geometry?

Answer 13-2

When the plane is perpendicular to the axis of the double cone, we get a circle, as shown in Fig. 20-5A. When the plane is not perpendicular to the axis of the cone but the intersection curve is closed, we get an ellipse (Fig. 20-5B). When we tilt the plane just enough to open up the curve, we get a parabola (Fig. 20-5C). When we tilt the plane still more, we get a hyperbola (Fig. 20-5D).

Question 13-3

What is meant by the term "eccentricity" with respect to a conic section? How do the eccentricity values compare for a circle, an ellipse, a parabola, and a hyperbola?

Answer 13-3

Eccentricity (symbolized e) is a nonnegative real number that defines the extent to which a conic section differs from a circle. Here's how the eccentricity values compare for the four types of conic section:

- A circle has $e = 0$
- An ellipse has $0 < e < 1$

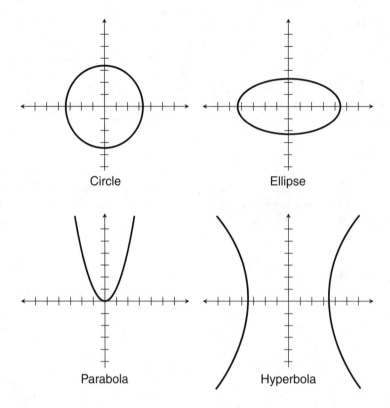

Figure 20-4   Illustration for Question and Answer 13-1.

- A parabola has $e = 1$
- A hyperbola has $e > 1$

Question 13-4
What's the focus of a parabola? What's the directrix of a parabola? How are they related?

Answer 13-4
The focus of a parabola is a point in the same plane as the parabola, and the directrix is a line in that plane that does not pass through the focus. On a parabola, every point is equidistant from a specific focus and a specific directrix, as shown in Fig. 20-6. For any particular focus and directrix in geometric space, there exists exactly one parabola. Conversely, for any particular parabola in space, there exists exactly one focus, and exactly one directrix.

Question 13-5
What's the focal length of a parabola?

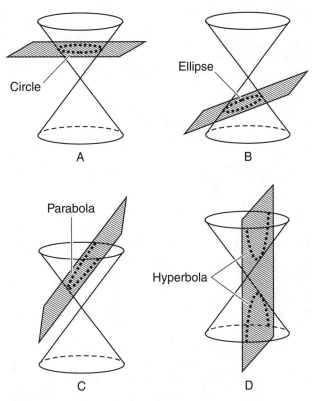

Figure 20-5 Illustration for Question and Answer 13-2.

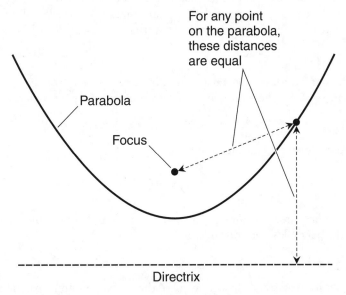

Figure 20-6 Illustration for Question and Answer 13-4.

Answer 13-5

The focal length of a parabola is the distance between the focus and the point on the parabola closest to the focus. The focal length is also equal to half the distance between the focus and the point on the directrix closest to the focus.

Question 13-6

What's the standard-form general equation for a circle in the Cartesian xy plane?

Answer 13-6

The standard-form general equation is

$$(x - x_0)^2 + (y - y_0)^2 = r^2$$

where x_0 and y_0 are real-number constants that tell us the coordinates (x_0, y_0) of the center of the circle, and r is a positive real-number constant that tells us the radius of the circle.

Question 13-7

What's the standard-form general equation for an ellipse in a Cartesian xy plane where the x axis is horizontal and the y axis is vertical?

Answer 13-7

The standard-form general equation is

$$(x - x_0)^2/a^2 + (y - y_0)^2/b^2 = 1$$

where x_0 and y_0 are real-number constants representing the coordinates (x_0, y_0) of the center of the ellipse, a is a positive real-number constant that tells us the length of the horizontal semi-axis, and b is a positive real-number constant that tells us the length of the vertical semi-axis.

Question 13-8

What's the standard-form general equation for a parabola that opens upward or downward in a Cartesian xy plane where the x axis is horizontal and the y axis is vertical?

Answer 13-8

The standard-form general equation is

$$y = ax^2 + bx + c$$

where a, b, and c are real-number constants, and $a \neq 0$. If $a > 0$, the parabola opens upward. If $a < 0$, the parabola opens downward.

Question 13-9

How can we locate the coordinates (x_0, y_0) of the vertex point on a parabola that opens upward or downward in a Cartesian xy plane where the x axis is horizontal and the y axis is vertical?

How can we tell whether that vertex point represents the absolute minimum value of y or the absolute maximum value of y?

Answer 13-9

We can find the coordinates (x_0, y_0) of the vertex point based on the constants in the standard-form equation of the parabola. The x value is

$$x_0 = -b/(2a)$$

The y value is

$$y_0 = -b^2/(4a) + c$$

If $a > 0$, the parabola opens upward, so the vertex represents the absolute minimum value of y on the curve. If $a < 0$, the parabola opens downward, so the vertex represents the absolute maximum value of y on the curve.

Question 13-10

What's the standard-form general equation for a hyperbola that opens toward the right and left in a Cartesian xy plane where the x axis is horizontal and the y axis is vertical?

Answer 13-10

The standard-form general equation is

$$(x - x_0)^2/a^2 - (y - y_0)^2/b^2 = 1$$

where x_0 and y_0 are real-number constants that tell us the coordinates (x_0, y_0) of the center of the hyperbola, a is a positive real-number constant that tells us the length of the horizontal semi-axis, and b is a positive real-number constant that tells us the length of the vertical semi-axis.

Chapter 14
Question 14-1

How can we informally describe the graph of the function

$$y = e^x$$

in the Cartesian xy plane?

Answer 14-1

The graph is a smooth, continually increasing curve that crosses the y axis at the point $(0,1)$. The domain is the set of all real numbers, and the range is the set of all positive real numbers. The curve is entirely contained within the first and second quadrants. As we move to the left (in the negative x direction), the curve approaches, but never reaches, the x axis. As we move to the right (in the positive x direction), the graph rises at an ever-increasing rate.

Question 14-2

How can we informally describe the graph of the function

$$y = e^{-x}$$

in the Cartesian xy plane?

Answer 14-2

The graph is a smooth, continually decreasing curve that crosses the y axis at $(0,1)$. The domain is the set of all real numbers, and the range is the set of all positive real numbers. The curve is entirely contained within the first and second quadrants. As we move to the right, the curve approaches the x axis but never quite reaches that axis. As we move to the left, the graph rises at an ever-increasing rate. In fact, the curve for the function

$$y = e^{-x}$$

has exactly the same shape as the curve for the function

$$y = e^{x}$$

but is reversed left-to-right around the y axis, so the two graphs are horizontal mirror images of each other.

Question 14-3

How can we informally describe the graphs of the functions

$$y = 10^{x}$$

and

$$y = 10^{-x}$$

in the Cartesian xy plane?

Answer 14-3

The graphs of these functions are curves that closely resemble the graphs of the functions

$$y = e^{x}$$

and

$$y = e^{-x}$$

respectively. Both base-10 graphs cross the y axis at $(0,1)$, just as the base-e graphs do. However, the contours differ. The base-10 curves are somewhat steeper than the base-e curves.

Question 14-4

How can we visually and qualitatively compare the graphs of the four functions described in Questions 14-1 through 14-3?

Answer 14-4

We can graph them all together on a generic rectangular-coordinate grid such as the one shown in Fig. 20-7.

Question 14-5

How can we informally describe the graph of the function

$$y = \ln x$$

in the Cartesian *xy* plane?

Answer 14-5

The graph is a smooth, continually increasing curve that crosses the *x* axis at the point (1,0). The domain is the set of positive real numbers, and the range is the set of all real numbers. The curve is entirely contained within the first and fourth quadrants. As we move to the left (in the negative *x* direction) from the point (1,0), the curve "blows up negatively," approaching the *y* axis but never reaching it. As we move to the right from (1,0), the graph rises at an ever-decreasing rate.

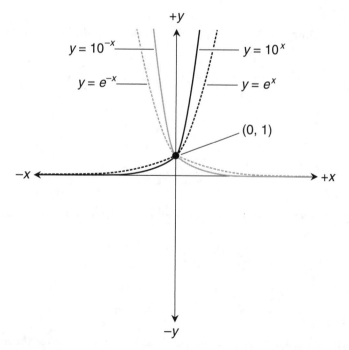

Figure 20-7   Illustration for Question and Answer 14-4.

Question 14-6

How can we verbally describe the graph of

$$y = \ln (1/x)$$

in the Cartesian xy plane?

Answer 14-6

The graph is a smooth, continually decreasing curve that crosses the y axis at $(1,0)$. The domain is the set of positive real numbers, and the range is the set of all real numbers. The curve is entirely contained within the first and fourth quadrants. As we move to the left from the point $(1,0)$, the curve "blows up positively," approaching the y axis but never reaching it. As we move to the right from $(1,0)$, the graph falls at an ever-decreasing rate. In fact, the curve representing

$$y = \ln (1/x)$$

has exactly the same shape as the curve for

$$y = \ln x$$

but is reversed top-to-bottom with respect to the x axis, so the two graphs are vertical mirror images of each other.

Question 14-7

How can we informally describe the graphs of the common-log functions

$$y = \log_{10} x$$

and

$$y = \log_{10} (1/x)$$

in the Cartesian xy plane?

Answer 14-7

The graphs of these functions closely resemble the graphs of the functions

$$y = \ln x$$

and

$$y = \ln (1/x)$$

respectively. Both common-log graphs cross the x axis at $(1,0)$, just as the natural-log graphs do. However, the contours differ. The natural-log curves are somewhat steeper than the common-log curves.

Question 14-8

How can we visually and qualitatively compare the graphs of the four functions described in Questions 14-5 through 14-7?

Answer 14-8

We can graph them all together on a generic rectangular-coordinate grid such as the one shown in Fig. 20-8.

Question 14-9

How can we plot the graph of the sum of two functions or the difference between two functions?

Answer 14-9

There are two ways in which this can be done. First, we can graph the original functions separately, and then add or subtract their values visually to infer the sum or difference graph. Second, we can calculate several outputs for each function using inputs that we've selected to get a good sampling. Then we can add or subtract these outputs arithmetically. Based on that data, we can graph the sum or difference function.

Question 14-10

Texts don't always agree in the denotation of logarithmic functions. How can we avoid confusion when we write our own papers?

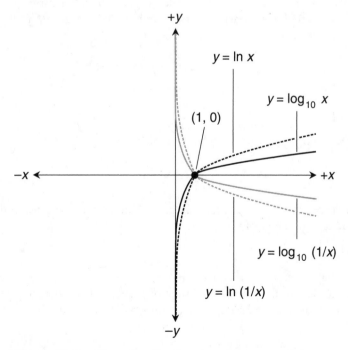

Figure 20-8   Illustration for Question and Answer 14-8.

Answer 14-10

We should always clarify the logarithmic base when we write "log" followed by anything. For example, we should write "$\log_{10}$" or "$\log_e$" instead of "log" (unless we can't portray the subscript within the constraints of a text-editing or Web site–building program). We don't need to write a subscript when we write "ln" to denote the natural logarithm, because "ln" means natural log or base-e log all the time.

Chapter 15

Special note

If you want to see graphical illustrations of the answers to the following 10 questions, feel free to look back at Chap. 15. Try to envision or draw the graphs yourself before you look back!

Question 15-1

Consider a function f of a real-number variable θ such that

$$f(\theta) = \sin \theta + \cos \theta$$

What are the period, the positive peak amplitude and the negative peak amplitude of f? What are the domain and range of f?

Answer 15-1

The period of f is 2π. That's the same as the period of the sine. It's also the same as the period of the cosine. The positive peak amplitude of f is $2^{1/2}$. The negative peak amplitude of f is $-2^{1/2}$. The domain of f is the set of all real numbers. The range of f is the set of all real numbers $f(\theta)$ such that

$$-2^{1/2} \leq f(\theta) \leq 2^{1/2}$$

Question 15-2

Consider a function f of a real-number variable θ such that

$$f(\theta) = \sin \theta \cos \theta$$

What are the period, the positive peak amplitude and the negative peak amplitude of f? What are the domain and range of f?

Answer 15-2

The period of f is π, which is equal to half the period of the sine, and is also half the period of the cosine. The positive peak amplitude of f is $1/2$. The negative peak amplitude of f is $-1/2$. The domain of f is the set of all real numbers. The range of f is the set of all real numbers $f(\theta)$ such that

$$-1/2 \leq f(\theta) \leq 1/2$$

Question 15-3

Consider a function f of a real-number variable θ such that

$$f(\theta) = \sin^2 \theta + \cos^2 \theta$$

What are the period, the positive peak amplitude and the negative peak amplitude of f? What are the domain and range of f?

Answer 15-3

In this case, the function f has a constant value of 1. The period is not defined, because the function's output value never changes, and is defined for all inputs. The positive peak amplitude of f is equal to 1. The negative peak amplitude of f is also equal to 1. The domain of f is the set of all real numbers. The range of f is the set containing the single number 1.

Question 15-4

Consider a function f of a real-number variable θ such that

$$f(\theta) = \sec \theta + \csc \theta$$

What are the period, the positive peak amplitude and the negative peak amplitude of f? What are the domain and range of f?

Answer 15-4

The period of f is 2π, which is the same as the period of the secant, and the same as the period of the cosecant. The positive and negative peak amplitudes of f are not defined, because f blows up in both the positive and negative directions whenever θ is an integer multiple of $\pi/2$. The domain of f is the set of all real numbers except the integer multiples of $\pi/2$. The range of f is the set of all real numbers.

Question 15-5

Consider a function f of a real-number variable θ such that

$$f(\theta) = \sec \theta \csc \theta$$

What are the period, the positive peak amplitude and the negative peak amplitude of f? What are the domain and range of f?

Answer 15-5

The period of f is π, which is half the period of the secant, and half the period of the cosecant. The positive and negative peak amplitudes of f are undefined, because f blows up both positively and negatively at all integer multiples of $\pi/2$. The domain of f is the set of all real numbers except the integer multiples of $\pi/2$. The range is the set of all real numbers $f(\theta)$ such that

$$f(\theta) \geq 2 \text{ or } f(\theta) \leq -2$$

Question 15-6

Consider a function f of a real-number variable θ such that

$$f(\theta) = \sec^2 \theta + \csc^2 \theta$$

What are the period, the positive peak amplitude and the negative peak amplitude of f? What are the domain and range of f?

Answer 15-6

The period of f is $\pi/2$, which is half the period of the secant squared, and half the period of the cosecant squared. The positive peak amplitude of f is undefined, because f "blows up" positively at all integer multiples of $\pi/2$. The negative peak amplitude of f is equal to 4, which occurs whenever θ is an odd-integer multiple of $\pi/4$. The domain of f is the set of all real numbers except the integer multiples of $\pi/2$. The range is of f the set of all real numbers $f(\theta)$ such that

$$f(\theta) \geq 4$$

Question 15-7

Consider a function f of a real-number variable θ such that

$$f(\theta) = \tan \theta + \cot \theta$$

What are the period, the positive peak amplitude and the negative peak amplitude of f? What are the domain and range of f?

Answer 15-7

The period of f is π, which is the same as that of the tangent, and the same as that of the cotangent. The positive and negative peak amplitudes of f are both undefined, because f blows up positively and negatively at all integer multiples of $\pi/2$. The domain of f is the set of all real numbers except the integer multiples of $\pi/2$. The range of f is the set of all real numbers $f(\theta)$ such that

$$f(\theta) \geq 2 \ \text{ or } \ f(\theta) \leq -2$$

Question 15-8

Consider a function f of a real-number variable θ such that

$$f(\theta) = \tan \theta \cot \theta$$

What are the period, the positive peak amplitude and the negative peak amplitude of f? What are the domain and range of f?

Answer 15-8

This particular function presents a strange situation. The graph of f is a horizontal, straight line with single-point gaps wherever θ is an integer multiple of $\pi/2$. The period of f is $\pi/2$, because the graph consists of infinitely many open line segments placed end to end, each of length $\pi/2$. The positive peak amplitude of f is equal to 1. The negative peak amplitude of f is also equal to 1. The domain of f is the set of all real numbers except the integer multiples of $\pi/2$. The range is the set containing the single element 1.

Question 15-9

Consider a function f of a real-number variable θ such that

$$f(\theta) = \tan^2 \theta + \cot^2 \theta$$

What are the period, the positive peak amplitude and the negative peak amplitude of f? What are the domain and range of f?

Answer 15-9

The period of f is $\pi/2$, which is half that of the tangent squared, and is also half that of the cotangent squared. The positive peak amplitude of f is undefined, because the function blows up positively at all integer multiples of $\pi/2$. The negative peak amplitude of f is equal to 2, which occurs whenever θ is an odd-integer multiple of $\pi/4$. The domain of f is the set of all real numbers except the integer multiples of $\pi/2$. The range of f is the set of real numbers $f(\theta)$ such that

$$f(\theta) \geq 2$$

Question 15-10

Consider a function f of a real-number variable θ such that

$$f(\theta) = (\tan^2 \theta)/(\cot^2 \theta)$$

What are the period, the positive peak amplitude and the negative peak amplitude of f? What are the domain and range of f?

Answer 15-10

The period of f is π, which is the same as the period of the tangent squared, and is also the same as the period of the cotangent squared. The positive peak amplitude of f is undefined, because the function blows up positively at all odd-integer multiples of $\pi/2$. The negative peak amplitude of f is undefined as well, although $f(\theta)$ approaches 0 whenever θ approaches any integer multiple of π from either side. (We can't say that the negative peak amplitude is 0, because the function never actually attains that value.) The domain of f is the set of all real numbers except the integer multiples of $\pi/2$. The range of f is the set of all positive real numbers.

Chapter 16

Question 16-1

What's a parameter? What's a set of parametric equations?

Answer 16-1

A parameter is an independent variable on which other variables depend. A set of parametric equations is a collection of equations, at least one of which has one or more variables that depend on the parameter. The parameter, which is often symbolized t, plays the role of master controller for the other variables in the system.

Question 16-2

Consider the pair of parametric equations

$$x = 3t$$

and

$$y = -t$$

where t is the parameter on which both x and y depend. How can we sketch a Cartesian graph of this system? How can we find an equivalent equation in terms of the variables x and y only, based on the graph?

Answer 16-2

We can input various values of t to both equations, and plot the ordered pairs (x,y) that come out of those equations. Following are some examples:

- When $t = -2$, we have $x = 3 \times (-2) = -6$ and $y = -1 \times (-2) = 2$.
- When $t = -1$, we have $x = 3 \times (-1) = -3$ and $y = -1 \times (-1) = 1$.
- When $t = 0$, we have $x = 3 \times 0 = 0$ and $y = -1 \times 0 = 0$.
- When $t = 1$, we have $x = 3 \times 1 = 3$ and $y = -1 \times 1 = -1$.
- When $t = 2$, we have $x = 3 \times 2 = 6$ and $y = -1 \times 2 = -2$.

When we plot the (x,y) ordered pairs based on this list as points on a Cartesian plane and then "connect the dots," we get a line through the origin with a slope of $-1/3$, as shown in Fig. 20-9. In slope-intercept form, the line can be represented as

$$y = (-1/3)x$$

Question 16-3

Consider again the pair of parametric equations

$$x = 3t$$

and

$$y = -t$$

How can we use algebra alone (without the help of a graph) to determine the equivalent equation in terms of x and y only?

Answer 16-3

We can take the first parametric equation

$$x = 3t$$

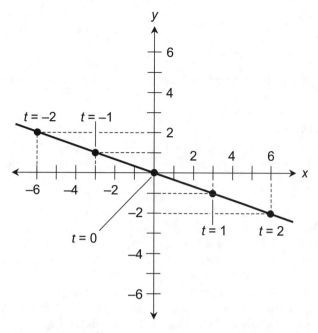

Figure 20-9 Illustration for Question and Answer 16-2.

and multiply it through by $-1/3$ to get

$$(-1/3)x = (-1/3)(3t) = -t$$

Deleting the middle portion, we get

$$(-1/3)x = -t$$

The second parametric equation tells us that $-t = y$, so we can substitute directly in the above equation to obtain

$$(-1/3)x = y$$

which is identical to the following slope-intercept equation:

$$y = (-1/3)x$$

Question 16-4

Consider the pair of parametric equations

$$\theta = 3t$$

and

$$r = -t$$

where t is the parameter on which both θ and r depend. How can we sketch a polar graph of this system?

Answer 16-4

We can input various values of t, restricting ourselves to values such that we see only the part of the graph corresponding to the first full counterclockwise rotation of the direction angle, where $0 \leq \theta \leq 2\pi$:

- When $t = 0$, we have $\theta = 3 \times 0 = 0$ and $r = -1 \times 0 = 0$.
- When $t = \pi/4$, we have $\theta = 3\pi/4$ and $r = -\pi/4 \approx -0.79$.
- When $t = \pi/2$, we have $\theta = 3\pi/2$ and $r = -\pi/2 \approx -1.57$.
- When $t = 2\pi/3$, we have $\theta = 3 \times 2\pi/3 = 2\pi$ and $r = -2\pi/3 \approx -2.09$.

Figure 20-10 illustrates this graph, based on these four points and the intuitive knowledge that the graph must be a spiral, starting at the origin and expanding as we rotate counterclockwise. The graph is a little tricky, because all of the radii are negative! Also, we should remember that the concentric circles represent radial divisions on the polar coordinate grid; the straight lines represent angular divisions.

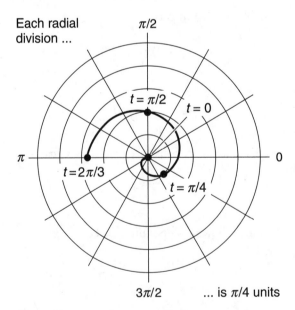

Figure 20-10   Illustration for Question and Answer 16-4. In this coordinate system, each radial division represents $\pi/4$ units.

Question 16-5

Consider again the pair of parametric equations

$$\theta = 3t$$

and

$$r = -t$$

How can we use algebra to determine the equivalent equation in terms of θ and r only?

Answer 16-5

The equation can be derived using the same algebraic process that we used in the Cartesian situation. We substitute θ in place of x, and we substitute r in place of y. When we do that, we get

$$r = (-1/3)\theta$$

Question 16-6

What are the parametric equations for a circle centered at the origin in the Cartesian xy plane?

Answer 16-6

The parametric equations are

$$x = a \cos t$$

and

$$y = a \sin t$$

where a is the radius of the circle and t is the parameter.

Question 16-7

What are the parametric equations for a circle centered at the origin in the polar coordinate plane?

Answer 16-7

Let the polar direction angle be θ, and let the polar radius be r. The parametric equations of a circle having radius a, and centered at the origin, are

$$\theta = t$$

and

$$r = a$$

where t is the parameter.

Question 16-8

Why does only one of the equations in Answer 16-7 contain the parameter *t*? Shouldn't both equations contain it?

Answer 16-8

The parameter *t* has no effect in the second equation, because the polar radius *r* of a circle centered at the origin is always the same, no matter how anything else varies.

Question 16-9

What are the parametric equations for an ellipse centered at the origin in the Cartesian *xy* plane?

Answer 16-9

The parametric equations are

$$x = a \cos t$$

and

$$y = b \sin t$$

where *a* is the length of the horizontal (*x*-coordinate) semi-axis, *b* is the length of the vertical (*y*-coordinate) semi-axis, and *t* is the parameter.

Question 16-10

Why is the passage of time a common parameter in science and engineering?

Answer 16-10

In the physical world, many effects and phenomena depend on elapsed time. If we find time acting as a mathematical variable, then that variable is almost always independent. We often come across situations where two or more factors fluctuate with the passage of time. An example is the variation of temperature, humidity, and barometric pressure versus time in a specific location. In a situation of this sort, time can be considered as the parameter on which the other three physical variables depend.

Chapter 17

Question 17-1

What information do we need to determine the equation of a plane in Cartesian *xyz* space?

Answer 17-1

We can find an equation for a plane in Cartesian *xyz* space if we know the direction of at least one vector that's perpendicular to the plane, and if we know the coordinates of at least one point in the plane. We don't have to know the magnitude of the vector, but only its direction. The point's coordinates don't have to tell us where the vector begins or ends; the point can be anywhere in the plane.

Question 17-2

Imagine a plane that passes through a point whose coordinates are (x_0,y_0,z_0) in Cartesian *xyz* space. Also suppose that we've found a vector $a\mathbf{i} + b\mathbf{j} + c\mathbf{k}$ that's normal (perpendicular) to the plane. Based on this information, how can we write down an equation that represents the plane?

Answer 17-2

We can write the plane's equation in the standard form

$$a(x - x_0) + b(y - y_0) + c(z - z_0) = 0$$

We can also write the equation as

$$ax + by + cz + d = 0$$

where d is a constant that works out to

$$d = -ax_0 - by_0 - cz_0$$

Question 17-3

What's the general equation for a sphere centered at the origin and having radius r in Cartesian *xyz* space?

Answer 17-3

The equation can be written in the standard form

$$x^2 + y^2 + z^2 = r^2$$

Question 17-4

What's the general equation for a sphere of radius r in Cartesian *xyz* space, centered at a point whose coordinates are (x_0,y_0,z_0)?

Answer 17-4

The equation can be written in the standard form

$$(x - x_0)^2 + (y - y_0)^2 + (z - z_0)^2 = r^2$$

Question 17-5

Can a sphere have a negative radius in Cartesian *xyz* space?

Answer 17-5

Normally, we define a sphere's radius as a positive real number. Nevertheless, spheres with negative radii can exist in theory. If we encounter a sphere whose radius happens to be defined as a negative real number, then that sphere has the same equation as it would if we defined the radius as the absolute value of that number. For all real numbers r, it's always true that $r^2 = |r|^2$, so the following two equations:

$$(x - x_0)^2 + (y - y_0)^2 + (z - z_0)^2 = r^2$$

and

$$(x - x_0)^2 + (y - y_0)^2 + (z - z_0)^2 = |r|^2$$

are equivalent, whether r is positive or negative.

Question 17-6

What's the general equation for a distorted sphere in Cartesian xyz space?

Answer 17-6

The equation can be written in the standard form

$$(x - x_0)^2/a^2 + (y - y_0)^2/b^2 + (z - z_0)^2/c^2 = 1$$

where (x_0, y_0, z_0) are the coordinates of the center, a is the is the axial radius in the x direction, b is the axial radius in the y direction, and c is the axial radius in the z direction. Normally, all three of the constants a, b, and c are positive reals.

Question 17-7

There are three distinct classifications of distorted sphere. What are they? How can we tell, from the standard-form equation, which of these three types we have?

Answer 17-7

We can have an oblate sphere, an ellipsoid, or an oblate ellipsoid. We can tell which of these three types a particular standard-form equation represents by comparing the values of the axial radii a, b, and c. We have an oblate sphere *if and only if* two of the positive real-number axial radii are equal, and the third is smaller. In that case, one of the following is true:

$$a < b = c$$
$$b < a = c$$
$$c < a = b$$

We have an ellipsoid *if and only if* two of the positive real-number axial radii are equal, and the third is larger. Then one of the following is true:

$$a > b = c$$
$$b > a = c$$
$$c > a = b$$

We have an oblate ellipsoid *if and only if* no two of the positive real-number axial radii are equal. In that scenario, all of the following are true:

$$a \neq b$$
$$b \neq c$$
$$a \neq c$$

Question 17-8

What's the general equation for a hyperboloid of one sheet in Cartesian *xyz* space?

Answer 17-8

The equation can be written in one of the following standard forms:

$$(x - x_0)^2/a^2 + (y - y_0)^2/b^2 - (z - z_0)^2/c^2 = 1$$
$$(x - x_0)^2/a^2 - (y - y_0)^2/b^2 + (z - z_0)^2/c^2 = 1$$
$$-(x - x_0)^2/a^2 + (y - y_0)^2/b^2 + (z - z_0)^2/c^2 = 1$$

where (x_0, y_0, z_0) are the coordinates of the center, the constants *a*, *b*, and *c* are positive real numbers that define the object's general shape, and the locations of the signs (plus and minus) define the object's orientation with respect to the coordinate axes.

Question 17-9

What's the general equation for a hyperboloid of two sheets in Cartesian *xyz* space?

Answer 17-9

The equation can be written in one of the following standard forms:

$$-(x - x_0)^2/a^2 + (y - y_0)^2/b^2 - (z - z_0)^2/c^2 = 1$$
$$(x - x_0)^2/a^2 - (y - y_0)^2/b^2 - (z - z_0)^2/c^2 = 1$$
$$-(x - x_0)^2/a^2 - (y - y_0)^2/b^2 + (z - z_0)^2/c^2 = 1$$

where (x_0, y_0, z_0) are the coordinates of the center, the constants *a*, *b*, and *c* are positive real numbers that define the object's general shape, and the locations of the signs (plus and minus) define the object's orientation with respect to the coordinate axes.

Question 17-10

What's the general equation for an elliptic cone in Cartesian *xyz* space?

Answer 17-10

The equation can be written in one of the following standard forms:

$$(x - x_0)^2/a^2 + (y - y_0)^2/b^2 - (z - z_0)^2/c^2 = 0$$
$$(x - x_0)^2/a^2 - (y - y_0)^2/b^2 + (z - z_0)^2/c^2 = 0$$
$$-(x - x_0)^2/a^2 + (y - y_0)^2/b^2 + (z - z_0)^2/c^2 = 0$$

where (x_0,y_0,z_0) are the coordinates of the point where the apexes of the two halves of the double cone meet, the constants a, b, and c are positive real numbers that define the object's general shape, and the locations of the signs (plus and minus) define the object's orientation with respect to the coordinate axes. Don't get confused by the similarity between these equations and those for hyperboloids of one sheet. The only difference is that the net values are all equal to 1 for the hyperboloids, and all equal to 0 for the cones.

Chapter 18

Question 18-1

What's the general symmetric equation for a straight line in Cartesian *xyz* space?

Answer 18-1

Imagine that (x_0,y_0,z_0) are the coordinates of a specific point. Suppose that a, b, and c are nonzero real-number constants. The general symmetric equation of a straight line passing through (x_0,y_0,z_0) is

$$(x - x_0)/a \ = \ (y - y_0)/b \ = \ (z - z_0)/c$$

The constants a, b, and c are called direction numbers. When considered all together as an ordered pair (a,b,c), these numbers define the direction or orientation of the line with respect to the coordinate axes.

Question 18-2

What are the general parametric equations for a straight line in Cartesian *xyz* space?

Answer 18-2

Let (x_0,y_0,z_0) be the coordinates of a specific point, and suppose that a, b, and c are nonzero real-number constants. The general parametric equations for a straight line passing through (x_0,y_0,z_0) are

$$x = x_0 + at$$
$$y = y_0 + bt$$
$$z = z_0 + ct$$

where the parameter t can range over the entire set of real numbers. As with the symmetric form, the constants a, b, and c are direction numbers that tell us how the line is orientated relative to the coordinate axes.

Question 18-3

What is meant by the expression "preferred direction numbers" when describing the orientation of a straight line in Cartesian *xyz* space?

Answer 18-3

For any line in Cartesian *xyz* space, there are infinitely many ordered triples that can define its orientation with respect to the coordinate axes. For example, if a line has the direction numbers (a,b,c), then we can multiply all three entries by a real number other than 0 or 1, and we'll get another valid ordered triple of direction numbers for that same line. For the sake of simplicity and elegance, mathematicians usually reduce the direction numbers so that their only common divisor is 1, and so that *at most* one of them is negative. Doing this produces a unique set of direction numbers "in lowest terms," and these are the preferred direction numbers for the line.

Question 18-4

What are the generalized parametric equations for a parabola in Cartesian *xyz* space where the value of *x* is constant, and the curve's axis is parallel to either the *y* axis or the *z* axis?

Answer 18-4

If *x* is constant and the axis of the parabola is parallel to the *y* axis, then the curve's parametric equations are

$$x = c$$
$$y = a_1 t^2 + a_2 t + a_3$$
$$z = t$$

where a_1, a_2, and a_3 are real-number coefficients, *c* is the real-number constant value to which *x* is held, and *t* is a parameter that can range over the set of all real numbers. If *x* is constant and the curve's axis is parallel to the *z* axis, then the parametric equations are

$$x = c$$
$$y = t$$
$$z = a_1 t^2 + a_2 t + a_3$$

In either case, the parabola lies in a plane parallel to the *yz* plane.

Question 18-5

What are the generalized parametric equations for a parabola in Cartesian *xyz* space where the value of *y* is constant, and the curve's axis is parallel to either the *x* axis or the *z* axis?

Answer 18-5

If *y* is constant and the axis of the parabola is parallel to the *x* axis, then the parametric equations are

$$x = a_1 t^2 + a_2 t + a_3$$
$$y = c$$
$$z = t$$

where a_1, a_2, and a_3 are real-number coefficients, *c* is the real-number constant value to which *y* is held, and *t* is a parameter that can range over the set of all real numbers. If *y* is constant and the curve's axis is parallel to the *z* axis, then the parametric equations are

$$x = t$$
$$y = c$$
$$z = a_1 t^2 + a_2 t + a_3$$

In either case, the parabola lies in a plane parallel to the *xz* plane.

Question 18-6

What are the generalized parametric equations for a parabola in Cartesian xyz space where the value of z is constant, and the curve's axis is parallel to either the x axis or the y axis?

Answer 18-6

If z is constant and the axis of the parabola is parallel to the x axis, then the parametric equations are

$$x = a_1 t^2 + a_2 t + a_3$$
$$y = t$$
$$z = c$$

where a_1, a_2, and a_3 are real-number coefficients, c is the real-number constant value to which z is held, and t is a parameter that can range over the set of all real numbers. If z is constant and the curve's axis is parallel to the y axis, then the parametric equations are

$$x = t$$
$$y = a_1 t^2 + a_2 t + a_3$$
$$z = c$$

In either case, the parabola lies in a plane parallel to the xy plane.

Question 18-7

What are the generalized parametric equations for a circle in Cartesian xyz space where the value of x is constant so the circle lies in a plane parallel to the yz plane, and the center of the circle lies on the x axis?

Answer 18-7

The parametric equations are

$$x = c$$
$$y = r \cos t$$
$$z = r \sin t$$

where r is the radius of the circle, c is the real-number constant value to which x is held, and t is a parameter that varies continuously over a real-number interval at least 2π units wide.

Question 18-8

What are the generalized parametric equations for a circle in Cartesian xyz space where the value of y is constant so the circle lies in a plane parallel to the xz plane, and the center of the circle lies on the y axis?

Answer 18-8

The parametric equations are

$$x = r \cos t$$
$$y = c$$
$$z = r \sin t$$

where r is the radius of the circle, c is the real-number constant value to which y is held, and t is a parameter that varies continuously over a real-number interval at least 2π units wide.

Question 18-9

What are the generalized parametric equations for a circle in Cartesian xyz space where the value of z is constant so the circle lies in a plane parallel to the xy plane, and the center of the circle lies on the z axis?

Answer 18-9

The parametric equations are

$$x = r \cos t$$
$$y = r \sin t$$
$$z = c$$

where r is the radius of the circle, c is the real-number constant value to which z is held, and t is a parameter that varies continuously over a real-number interval at least 2π units wide.

Question 18-10

Imagine a circle in Cartesian xyz space whose sets of parametric equations have one of the forms described in Answer 18-7 through 18-9. Consider the variable that's held constant. Suppose that, instead of insisting that it always keep the same value, we allow that variable to follow a constant multiple of the parameter t. What sort of curve will we get under these conditions?

Answer 18-10

We'll get a circular helix centered on the axis for whichever variable follows the constant multiple of t.

Chapter 19
Question 19-1

What's the difference between a sequence (also called a progression) and a series?

Answer 19-1

A sequence is a list of numbers or variables. Such a list can contain anywhere from two to infinitely many elements. A series is the sum of the elements in a specific sequence.

Question 19-2

What's an arithmetic sequence?

Answer 19-2

An arithmetic sequence is a list of numbers that starts at a certain value, and then increases or decreases at a constant rate after that. Therefore, each element is larger or smaller than its predecessor by a certain fixed amount.

Question 19-3

What's the general form of a finite arithmetic sequence of real numbers? What's the general form of an infinite arithmetic sequence of real numbers?

Answer 19-3

The general form of a finite arithmetic sequence S_{fin} is

$$S_{fin} = s_0, (s_0 + c), (s_0 + 2c), (s_0 + 3c), ..., (s_0 + nc)$$

where s_0 is a real number representing the first element, c is a real number representing the sequence constant, and n is a positive integer. In this case, the sequence has $n + 1$ elements. The general form of an infinite arithmetic sequence S_{inf} is

$$S_{inf} = s_0, (s_0 + c), (s_0 + 2c), (s_0 + 3c), ...$$

where s_0 is a real number representing the first element, and c is a real number representing the sequence constant. The ellipsis (...) tells us that the sequence continues without end.

Question 19-4

What are the partial sums of an infinite arithmetic sequence?

Answer 19-4

When we add up the elements of a numeric sequence, we get another list of numbers called the sequence of partial sums. For S_{inf} described in Answer 19-3, the first five partial sums are

$$s_0$$
$$s_0 + s_0 + c$$
$$s_0 + s_0 + c + s_0 + 2c$$
$$s_0 + s_0 + c + s_0 + 2c + s_0 + 3c$$
$$s_0 + s_0 + c + s_0 + 2c + s_0 + 3c + s_0 + 4c$$

which can be simplified to

$$s_0$$
$$2s_0 + c$$
$$3s_0 + 3c$$

$$4s_0 + 6c$$
$$5s_0 + 10c$$

Question 19-5

What's a geometric sequence?

Answer 19-5

A geometric sequence is a list of numbers with a starting value that's repeatedly multiplied by a constant factor. If we take any element in the sequence (except the first one) and divide it by its predecessor, we always get the same constant.

Question 19-6

What's the general form of a finite geometric sequence of real numbers? What's the general form of an infinite geometric sequence of real numbers?

Answer 19-6

The general form of a finite geometric sequence T_{fin} is

$$T_{\text{fin}} = t_0, t_0 k, t_0 k^2, t_0 k^3, t_0 k^4, ..., t_0 k^n$$

where t_0 is a real number representing the first element, k is a real number representing the sequence constant, and n is a positive integer. In this case, the sequence has $n + 1$ elements. The general form of an infinite geometric sequence T_{inf} is

$$T_{\text{inf}} = t_0, t_0 k, t_0 k^2, t_0 k^3, t_0 k^4, ...$$

where t_0 is a real number representing the first element, and k is a real number representing the sequence constant.

Question 19-7

What are the partial sums of an infinite geometric sequence?

Answer 19-7

For T_{inf} described in Answer 19-6, the first five partial sums are

$$t_0$$
$$t_0 + t_0 k$$
$$t_0 + t_0 k + t_0 k^2$$
$$t_0 + t_0 k + t_0 k^2 + t_0 k^3$$
$$t_0 + t_0 k + t_0 k^2 + t_0 k^3 + t_0 k^4$$

which can be simplified to

$$t_0$$
$$t_0 \, (1 + k)$$
$$t_0 \, (1 + k + k^2)$$
$$t_0 \, (1 + k + k^2 + k^3)$$
$$t_0 \, (1 + k + k^2 + k^3 + k^4)$$

Question 19-8

How can summation notation be used to symbolize a finite series? How can summation notation be used to symbolize an infinite series?

Answer 19-8

Suppose we have a series with n terms

$$a_1 + a_2 + a_3 + \cdots + a_{n-2} + a_{n-1} + a_n$$

We can symbolize it by writing

$$\sum_{i=1}^{n} a_i$$

and read it as, "The summation of the terms a_i, from $i = 1$ to n." If we have an infinite series

$$a_1 + a_2 + a_3 + a_4 + a_5 + \cdots$$

then we can symbolize it by writing

$$\sum_{i=1}^{\infty} a_i$$

and read it as, "The summation of the terms a_i, from $i = 1$ to infinity."

Question 19-9

What is the limit of an infinite sequence, an infinite series, a relation, or a function? How can we symbolize the fact that the limit of x^{-5}, as a real number x approaches infinity, is equal to 0? What does the term "convergent" mean in relation to an infinite sequence or series?

Answer 19-9

A limit is a value that an infinite sequence, an infinite series, a relation, or a function approaches, but does not necessarily reach. We can symbolize the fact that the limit of x^{-5}, as a real-number variable x approaches infinity, is equal to 0 by writing

$$\lim_{x \to \infty} x^{-5} = 0$$

An infinite sequence is convergent *if and only if*, as we move along the sequence from term to term, the values of the terms approach a definable limit. An infinite series is convergent *if and only if,* as we move along the series from term to term, the values of the partial sums approach a definable limit.

Question 19-10

What is the right-hand limit of a function at a point? What is the left-hand limit of a function at a point?

Answer 19-10

The right-hand limit of a function at a point is the value that the function approaches as we move toward the point from the positive direction. We denote a right-hand limit by writing a small plus sign at the end of the subscript. For example, if we approach the point where $x = 0$ along the x axis from the positive side, then the expression

$$\underset{x \to 0+}{Lim} \ln x$$

refers to the limit of the natural logarithm of x, as x approaches 0 from the right. (You might immediately see that this particular limit is not defined.) The left-hand limit of a function at a point is the value that the function approaches as we move toward the point from the negative direction. We denote a left-hand limit by writing a small minus sign at the end of the subscript. For example, if we approach the point where $x = 3$ along the x axis from the negative side, then the expression

$$\underset{x \to 3-}{Lim} e^x$$

refers to the limit of the natural exponential of x, as x approaches 3 from the left. (This particular limit happens to be defined, and is equal to e^3.)

Final Exam

This exam is designed to test your general knowledge, not to measure how fast you can perform calculations. A good score is at least 80 correct answers. The answers are listed in App. C. This test is long, so don't try to take it in a single session. Feel free to draw diagrams, sketch graphs, or use a calculator. But don't look back at the text or refer to outside information sources.

1. Under what conditions is the dot product of two *nonzero* polar-plane vectors equal to 0?
   (a) When the two vectors point in the same direction.
   (b) When the two vectors point in opposite directions.
   (c) When the two vectors have equal magnitude.
   (d) When the two vectors are mutually perpendicular.
   (e) Under more than one of the above conditions (a), (b), (c), and (d).

2. The cosecant function is singular when the input, in radians, is equal to
   (a) 0.
   (b) $\pi/6$.
   (c) $\pi/4$.
   (d) $\pi/3$.
   (e) $\pi/2$.

3. The conjugate of $8 - j6$ is
   (a) undefined.
   (b) $8 + j6$.
   (c) $6 - j8$.
   (d) $6 + j8$.
   (e) the pure real number 10.

4. Consider two vectors **a** and **b** in the Cartesian *xy* plane, both of which originate at (−1,4). Suppose that vector **a** terminates at (0,5) and vector **b** terminates at (0,3). If we add these vectors and express the result in standard form, what do we get?

(a) $\mathbf{a} + \mathbf{b} = (3^{1/2}, 5^{1/2})$

(b) $\mathbf{a} + \mathbf{b} = (-1, 1)$

(c) $\mathbf{a} + \mathbf{b} = (-1, 9)$

(d) $\mathbf{a} + \mathbf{b} = (2, 0)$

(e) $\mathbf{a} + \mathbf{b} = (0, 2^{1/2})$

5. Suppose that we have a complex number *c* in polar form, such that

$$c = r \cos \theta + j(r \sin \theta)$$

where *r* is the real-number polar vector magnitude and θ is the real-number polar vector angle. Also suppose that *n* is an integer. DeMoivre's theorem tells us that the *n*th power of this complex number is equal to

(a) $rn \cos (n\theta) + j[rn \sin (n\theta)]$.

(b) $(r + n) \cos (n\theta) + j[(r + n) \sin (n\theta)]$.

(c) $r^n \cos (n\theta) + j[r^n \sin (n\theta)]$.

(d) $rn \cos (\theta) + j[rn \sin (\theta)]$.

(e) $r^n \cos (\theta) + j[r^n \sin (\theta)]$.

6. The point $(x,y) = (0,-3)$ is

(a) in the first quadrant of the Cartesian plane.

(b) in the second quadrant of the Cartesian plane.

(c) in the third quadrant of the Cartesian plane.

(d) in the fourth quadrant of the Cartesian plane.

(e) not in any quadrant of the Cartesian plane.

7. Under what conditions is the cross product of two *nonzero* polar-plane vectors equal to the zero vector?

(a) When the two vectors point in the same direction.

(b) When the two vectors point in opposite directions.

(c) When the two vectors have equal magnitude.

(d) When the two vectors are mutually perpendicular.

(e) Under more than one of the above conditions (a), (b), (c), and (d).

8. In polar coordinates, which, if any, of the following equations can represent the dashed line graphed in Fig. FE-1?

(a) $\theta = 0$

(b) $\theta = r$

(c) $\theta = 2r\pi/3$

(d) $\theta = 2\pi/3$

(e) We can't say without knowing the size of each radial increment.

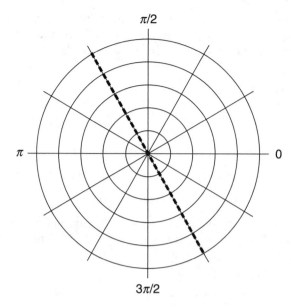

Figure FE-1   Illustration for Question 8.

9. The intersection of the sets of real and imaginary numbers is the set
   (a) $\{j\}$.
   (b) $\{1\}$.
   (c) $\{-j\}$.
   (d) $\{-1\}$.
   (e) $\{0\}$.

10. Which, if any, of these geometric figures would be the graph of a true function of x if drawn on the Cartesian xy plane, and the graph of a true function of θ if drawn on the polar coordinate plane?
    (a) A circle centered at the origin.
    (b) A straight line passing through the origin.
    (c) A straight line parallel to the Cartesian x axis, but not passing through the origin.
    (d) A straight line parallel to the Cartesian y axis, but not passing through the origin.
    (e) None of the above

11. When we multiply a polar vector by a negative scalar, what restrictions, if any, should we put on the direction angle of the product?
    (a) It should be nonnegative, but less than $\pi/2$.
    (b) It should be nonnegative, but less than π.
    (c) It should be nonnegative, but less than 2π.
    (d) It should be at least $-\pi$, but less than π.
    (e) We don't have to take any precautions concerning the direction angle.

12. If we quadruple the value of each coordinate of a point in Cartesian three-space, then its distance from the origin increases by a factor of

   (a) 2.

   (b) 4.

   (c) 8.

   (d) 16.

   (e) 64.

13. If we graph the unit circle in the Cartesian xy plane and then pick a point (x_0, y_0) on that circle such that $x_0 \neq 0$, then $1/x_0$ is equal to

   (a) the cosine of the counterclockwise angle between the positive x axis and a ray going out from the origin through (x_0, y_0).

   (b) the Arccosine of the counterclockwise angle between the positive x axis and a ray going out from the origin through (x_0, y_0).

   (c) the tangent of the counterclockwise angle between the positive x axis and a ray going out from the origin through (x_0, y_0).

   (d) the Arctangent of the counterclockwise angle between the positive x axis and a ray going out from the origin through (x_0, y_0).

   (e) the secant of the counterclockwise angle between the positive x axis and a ray going out from the origin through (x_0, y_0).

14. Suppose that we have a vector in Cartesian xyz space whose originating point is $(1, -1, 3)$ and whose terminating point is $(7, -4, -3)$. What is the ordered triple representing the Cartesian standard form of this vector?

   (a) We need more information to answer this question.

   (b) $(8, -5, 0)$

   (c) $(6, -3, -6)$

   (d) $(7, 4, -9)$

   (e) $(4, -5/2, 0)$

15. What are the Cartesian xy plane coordinates of the point P plotted in Fig. FE-2? Assume that each concentric-circle radial division represents 2 units.

   (a) $(2^{1/2}, 6)$

   (b) $(2^{1/2}, -6)$

   (c) $(-2^{1/2}, 6)$

   (d) $(-2^{1/2}, -6)$

   (e) None of the above

16. What is the range of values for x in the interval $(-4, 0]$?

   (a) $-4 < x < 0$

   (b) $-4 \leq x < 0$

   (c) $-4 < x \leq 0$

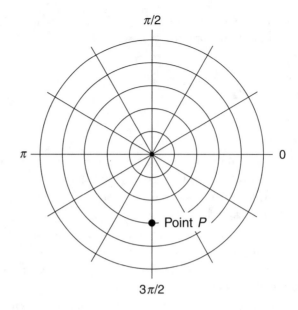

Figure FE-2   Illustration for Question 15.

(d) $-4 \le x \le 0$

(e) We can't say, because the notation (–4,0] is meaningless.

17. The direction of a standard-form vector in Cartesian *xyz* space can be uniquely defined by
    (a) the angle that the vector subtends with respect to the positive *x* axis.
    (b) the angle that the vector subtends with respect to the positive *y* axis.
    (c) the angle that the vector subtends with respect to the positive *z* axis.
    (d) the sum of the angles that the vector subtends with respect to the positive *x*, *y*, and *z* axes.
    (e) None of the above

18. In Cartesian *xyz* space, the point (2,3,4) is
    (a) 3 units from the origin.
    (b) 9 units from the origin.
    (c) $24^{1/2}$ units from the origin.
    (d) $29^{1/2}$ units from the origin.
    (e) 24 units from the origin.

19. The Cartesian negative of a vector in *xyz* space
    (a) points in the same direction as the original vector.
    (b) has the same magnitude as the original vector.
    (c) has the same direction angles as the original vector.
    (d) always has negative coordinates.
    (e) always has coordinates whose absolute values are negative.

20. The radian is an angle whose measure is precisely equivalent to
    (a) $\pi/2$ of a full circle.
    (b) $2/\pi$ of a full circle.
    (c) $4/\pi$ of a full circle.
    (d) $1/(2\pi)$ of a full circle.
    (e) $3\pi/2$ of a full circle.

21. When we multiply a vector in two-space by a positive scalar k_+, the magnitude
    (a) changes by a factor of k_+, while the direction angle stays the same.
    (b) changes by a factor of k_+, while the direction angle reverses.
    (c) stays the same, but the direction angle changes by a factor of k_+.
    (d) stays the same, but the direction angle becomes k_+ radians larger.
    (e) and direction angle both change by a factor of k_+.

22. Suppose we see the ordered pair $(5\pi/2,-2)$ as the representation for a point in the polar coordinate plane. If we want to keep the direction angle nonnegative but less than 2π, and if we want to keep the radius nonnegative, we should rewrite this ordered pair as
    (a) $(3\pi/2,2)$.
    (b) $(\pi/2,2)$.
    (c) $(2\pi/5,2)$.
    (d) $(2\pi/5,1/2)$.
    (e) $(\pi/2,1/2)$.

23. In Cartesian xyz space, the distance between the points $(-1,-2,-3)$ and $(3,2,1)$ is
    (a) 6 units.
    (b) 8 units.
    (c) 9 units.
    (d) 12 units.
    (e) None of the above

24. In Cartesian two-space, a line segment connecting the points $(-3,10)$ and $(5,16)$ is exactly
    (a) $145^{1/2}$ units long.
    (b) 10 units long.
    (c) $97^{1/2}$ units long.
    (d) 9 units long.
    (e) $79^{1/2}$ units long.

25. What are the polar coordinates of the point plotted in Fig. FE-3?
    (a) $(3\pi/4,50^{1/2})$
    (b) $(5\pi/4,-50^{1/2})$
    (c) $(7\pi/4,50^{1/2})$

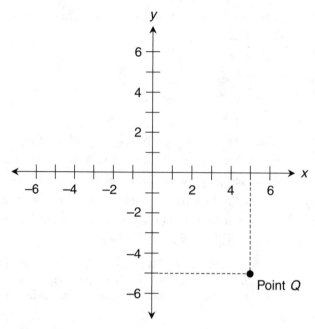

Figure FE-3 Illustration for Question 25.

(d) $(5\pi/2,-50^{1/2})$

(e) None of the above

26. Imagine a polar complex vector **p**, as follows:

$$\mathbf{p} = (\theta,r) = (3\pi/4,72^{1/2})$$

What complex number does this vector represent?

(a) $6+j6$

(b) $-6+j6$

(c) $-6-j6$

(d) $6-j6$

(e) More than one of the above

27. What's the ordered pair representing the standard form of a vector in the Cartesian *xy* plane whose originating point is $(-2,-2)$ and whose ending point is $(3,3)$?

(a) $(5,5)$

(b) $(-5,-5)$

(c) $(1,1)$

(d) $(-1,-1)$

(e) $(1/2,1/2)$

28. How is the cross product **b** × **a** oriented in the situation of Fig. FE-4?
    (a) It points in the direction bisecting the smaller angle between **a** and **b**.
    (b) It points in the direction bisecting the larger angle between **a** and **b**.
    (c) It points straight out of the page toward us.
    (d) It points straight out of the page away from us.
    (e) It has no orientation, because it's a scalar, not a vector!

29. How is the dot product **b** • **a** oriented in the situation of Fig. FE-4?
    (a) It points in the direction bisecting the smaller angle between **a** and **b**.
    (b) It points in the direction bisecting the larger angle between **a** and **b**.
    (c) It points straight out of the page toward us.
    (d) It points straight out of the page away from us.
    (e) It has no orientation, because it's a scalar, not a vector!

30. The midpoint coordinates of a line segment in Cartesian two-space can be found by
    (a) adding the coordinates of the endpoints.
    (b) multiplying the coordinates of the endpoints.
    (c) multiplying the distances of the endpoints from the origin.
    (d) averaging the coordinates of the endpoints.
    (e) averaging the distances of the endpoints from the origin.

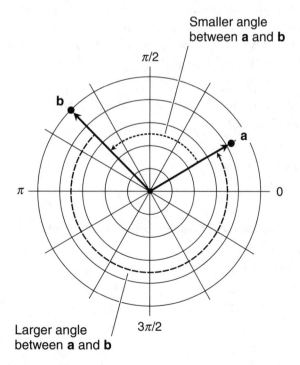

Figure FE-4   Illustration for Questions 28 and 29.

31. Suppose we're given two points P and Q in the Cartesian xy plane, such that their y values are negatives of each other. Based on our knowledge of the midpoint formula for Cartesian two-space, we can be absolutely certain that if we connect P and Q with a line segment, the midpoint of that line segment lies

    (a) on the x axis.

    (b) at the origin.

    (c) on the y axis.

    (d) in either the first quadrant or the third quadrant.

    (e) in either the second quadrant or the fourth quadrant.

32. In Cartesian xyz space, the point midway between $(-1,-2,-3)$ and $(3,2,1)$ is

    (a) $(0,0,0)$.

    (b) $(1,0,-1)$.

    (c) $(2,2,2)$.

    (d) $(2,-2,2)$.

    (e) $(-2,0,2)$.

33. How do we find the negative of a vector in polar coordinates?

    (a) We negate the magnitude, but leave the direction angle unchanged.

    (b) We negate the direction angle, but leave the magnitude unchanged.

    (c) We add or subtract π to or from the direction angle, keeping the angle nonnegative but less than 2π, but leave the magnitude unchanged.

    (d) We add or subtract π to or from the direction angle, keeping the angle nonnegative but less than 2π, and negate the magnitude.

    (e) We don't, because we can't!

34. Fill in the blank to make the following sentence true: "A _____ exists between the set of all polar-plane vectors and the set of all Cartesian-plane vectors."

    (a) circular relation

    (b) linear function

    (c) quadratic function

    (d) bijection

    (e) trijection

35. In Fig. FE-5, what complex number does the longer vector represent?

    (a) $5\pi/4 + j14$

    (b) $-5\pi/4 - j14$

    (c) $70^{1/2}\pi/4 + j70^{1/2}\pi/4$

    (d) $-70^{1/2}\pi/4 - j70^{1/2}\pi/4$

    (e) $-98^{1/2} - j98^{1/2}$

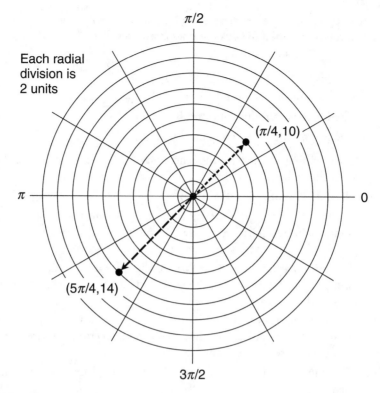

Figure FE-5   Illustration for Questions 35 and 36.

36. In Fig. FE-5, what is the magnitude of the cross product of the two vectors?
    (a) 0
    (b) $70^{1/2}$
    (c) $98^{1/2}$
    (d) 24
    (e) 140

37. Imagine two generic standard-form vectors in *xyz* space, defined by ordered triples as

$$\mathbf{a} = (x_a, y_a, z_a)$$

and

$$\mathbf{b} = (x_b, y_b, z_b)$$

Now consider the quantity

$$k = [(x_a^2 + y_a^2 + z_a^2)(x_b^2 + y_b^2 + z_b^2)]^{1/2} \cos \theta_{ab}$$

where θ_{ab} is the angle between **a** and **b** as determined in the plane containing them both, rotating from **a** to **b**. What does k represent?

(a) The dot product of **a** and **b**.

(b) The product of the magnitudes of **a** and **b**.

(c) The ratio of the magnitudes of **a** and **b**.

(d) The Cartesian product of **a** and **b**.

(e) The magnitude of the cross product of **a** and **b**.

38. Is there anything wrong with the rendition of Cartesian *xyz* space in Fig. FE-6? If so, how can things be made right?

(a) Nothing is wrong with Fig. FE-6.

(b) The axis polarities do not conform to the rules for Cartesian *xyz* space. To make things right, the polarity of the *x* axis can be reversed, while leaving the polarities of the other two axes as they are.

(c) The axis polarities do not conform to the rules for Cartesian *xyz* space. To make things right, the polarity of the *y* axis can be reversed, while leaving the polarities of the other two axes as they are.

(d) The axis polarities do not conform to the rules for Cartesian *xyz* space. To make things right, the polarity of the *z* axis can be reversed, while leaving the polarities of the other two axes as they are.

(e) Any single one of the above actions (b), (c), or (d) can be taken, and things will be made right.

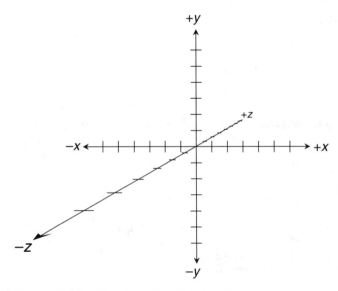

Figure FE-6   Illustration for Question 38.

39. The square of the sine of an angle plus the square of the cosine of the same angle is always equal to
    (a) 0.
    (b) 1.
    (c) $\pi/2$.
    (d) π.
    (e) 2π.

40. Imagine two generic standard-form vectors in xyz space, defined by ordered triples as

$$\mathbf{a} = (x_a, y_a, z_a)$$

and

$$\mathbf{b} = (x_b, y_b, z_b)$$

Now consider the quantity

$$n = x_a x_b + y_a y_b + z_a z_b$$

What does n represent?
    (a) The dot product of $\mathbf{a}$ and $\mathbf{b}$.
    (b) The product of the magnitudes of $\mathbf{a}$ and $\mathbf{b}$.
    (c) The sum of the magnitudes of $\mathbf{a}$ and $\mathbf{b}$.
    (d) The arithmetic mean of $\mathbf{a}$ and $\mathbf{b}$.
    (e) The magnitude of the cross product of $\mathbf{a}$ and $\mathbf{b}$.

41. Here's a claim concerning coordinate conversions. Suppose we have a point (θ, r, h) in cylindrical coordinates. We can find the Cartesian x value of this point using the formula

$$x = r \cos \theta$$

The Cartesian y value is

$$y = r \sin \theta$$

The Cartesian z value is

$$z = h$$

What, if anything, is wrong with this claim as stated? If anything is wrong with it, how can it be made right?
    (a) The x and y conversions are wrong. It should say $x = r \sin \theta$ and $y = r \cos \theta$.
    (b) The x and y conversions are wrong. It should say $x = h \cos \theta$ and $y = h \sin \theta$.

(c) The z conversion is wrong. It should say $z = (r^2 + h^2)^{1/2}$.

(d) The z conversion is wrong. It should say $z = h \tan \theta$.

(e) Nothing is wrong with the claim as stated.

42. If we graph the unit circle in the Cartesian xy plane and then pick a point (x_0, y_0) on that circle such that $x_0 \neq 0$, then y_0/x_0 is equal to

(a) the cosine of the counterclockwise angle between the positive x axis and a ray going out from the origin through (x_0, y_0).

(b) the Arccosine of the counterclockwise angle between the positive x axis and a ray going out from the origin through (x_0, y_0).

(c) the tangent of the counterclockwise angle between the positive x axis and a ray going out from the origin through (x_0, y_0).

(d) the Arctangent of the counterclockwise angle between the positive x axis and a ray going out from the origin through (x_0, y_0).

(e) the secant of the counterclockwise angle between the positive x axis and a ray going out from the origin through (x_0, y_0).

43. Figure FE-7 illustrates a general cylindrical coordinate system. Note that the line segment connecting the origin and point P' is always perpendicular to the line segment connecting points P' and P. Based on this knowledge and the information in the diagram, the straight-line distance d between the origin and point P is

(a) $(r^2 + h^2)^{1/2}$.

(b) $r \sin \theta$.

(c) $r \cos \theta$.

(d) $rh \cos \theta$.

(e) impossible to determine unless we have more information.

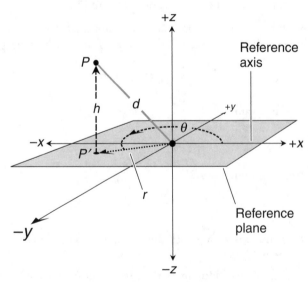

Figure FE-7   Illustration for Question 43.

44. The Arctangent function
    (a) reverses the work of the tangent function.
    (b) is the reciprocal of the tangent function.
    (c) tells us the tangent of an angle measuring an integer multiple of π radians.
    (d) tells us the tangent of an angle measuring an odd-integer multiple of $\pi/2$ radians.
    (e) tells us the length of an arc having a measure of a given angle.

45. Which of the three variables portray the same geometric dimension in both spherical and cylindrical coordinates?
    (a) The radius.
    (b) The vertical direction angle.
    (c) The radius and the vertical direction angle.
    (d) The horizontal direction angle.
    (e) The horizontal and vertical direction angles.

46. In Cartesian coordinates, the point $(-5,-12)$ is the same distance from the origin as the point
    (a) $(0,17)$.
    (b) $(10,7)$.
    (c) $(0,-13)$.
    (d) $(6,11)$.
    (e) All of the above.

47. Imagine two generic standard-form vectors in *xyz* space, defined by ordered triples as

$$\mathbf{a} = (x_a, y_a, z_a)$$

and

$$\mathbf{b} = (x_b, y_b, z_b)$$

Now consider the quantity

$$q = [(x_a^2 + y_a^2 + z_a^2)(x_b^2 + y_b^2 + z_b^2)]^{1/2} \sin \theta_{ab}$$

where θ_{ab} is the smaller angle between $\mathbf{a}$ and $\mathbf{b}$ as determined in the plane containing them both. What does q represent?
    (a) The dot product of $\mathbf{a}$ and $\mathbf{b}$.
    (b) The product of the magnitudes of $\mathbf{a}$ and $\mathbf{b}$.
    (c) The ratio of the magnitudes of $\mathbf{a}$ and $\mathbf{b}$.
    (d) The Cartesian product of $\mathbf{a}$ and $\mathbf{b}$.
    (e) The magnitude of the cross product of $\mathbf{a}$ and $\mathbf{b}$.

48. In spherical coordinates, the graph of the equation $r = 0$ is
     (a) an infinitely tall vertical cylinder.
     (b) an infinitely long vertical line.
     (c) a sphere.
     (d) a point.
     (e) undefined.

49. How can we add two polar-coordinate vectors?
     (a) Convert them to standard form in Cartesian coordinates, add the Cartesian vectors, and then convert the Cartesian sum back to polar form.
     (b) Convert them to standard form in cylindrical coordinates, add the cylindrical vectors, and then convert the cylindrical sum back to polar form.
     (c) Convert them to standard form in spherical coordinates, add the spherical vectors, and then convert the spherical sum back to polar form.
     (d) Add the direction angles of the addend vectors to get the direction angle of the sum vector, and add the magnitudes of the addend vectors to get the magnitude of the sum vector.
     (e) We can't! Addition of polar vectors is not defined.

50. In a system of spherical coordinates, the constant-radius increments appear as
     (a) concentric circles.
     (b) concentric cylinders.
     (c) concentric spheres.
     (d) straight lines passing through the origin.
     (e) parallel planes.

51. Consider the sum of the tangent and the cotangent. Let

     $$f(\theta) = \tan \theta + \cot \theta$$

     What's the positive peak amplitude in the graph of f?
     (a) 1
     (b) 2
     (c) π
     (d) 2π
     (e) It's not defined.

52. Consider the following system of parametric equations representing a straight line in Cartesian xyz space:

     $$x = -4 + 7t$$
     $$y = 3 - 5t$$
     $$z = 2 + 6t$$

Which of the following is a valid expression of the line's direction numbers?

(a) $(-4,3,2)$

(b) $(3,-2,8)$

(c) $(-11,8,-4)$

(d) $(11,-8,4)$

(e) $(-14,10,-12)$

53. When we add, subtract, or multiply one function by another function of the same variable, the domain of the resultant function is

(a) the intersection of the ranges of the two functions.

(b) the union of the ranges of the two functions.

(c) the intersection of the domains of the two functions.

(d) the union of the domains of the two functions.

(e) None of the above.

54. Consider the following equation in three variables x, y, and z:

$$(x+2)^2/4 + (x+3)^2/9 + (x+4)^2/16 = 1$$

In Cartesian xyz-space, this equation represents

(a) an elliptic cone.

(b) an oblate ellipsoid.

(c) a cylinder.

(d) a hyperboloid.

(e) a paraboloid.

55. The solid black curve in Fig. FE-8 shows the graph of a relation f in the Cartesian xy plane. Which, if any, of the four dashed gray curves portrays the graph of f^{-1}?

(a) Curve A

(b) Curve B

(c) Curve C

(d) Curve D

(e) None of the above

56. Which, if any, of the curves in Fig. FE-8 represents a relation that's *not* a true function of either x or y?

(a) Curve A

(b) Curve B

(c) Curve C

(d) Curve D

(e) The solid black curve

57. In the Cartesian xy plane, the unit hyperbola intersects the line $x = 3$ at

(a) no points.

(b) one point.

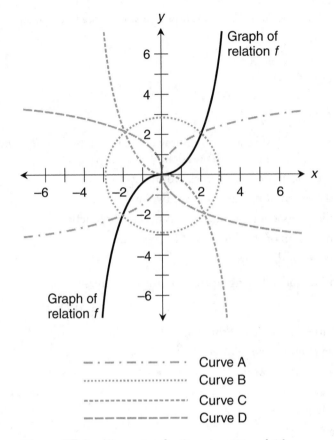

y

6

4

2

Graph of
relation *f*

−6   −4   −2   2   4   6   → x

−2

−4

Graph of
relation *f*

−6

— · — · — · — · — Curve A
·················· Curve B
— — — — — — Curve C
— — — — — — Curve D

Figure FE-8   Illustration for Questions 55 and 56.

(c) two points.

(d) four points.

(e) infinitely many points.

58. Which of the following is an arithmetic sequence and also a geometric sequence?

(a) 4, 4, 4, 4, 4, 4, ...

(b) 4, 2, 4, 2, 4, 2, ...

(c) 1, −1, 1, −1, 1, −1, ...

(d) 3, 2, 1, 0, −1, −2, ...

(e) 1, 1/2, 1/3, 1/4, 1/5, 1/6, ...

59. What does the following expression represent?

$$\sum_{i=0}^{\infty} 2^{-i}$$

(a) A divergent harmonic series

(b) A divergent geometric series

(c) A convergent harmonic series

(d) A convergent geometric series

(e) An undefined series

60. What's the value of the following limit?

$$\mathop{Lim}_{x \to 0-} \ln(-x)$$

(a) 1

(b) −1

(c) e

(d) −e

(e) It's not defined.

61. Consider a relation whose graph in the Cartesian xy plane looks like Fig. FE-9. What can we say about this relation?

(a) It has no inverse.

(b) It's a function of x, but not y.

(c) It's a function of y, but not x.

(d) It's a function of both x and y.

(e) It's identical to its inverse.

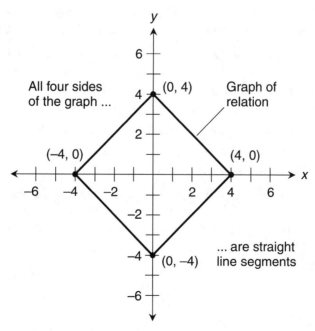

Figure FE-9   Illustration for Question 61.

62. We can uniquely identify a plane in Cartesian three-space if we know
    (a) the locations of two points in the plane.
    (b) the direction of a vector normal to the plane.
    (c) the location of a point in the plane, and the direction of a vector normal to the plane.
    (d) the direction of a vector in the plane.
    (e) the direction numbers of a line that passes through the plane and the coordinate origin.

63. Which of the following functions has an inverse that's also a function when we allow x to span the entire set of real numbers?
    (a) $f_1(x) = x^5 + 2$
    (b) $f_2(x) = 3x^4 - 1$
    (c) $f_3(x) = \sin x$
    (d) $f_4(x) = x^2 + 10$
    (e) $f_5(x) = -\tan x + 3$

64. Consider the sum of twice the secant and twice the cosecant. Let

$$f(\theta) = 2 \sec \theta + 2 \csc \theta$$

    The range of f is
    (a) the set of all real numbers.
    (b) the set of all positive real numbers.
    (c) the set of all real numbers except those in the interval $[-1,1]$.
    (d) the set of all real numbers except those in the interval $[-2,2]$.
    (e) the set of all real numbers except those in the interval $[-\pi,\pi]$.

65. Figure FE-10 illustrates a parabola in the Cartesian xy plane, along with the generalized standard equation for that type of curve. Based on the information shown, we know that
    (a) $c > 0$, because the curve has an absolute maximum.
    (b) $b = 0$, because the curve's axis is vertical.
    (c) $c < 0$, because the x-value of the curve's vertex point is negative.
    (d) $a < 0$, because the curve opens downward.
    (e) $a = 0$, because the curve doesn't turn any sharp corners.

66. Imagine a double right circular cone, through which a flat plane passes. Suppose that the plane has a Cartesian coordinate xy coordinate grid drawn on it. Which of the following equations cannot, under any circumstances, represent the intersection of the plane and the cone?
    (a) $x^2 + y^2 = 4$
    (b) $3x^2 - 4y^2 = 12$
    (c) $y = x^3 - 2x^2 + 1$
    (d) $y = 3x$
    (e) $x^2/4 + y^2/9 = 1$

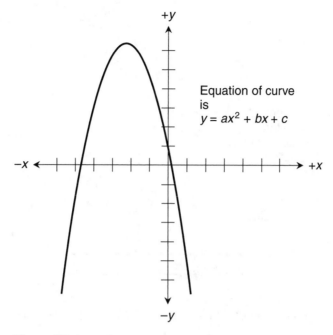

Equation of curve
is
$y = ax^2 + bx + c$

Figure FE-10 Illustration for Question 65.

67. In Fig. FE-11, line L is a good portrayal of the graph of
    (a) the product of the natural exponential function and its reciprocal.
    (b) the sum of the natural exponential function and its reciprocal.
    (c) the difference between the natural exponential function and its reciprocal.
    (d) the natural exponential function divided by its reciprocal.
    (e) the reciprocal of the natural exponential function divided by the natural exponential function.

68. In Fig. FE-11, curve C is a good portrayal of the graph of
    (a) the product of the natural exponential function and its reciprocal.
    (b) the sum of the natural exponential function and its reciprocal.
    (c) the difference between the natural exponential function and its reciprocal.
    (d) the natural exponential function divided by its reciprocal.
    (e) the reciprocal of the natural exponential function divided by the natural exponential function.

69. A relation that's both one-to-one and onto is known as
    (a) a surjection.
    (b) a bijection.
    (c) a monojection.
    (d) an injection.
    (e) a superjection.

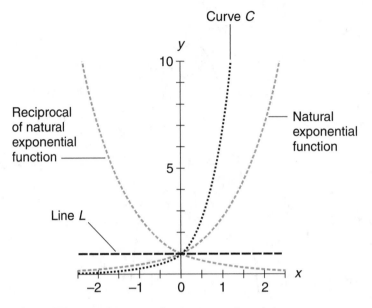

Figure FE-11 Illustration for Questions 67 and 68.

70. Which of the following pairs of graphs have identical asymptotes in the Cartesian xy plane?

   (a) The graphs of $y = \log_{10} x$ and $y = \tan x$

   (b) The graphs of $y = \log_{10} x$ and $y = \ln x$

   (c) The graphs of $y = \ln x$ and $x^2 - y^2 = 1$

   (d) The graphs of $y = \ln x$ and $y = \csc x$

   (e) The graphs of $x^2 - y^2 = 1$ and $x^2 + y^2 = 1$

71. Consider the relation in the Cartesian xy plane represented by

$$9(x - 2)^2 + 4(y + 3)^2 = 36$$

   The graph of this relation intersects the y axis at

   (a) no points.

   (b) one point.

   (c) two points.

   (d) four points.

   (e) infinitely many points.

72. Consider the function that we get when we multiply the square of the tangent by the square of the cotangent. Let

$$f(\theta) = \tan^2 \theta \cot^2 \theta$$

What's the range of f?

(a) $\{-\pi\}$

(b) $\{\pi\}$

(c) $\{-1\}$

(d) $\{1\}$

(e) $\{0\}$

73. Which of the following statements is true of all conic sections?

(a) For a circle, the eccentricity is 0; for an ellipse, the eccentricity is positive but less than 1; for a parabola, the eccentricity is equal to 1; for a hyperbola, the eccentricity is greater than 1.

(b) For a parabola, the eccentricity is 0; for an ellipse, the eccentricity is positive but less than 1; for a circle, the eccentricity is equal to 1; for a hyperbola, the eccentricity is greater than 1.

(c) For a hyperbola, the eccentricity is 0; for a parabola, the eccentricity is positive but less than 1; for an ellipse, the eccentricity is equal to 1; for a circle, the eccentricity is greater than 1.

(d) For an ellipse, the eccentricity is 0; for a parabola, the eccentricity is positive but less than 1; for a circle, the eccentricity is equal to 1; for a hyperbola, the eccentricity is greater than 1.

(e) For an ellipse, the eccentricity is 0; for a hyperbola, the eccentricity is positive but less than 1; for a circle, the eccentricity is equal to 1; for a parabola, the eccentricity is greater than 1.

74. Which of the following functions appears as a straight line in log-log coordinates, meaning that both axes are graduated according to the common logarithm of the displacement?

(a) $y = 3$

(b) $y = \ln x$

(c) $y = e^x$

(d) $y = \log_{10} x$

(e) $y = 10^x$

75. Suppose that t is a parameter on which two variables x and y depend. Which of the following pairs of parametric equations can represent the straight line graphed in Fig. FE-12?

(a) $x = t - 3$ and $y = t - 4$

(b) $x = t + 3$ and $y = t + 4$

(c) $x = 3t/4$ and $y = 4t/3$

(d) $x = t$ and $y = 3t/4$

(e) $x = 3t$ and $y = 4t$

76. Consider the following pair of parametric equations:

$$x = 3 \cos t$$

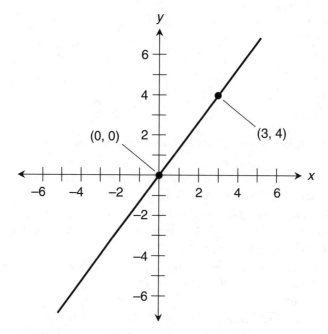

Figure FE-12   Illustration for Question 75.

and

$$y = 4 \sin t$$

In the Cartesian xy plane, the graph of this system appears as
(a) an ellipse centered at the origin.
(b) an ellipse passing through the origin.
(c) a hyperbola centered at the origin.
(d) a hyperbola passing through the origin.
(e) a parabola whose focus is at the origin.

77. The peak-to-peak amplitude of the wave representing the function $y = \sin \pi x$ is
    (a) 2π.
    (b) $\pi/2$.
    (c) 2.
    (d) 1/2.
    (e) impossible to determine without more information.

78. If we divide a function by another function, the resultant function is undefined for any value of the independent variable where
    (a) the value of the numerator function is negative.
    (b) the value of the denominator function is equal to 0.

(c) the value of the denominator function is negative.

(d) the values of both the numerator and denominator function are negative.

(e) the values of the numerator and denominator function have opposite signs.

79. Consider the following system of parametric equations in Cartesian *xyz* space:

$$x = t$$
$$y = -1$$
$$z = -2t^2 - 7t$$

This system represents

(a) a straight line in a plane parallel to the *xy* plane.

(b) a hyperbola in a plane parallel to the *xy* plane.

(c) a circle in a plane parallel to the *yz* plane.

(d) a parabola in a plane parallel to the *xz* plane.

(e) an ellipse that intersects the *z* axis at two points.

80. What does the following expression portray?

$$S = -1/2 + 1/2 - 1/2 + 1/2 - 1/2 + 1/2 - 1/2 + \cdots$$

(a) An infinite arithmetic series

(b) An infinite harmonic series

(c) An infinite geometric series

(d) An infinite hyperbolic series

(e) An infinite circular series

81. Suppose that *t* is a parameter on which two variables *x* and *y* depend. Which of the following pairs of parametric equations can represent the straight line graphed in Fig. FE-13?

(a) $x = t - 5$ and $y = t - 3$

(b) $x = 5t$ and $y = 3t$

(c) $x = 3t/5$ and $y = 5t/3$

(d) $x = t$ and $y = 3 - 3t/5$

(e) $x = 3t$ and $y = 5t$

82. Consider an object in Cartesian *xyz* space whose parametric equations are

$$x = (\cos t)/\pi$$
$$y = \pi \sin t$$
$$z = \pi t$$

The graph of this object is an elliptical

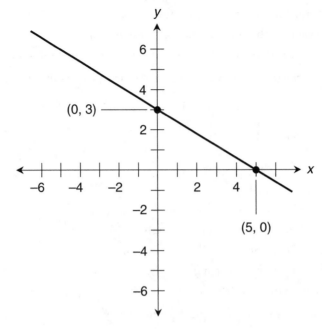

Figure FE-13   Illustration for Question 81.

(a) hyperboloid.

(b) helix.

(c) paraboloid.

(d) cylinder.

(e) cone.

83. In an infinite arithmetic sequence, we can find a constant that determines

(a) the ratio of any term to its successor.

(b) the difference between any term and its successor.

(c) the product of any term and its successor.

(d) the sum of all the terms.

(e) the product of all the terms.

84. If the ordered pair (−1,8) represents a point that lies on the graph of a relation f in the Cartesian xy plane, then its counterpoint on the graph of the inverse relation f^{-1} is represented by the ordered pair

(a) (1,−8).

(b) (−1,1/8).

(c) (8,−1).

(d) (1/8,−1).

(e) (−1/8,1).

85. In the Cartesian *xy* plane, a unit hyperbola can be represented by which one of the following pairs of parametric equations, where *t* is the parameter?

(a) $x = t^2$ and $y = -t^2$

(b) $x = \cos t$ and $y = -\sin t$

(c) $x = \sin t$ and $y = -\sin t$

(d) $x = \cos t$ and $y = -\cos t$

(e) $x = t$ and $y = \pm(t^2 - 1)^{1/2}$

86. Which of the following equations might describe the hyperboloid of two sheets shown in Fig. FE-14? Assume the center of the entire object is at the origin.

(a) $-x^2/2 + y^2/2 - z^2/3 = 1$

(b) $x^2/2 + y^2/2 + z^2/3 = 1$

(c) $x^2/2 + y^2/2 + z^2/3 = 0$

(d) $x/2 + y/2 + z/3 = 1$

(e) $(x - 2)(y + 2)(z - 3) = 1$

87. If we move the entire object in Fig. FE-14 to place its center at the point (5,5,5) rather than at the origin, its equation becomes which one of the following?

(a) $x^2/2 + y^2/2 + z^2/3 = 5$

(b) $5x^2/2 + 5y^2/2 + 5z^2/3 = 0$

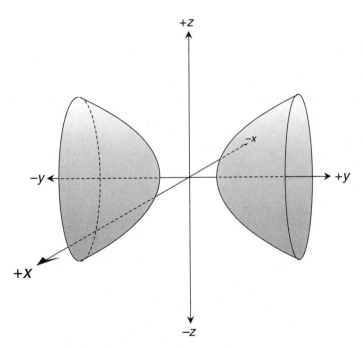

Figure FE-14   Illustration for Questions 86 and 87.

(c) $x/10 + y/10 + z/15 = 1$

(d) $(5x - 10)(5y + 10)(5z - 15) = 1$

(e) $-(x - 5)^2/2 + (y - 5)^2/2 - (z - 5)^2/3 = 1$

88. Consider the graph of the pair of parametric equations

$$\theta = \pi/3$$

and

$$r = -3t$$

In the polar coordinate plane, the graph of this system is a

(a) spiral that expands as we rotate clockwise.

(b) circle passing through the origin.

(c) circle centered at the origin.

(d) straight line that doesn't pass through the origin.

(e) straight line passing through the origin.

89. The period of the graph representing the function $y = 2\pi \csc x$ is

(a) 2π.

(b) $\pi/2$.

(c) 2.

(d) 1/2.

(e) undefined.

90. In two-space, a relation can always be represented as a set of

(a) lines.

(b) circles.

(c) parabolas.

(d) closed curves.

(e) ordered pairs.

91. Consider the two functions

$$y = \ln x$$

and

$$y = \ln (1/x)$$

When we graph the sum of these two functions in the Cartesian xy plane, we get

(a) an open-ended ray corresponding to the line $y = x$ in the first quadrant.

(b) an open-ended ray corresponding to the positive y axis.

(c) an open-ended ray corresponding to the positive x axis.

(d) an open-ended ray corresponding to the line $y = -x$ in the fourth quadrant.

(e) a circle centered at the origin and having a radius of e units.

92. Which of the following relations is *not* a true function of x?

(a) $y = 2x + 3$

(b) $x = 2y + 3$

(c) $y = x^2 + 3$

(d) $x = y^2 + 3$

(e) $y = x/2 + 3$

93. Consider the following equation in three variables x, y, and z:

$$-2x + y + 7z = 3$$

In Cartesian xyz space, this equation represents a

(a) straight line.

(b) circle.

(c) sphere.

(d) hyperboloid.

(e) plane.

94. Consider again the equation in Question 93, and the geometric figure it represents in Cartesian xyz space. At what coordinates, if any, does the figure cross the z axis?

(a) $(0,0,-3/2)$

(b) $(0,0,1/3)$

(c) $(0,0,3/7)$

(d) $(0,0,1/2)$

(e) The figure doesn't cross the z axis anywhere.

95. Which of the following types of graphs is not a conic section?

(a) A circular curve

(b) An elliptical curve

(c) A logarithmic curve

(d) A parabolic curve

(e) A hyperbolic curve

96. Consider the following equation in three variables x, y, and z:

$$x^2 + y^2 + z^2 + 2z + 1 = 20$$

In Cartesian *xyz*-space, this equation represents a
- (a) plane.
- (b) cone.
- (c) cylinder.
- (d) hyperboloid.
- (e) sphere.

97. In two-space, a function is a relation that
    - (a) never maps any specific value of the independent variable to more than one value of the dependent variable.
    - (b) never maps more than one value of the independent variable to any specific value of the dependent variable.
    - (c) has a domain and range that both encompass the entire set of real numbers.
    - (d) has a domain that's a proper subset of the range.
    - (e) has a range that's a proper subset of the domain.

98. Consider the following three-way equation:

$$x/3 = y/4 = -z/2$$

In Cartesian *xyz* space, this equation represents a
- (a) straight line through the origin.
- (b) circle centered at the origin.
- (c) circular cone centered at the origin.
- (d) circle centered at the point (3,4,–2).
- (e) circular cone whose apex is at the point (3,4,–2).

99. Consider the following system of parametric equations:

$$x = \pi \cos t$$

$$y = -\pi$$

$$z = 2\pi \sin t$$

The graph of this system in Cartesian *xyz* space is
- (a) an ellipse that lies in a plane perpendicular to the *y* axis.
- (b) a parabola that lies in the *xz* plane.
- (c) a circle that passes through all three axes.
- (d) a hyperbola that's centered on the *y* axis.
- (e) impossible to figure out based on the information given here.

100. The period of the wave representing the function $y = \sin \pi x$ is equal to

   (a) 2π.

   (b) $\pi/2$.

   (c) $1/2$.

   (d) 2.

   (e) impossible to determine without more information.

A

Worked-Out Solutions to Exercises:
Chapter 1-9

These solutions do not necessarily represent the only ways the chapter-end problems can be figured out. If you think you can solve a particular problem in a quicker or better way than you see here, by all means go ahead! But always check your work to be sure your alternative answer is correct.

Chapter 1

1. As shown in the graph of Fig. 1-10, the x axis is horizontal and the y axis is vertical. Unless otherwise stated, the horizontal axis represents the independent variable in Cartesian coordinates, and the vertical axis represents the dependent variable. The independent variable is listed first in an ordered pair, and the dependent variable is listed second. According to the following rules:

   - The point (0,0) has $x = 0$ and $y = 0$
   - The point (−4,5) has $x = −4$ and $y = 5$
   - The point (−5,−3) has $x = −5$ and $y = −3$
   - The point (1,−6) has $x = 1$ and $y = −6$

2. Let's call our point P, so we have $P = (−4,5)$. That means $x_p = −4$ and $y_p = 5$. When we plug these values into the formula for the distance d of a point from the origin, we get

$$d = (x_p^2 + y_p^2)^{1/2} = [(−4)^2 + 5^2]^{1/2} = (16 + 25)^{1/2} = 41^{1/2}$$

That's an irrational number. We can use a calculator to approximate its value to three decimal places, getting

$$d \approx 6.403$$

The "wavy" or "squiggly" equals sign means "is approximately equal to."

3. This time, let's say that $P = (-5,-3)$, so $x_p = -5$ and $y_p = -3$. Plugging these values into the formula for d gives us

$$d = (x_p^2 + y_p^2)^{1/2} = [(-5)^2 + (-3)^2]^{1/2} = (25 + 9)^{1/2} = 34^{1/2}$$

Once again, we have an irrational number. Using a calculator, we can approximate it to three decimal places as

$$d \approx 5.831$$

Note that we've *rounded off* the value here, because that's what we were asked to do. Remember that rounding is not the same thing as *truncation*, where we simply delete all the digits after a certain place. Whenever we want to approximate a value to a certain number of decimal places or significant figures, we should round it either up or down as necessary, not truncate it, unless we're specifically told to truncate it. (If you've forgotten the rules for rounding, this is a good time to review your pre-algebra book!)

4. We can call $P = (1,-6)$, so we have $x_p = 1$ and $y_p = -6$. Plugging these values into the formula, we obtain

$$d = (x_p^2 + y_p^2)^{1/2} = [1^2 + (-6)^2]^{1/2} = (1 + 36)^{1/2} = 37^{1/2}$$

Our answer is irrational again. Approximating to three decimal places, we get

$$d \approx 6.083$$

5. Let's call the points P and Q, and assign them the ordered pairs

$$P = (-4,5)$$

and

$$Q = (-5,-3)$$

The values of the coordinates are

$$x_p = -4$$
$$y_p = 5$$
$$x_q = -5$$
$$y_q = -3$$

Plugging these numbers into the formula for the distance d between two points, we get

$$d = [(x_p - x_q)^2 + (y_p - y_q)^2]^{1/2} = \{[-4 - (-5)]^2 + [5 - (-3)]^2\}^{1/2}$$
$$= [1^2 + 8^2]^{1/2} = (1 + 64)^{1/2} = 65^{1/2}$$

When we use a calculator to work this out and round d off to three decimal places, we get

$$d \approx 8.062$$

6. This time, let's call the points

$$P = (-5,-3)$$

and

$$Q = (1,-6)$$

The individual coordinates are

$$x_p = -5$$
$$y_p = -3$$
$$x_q = 1$$
$$y_q = -6$$

Plugging these numbers into the formula, we get

$$d = [(x_p - x_q)^2 + (y_p - y_q)^2]^{1/2} = \{(-5 - 1)^2 + [-3 - (-6)]^2\}^{1/2}$$

$$= [(-6)^2 + 3^2]^{1/2} = (36 + 9)^{1/2} = 45^{1/2}$$

Using a calculator and rounding to three decimal places, we get

$$d \approx 6.708$$

7. Let's call the points P and Q once again, and give them the ordered pairs

$$P = (1,-6)$$

and

$$Q = (-4,5)$$

The coordinates are

$$x_p = 1$$
$$y_p = -6$$
$$x_q = -4$$
$$y_q = 5$$

Plugging these numbers into the formula yields

$$d = [(x_p - x_q)^2 + (y_p - y_q)^2]^{1/2} = \{[1 - (-4)]^2 + (-6 - 5)^2\}^{1/2}$$
$$= [3^2 + (-11)^2]^{1/2} = (9 + 121)^{1/2} = 130^{1/2}$$

Using a calculator and rounding off to three decimal places, we get

$$d \approx 11.402$$

8. Let's call the endpoints of our line segment L by the names P and Q, such that

$$P = (-4,5)$$

and

$$Q = (-5,-3)$$

The coordinate values of these points are

$$x_p = -4$$
$$y_p = 5$$
$$x_q = -5$$
$$y_q = -3$$

Using the formula to find the midpoint (x_m,y_m), we obtain

$$(x_m,y_m) = [(x_p + x_q)/2,(y_p + y_q)/2] = \{[-4 + (-5)]/2,[5 + (-3)]/2\}$$

$$= (-9/2,2/2) = (-9/2,1)$$

In decimal form, the ordered pair is exactly

$$(x_m,y_m) = (-4.5,1)$$

9. Let's call the endpoints of M by the names P and Q, such that

$$P = (-5,-3)$$

and

$$Q = (1,-6)$$

The coordinates are

$$x_p = -5$$
$$y_p = -3$$
$$x_q = 1$$
$$y_q = -6$$

Plugging these numbers into the midpoint formula, we get

$$(x_m,y_m) = [(x_p + x_q)/2,(y_p + y_q)/2] = \{(-5 + 1)/2,[-3 + (-6)]/2\}$$
$$= (-4/2,-9/2) = (-2,-9/2)$$

When we express this ordered pair in decimal form, we have exactly

$$(x_m,y_m) = (-2,-4.5)$$

10. We can call the endpoints of N by the names P and Q, such that

$$P = (1,-6)$$

and

$$Q = (-4,5)$$

This time, we have

$$x_p = 1$$
$$y_p = -6$$
$$x_q = -4$$
$$y_q = 5$$

When we put these values into the formula for the midpoint, we come up with

$$(x_m,y_m) = [(x_p + x_q)/2,(y_p + y_q)/2] = \{[1 + (-4)]/2,(-6 + 5)/2\}$$
$$= (-3/2,-1/2)$$

In decimal form, this is exactly

$$(x_m,y_m) = (-1.5,-0.5)$$

Chapter 2

1. There are 2π radians in a full circle of 360°. If we assume that $\pi \approx 3.14159$, then a full circle has a radian measure of

$$2\pi \approx 2 \times 3.14159 \approx 6.28318$$

To get the radian measure in 1°, we divide 2π by 360. That gives us

$$1° \approx 6.28318/360 \approx 0.0175$$

rounded off to four decimal places.

2. If we go 7/8 of the way around a circle counterclockwise, we rotate thorough an angle of 7/8 × 2π, or 7π/4.

3. An angle of 120° is 1/3 of a circular rotation, because 120° is 1/3 of 360°. If we go 1/3 of the way around a circle counterclockwise, that's an angle of 1/3 × 2π, or 2π/3.

4. Imagine that we travel over the earth in a *great circle* (the shortest path between two points on the surface of a sphere, as measured on that surface) for 1000 /π km. If the earth's circumference is 40,000 km and the planet is a perfectly smooth sphere, then our distance traveled is

$$(1,000 /\pi)/40,000 = 1000/(40,000\pi) = 1/(40\pi)$$

of a complete circumnavigation. If we travel exactly once around the earth along a great circle, we go through an angle of 2π. The angular separation, in radians, of two points located 1000 /π km apart on the surface is therefore

$$[1/(40\pi)] \times 2\pi = (2\pi)/(40\pi) = 2/40 = 1/20$$

5. Figure A-1 shows the graphs of $y = 2 \sin x$ (solid curve) and $y = \sin x$ (dashed curve). The graph of $y = 2 \sin x$ resembles the graph of $y = \sin x$, but the amplitude is doubled.

6. Figure A-2 shows the graphs of $y = \sin 2x$ (solid curve) and $y = \sin x$ (dashed curve). The graph of $y = \sin 2x$ resembles the graph of $y = \sin x$, but the frequency is doubled.

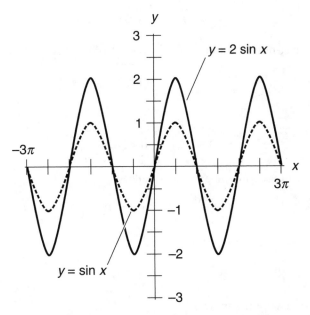

Figure A-1   Illustration for the solution to Problem 5 in Chap. 2.

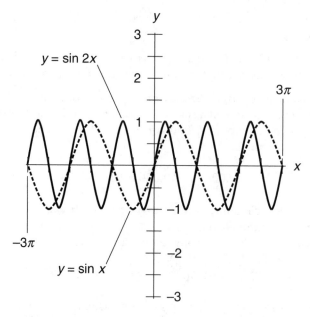

Figure A-2   Illustration for the solution to Problem 6 in Chap. 2.

7. The secant is the reciprocal of the cosine. The cosine has a range of output values covering the closed interval [−1,1]. That means

$$-1 \leq \cos x \leq 1$$

for all real-number input values x. We can break this fact down into the two statements

$$-1 \leq \cos x \leq 0$$

and

$$0 \leq \cos x \leq 1$$

The reciprocals are the one-ended ranges

$$1/\cos x \leq -1$$

and

$$1 \leq 1/\cos x$$

We can rewrite the above as

$$\sec x \leq -1$$

and

$$1 \leq \sec x$$

These two inequalities tell us that the secant function never attains any values in the open interval $(-1,1)$.

8. The cosecant is the reciprocal of the sine. The sine has a range of output values covering the closed interval $[-1,1]$. In other words, no matter what the real-number input x, we always have

$$-1 \leq \sin x \leq 1$$

We can split this into the statements

$$-1 \leq \sin x \leq 0$$

and

$$0 \leq \sin x \leq 1$$

Therefore,

$$1/\sin x \leq -1$$

and

$$1 \leq 1/\sin x$$

We can rewrite the above as

$$\csc x \leq -1$$

and

$$1 \leq \csc x$$

The output of the cosecant function, like the output of the secant, is never equal to anything in the open interval $(-1,1)$.

9. We start with the Pythagorean theorem for the sine and cosine, which is

$$\sin^2 \theta + \cos^2 \theta = 1$$

When we subtract $\sin^2 \theta$ from either side, we get

$$\cos^2 \theta = 1 - \sin^2 \theta$$

We can divide through by the square of the cosine, as long as we don't allow θ to be an odd-integer multiple of $\pi/2$. (If it is, then $\cos \theta = 0$, which means that $\cos^2 \theta = 0$ and we end up dividing by 0.) Performing the division, we get

$$\cos^2 \theta / \cos^2 \theta = 1/\cos^2 \theta - \sin^2 \theta / \cos^2 \theta$$

The left-hand side of this equation is equal to 1 regardless of the value of θ, as long as it's not one of the forbidden values. The first term on the right-hand side is the reciprocal of the cosine squared, which is the same as the secant squared. The second term on the right-hand side is the ratio of the sine squared to the cosine squared, which is same as the tangent squared. We can therefore simplify the above equation to

$$1 = \sec^2 \theta - \tan^2 \theta$$

which is, of course, the same as

$$\sec^2 \theta - \tan^2 \theta = 1$$

10. Again, we start with the Pythagorean theorem for the sine and cosine

$$\sin^2 \theta + \cos^2 \theta = 1$$

This derivation goes a lot like the solution to Problem 9. Let's subtract $\cos^2 \theta$ from either side to get

$$\sin^2 \theta = 1 - \cos^2 \theta$$

We can divide through by the square of the sine, provided that we don't allow θ to be an integer multiple of π. (If it is, then we end up dividing by 0.) This gives us

$$\sin^2 \theta / \sin^2 \theta = 1/\sin^2 \theta - \cos^2 \theta / \sin^2 \theta$$

The left-hand side of the above equation is always equal to 1, as long as θ is not one of the forbidden values. The first term on the right-hand side is the reciprocal of the sine squared; that's the same as the cosecant squared. The second term on the right-hand side is the ratio of the cosine squared to the sine squared. That's the same as the cotangent squared. We can therefore simplify the above equation to

$$1 = \csc^2 \theta - \cot^2 \theta$$

which can be rearranged to

$$\csc^2 \theta - \cot^2 \theta = 1$$

Chapter 3

1. Figure A-3 shows the graphs of the equations $\theta = \pi/4$ and $\theta = \pi/2$ in polar coordinates, where θ is the independent variable and r is the dependent variable. Neither of these are functions of θ. In the first case, r can be any real number when $\theta = \pi/4$. In the second case, r can be any real number when $\theta = \pi/2$.

2. The graph of $\theta = \pi/4$ is a sloping line through the origin in the Cartesian xy plane. The graph of $\theta = \pi/2$ is a vertical line that coincides with the y axis. Figure A-4 shows both graphs. The line representing $\theta = \pi/4$ portrays a function of x in the Cartesian xy plane, because there is never more than one value of y for any value of x. But the line representing $\theta = \pi/2$ does not portray a function of x in the Cartesian xy plane, because when $x = 0$, y can be any real number.

Figure A-3 Illustration for the solution to Problem 1 in Chap. 3.

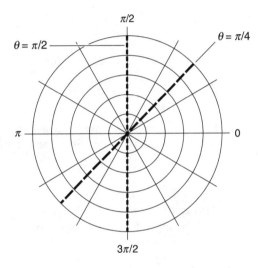

Figure A-4 Illustration for the solution to Problem 2 in Chap. 3.

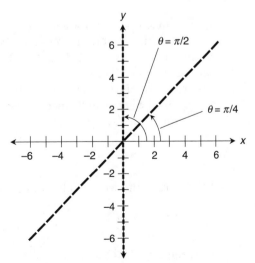

3. The equation $r = -a$ represents the same circle as the equation $r = a$.

4. Imagine a ray that points straight to the right along the reference axis labeled 0. As the ray rotates counterclockwise so that θ starts out at 0 and increases positively, the corresponding radius r starts out at 0 and increases negatively. This tells us that the constant a is negative. When the ray has turned through 1/2 rotation so that $\theta = \pi$, the radius of the solid spiral reaches the value $r = -2\pi$. (Don't get this confused with the apparent radius of $r = 4\pi$ on the solid spiral! The larger value is actually $r = -4\pi$, which we get when the ray has rotated through a complete circle so that $\theta = 2\pi$.) We can solve for a by substituting the number pair $(\theta,r) = (\pi,-2\pi)$ in the general spiral equation

$$r = a\theta$$

This gives us

$$-2\pi = a\pi$$

which solves to $a = -2$. Therefore, the equation of the pair of spirals is

$$r = -2\theta$$

5. Line L runs through the origin and up to the left at an angle halfway between the $\pi/2$ axis and the π axis. That direction is represented by

$$\theta = 3\pi/4$$

This is the equation of L. But we can also imagine that line L runs down and to the right at an angle corresponding to

$$\theta = 7\pi/4$$

so this can also serve as the equation of L. Theoretically, we can add or subtract any integer multiple of π from $3\pi/4$ and get a valid equation for L. By convention, we stick to the range of angles $0 \leq \theta < 2\pi$, so the above two equations are preferred over any others.

Circle C is centered at the origin and has a radius of 3 units, as we can see by inspecting the graph and remembering that each radial division equals 1 unit. Therefore, C can be represented by

$$r = 3$$

We can also consider the radius to be -3 units, so

$$r = -3$$

is an equally valid equation for C.

6. Based on the solution to Problem 5, we can represent the intersection point at the upper left as either

$$P = (3\pi/4, 3)$$

or

$$P = (7\pi/4, -3)$$

We can represent the intersection point at the lower right as either

$$Q = (7\pi/4, 3)$$

or

$$Q = (3\pi/4, -3)$$

The more intuitive representations are the coordinates with positive radii, which are

$$P = (3\pi/4, 3)$$

and

$$Q = (7\pi/4, 3)$$

7. Before we can solve the system of equations for L and C as they are shown in Fig. 3-8, we must be certain that we've completely identified the system. For L, we have

$$\theta = 3\pi/4$$

or

$$\theta = 7\pi/4$$

and for C, we have

$$r = 3$$

or

$$r = -3$$

Solving this system is deceptively simple. It doesn't require algebra at all! We merely combine all the possible combinations of angles and radii we've listed above to get the following four ordered pairs:

$$(\theta, r) = (3\pi/4, 3)$$
$$(\theta, r) = (3\pi/4, -3)$$
$$(\theta, r) = (7\pi/4, 3)$$
$$(\theta, r) = (7\pi/4, -3)$$

Using plus-and-minus notation for the radii, we can reduce this list to two items:

$$(\theta, r) = (3\pi/4, \pm 3)$$

and

$$(\theta, r) = (7\pi/4, \pm 3)$$

That's redundant, but it's valid. If we want to be more elegant, we can get rid of the redundancy and list the solutions as

$$(\theta, r) = (3\pi/4, 3)$$

and

$$(\theta, r) = (7\pi/4, 3)$$

We can tell which ordered pair represents P and which one represents Q by looking again at Fig. 3-8. It's obvious that

$$P = (3\pi/4, 3)$$

and

$$Q = (7\pi/4, 3)$$

8. Let's take away the polar grid in Fig. 3-8 and put a Cartesian grid in its place, as shown in Fig. A-5. Because we've been told that line L is equally distant from the vertical and

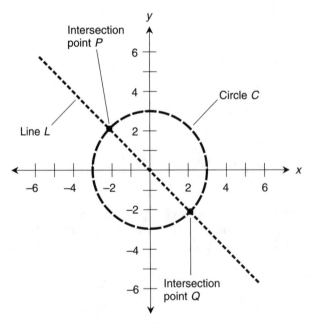

Figure A-5 Illustration for the solutions to Problems 8 through 10 in Chap. 3.

horizontal axes, we know that its slope is −1. Because we've been told that line L passes through the origin, we know that its y-intercept is 0. From algebra, we remember that the slope-intercept form of the Cartesian equation for a straight line is

$$y = mx + b$$

where m is the slope and b is the y-intercept. Plugging in −1 for m and 0 for b, we find that the Cartesian equation for line L is

$$y = -x$$

We've been told that circle C is centered at the origin and has a radius of 3 units. From algebra, we recall that the general form for the equation of a circle centered at the origin is

$$x^2 + y^2 = r^2$$

where r is the radius. When we plug in either 3 or −3 for r, we find that the Cartesian equation for circle C is

$$x^2 + y^2 = 9$$

9. Here's the system of Cartesian equations that we've found, representing line L and circle C as shown in Figs. 3-8 and A-5:

$$y = -x$$

and

$$x^2 + y^2 = 9$$

Let's replace y in the second equation by −x, so we get

$$x^2 + (-x)^2 = 9$$

Because $(-x)^2 = x^2$ for any real number x, we can rewrite the above equation as

$$x^2 + x^2 = 9$$

which simplifies to

$$2x^2 = 9$$

and further to

$$x^2 = 9/2$$

The solutions to this equation are

$$x = (9/2)^{1/2}$$

or

$$x = -(9/2)^{1/2}$$

To solve for y, we must plug in these values of x to either of the equations in our original system. Let's use $y = -x$. When we put the first of these solutions into that equation, we obtain

$$y = -(9/2)^{1/2}$$

which tells us that one of the points is $(x,y) = [(9/2)^{1/2},-(9/2)^{1/2}]$. When we plug the second solution for x into the equation $y = -x$, we get

$$y = -[-(9/2)^{1/2}] = (9/2)^{1/2}$$

so we know that the other point is $(x,y) = [-(9/2)^{1/2},(9/2)^{1/2}]$. By inspecting Fig. A-5, we can see that the points must be

$$P = [-(9/2)^{1/2},(9/2)^{1/2}]$$

and

$$Q = [(9/2)^{1/2},-(9/2)^{1/2}]$$

10. To get the Cartesian equivalents of the points we found when we solved Problems 6 and 7, we use the conversion formulas

$$x = r \cos \theta$$

and

$$y = r \sin \theta$$

The polar form of point P is

$$(\theta,r) = (3\pi/4,3)$$

In this case, we have

$$x = 3 \cos (3\pi/4) = 3 \times (-2^{1/2})/2 = -(9/2)^{1/2}$$

and

$$y = 3 \sin (3\pi/4) = 3 \times 2^{1/2}/2 = (9/2)^{1/2}$$

so the ordered pair is

$$(x,y) = [-(9/2)^{1/2},(9/2)^{1/2}]$$

The polar form of point Q is

$$(\theta,r) = (7\pi/4,3)$$

In this case, we have

$$x = 3 \cos (7\pi/4) = 3 \times 2^{1/2}/2 = (9/2)^{1/2}$$

and

$$y = 3 \sin (7\pi/4) = 3 \times (-2^{1/2})/2 = -(9/2)^{1/2}$$

so the ordered pair is

$$(x,y) = [(9/2)^{1/2},-(9/2)^{1/2}]$$

We have found that

$$P = [-(9/2)^{1/2},(9/2)^{1/2}]$$

and

$$Q = [(9/2)^{1/2},-(9/2)^{1/2}]$$

These results agree with what we got when we solved Problem 9. They are the Cartesian coordinates of points P and Q as shown in Figs. 3-8 and A-5.

Chapter 4

1. Here are the two vectors we've been told to work with:

$$\mathbf{a} = (-3,6)$$

and

$$\mathbf{b} = (2,5)$$

In this situation, $x_a = -3$, $x_b = 2$, $y_a = 6$, and $y_b = 5$. The Cartesian sum $\mathbf{a} + \mathbf{b}$ is

$$\mathbf{a} + \mathbf{b} = [(x_a + x_b),(y_a + y_b)] = [(-3 + 2),(6 + 5)] = (-1,11)$$

Reversing the order of the sum, we get

$$\mathbf{b} + \mathbf{a} = [(x_b + x_a),(y_b + y_a)] = [2 + (-3),(5 + 6)] = (-1,11)$$

The Cartesian difference $\mathbf{a} - \mathbf{b}$ is

$$\mathbf{a} - \mathbf{b} = [(x_a - x_b),(y_a - y_b)] = [(-3 - 2),(6 - 5)] = (-5,1)$$

Reversing the order of the difference, we obtain

$$\mathbf{b} - \mathbf{a} = [(x_b - x_a),(y_b - y_a)] = \{[2 - (-3)],(5 - 6)\}$$
$$= [(2 + 3),(5 - 6)] = (5,-1)$$

2. Imagine that we have an arbitrary Cartesian vector

$$\mathbf{a} = (x_a,y_a)$$

Its Cartesian negative is

$$-\mathbf{a} = (-x_a,-y_a)$$

By definition, the Cartesian sum vector $\mathbf{a} + (-\mathbf{a})$ is

$$\mathbf{a} + (-\mathbf{a}) = \{[x_a + (-x_a)],[y_a + (-y_a)]\} = [(x_a - x_a),(y_a - y_a)] = (0,0) = \mathbf{0}$$

Reversing the order of the sum, we get

$$-\mathbf{a} + \mathbf{a} = [(-x_a + x_a),(-y_a + y_a)] = \{[x_a + (-x_a)],[y_a + (-y_a)]\}$$
$$= [(x_a - x_a),(y_a - y_a)] = (0,0) = \mathbf{0}$$

3. As with the solutions to Problems 1 and 2, demonstrating this fact is a mere exercise in arithmetic. Nevertheless, we can get some practice in mathematical rigor by carefully working our way through each step in the process. According to the formula for the Cartesian difference between two vectors from the chapter text, we have

$$\mathbf{a} - \mathbf{b} = [(x_a - x_b),(y_a - y_b)]$$

and

$$\mathbf{b} - \mathbf{a} = [(x_b - x_a),(y_b - y_a)]$$

Now let's look closely at the coordinates for these two vectors, and compare them. The x coordinate of $\mathbf{a} - \mathbf{b}$ is the real number $x_a - x_b$, while the x coordinate of $\mathbf{b} - \mathbf{a}$ is the real number $x_b - x_a$. From pre-algebra, we remember that when we reverse the order of the difference between two numbers, we get the negative. In this case, it means

$$x_b - x_a = -(x_a - x_b)$$

The same thing happens with the other elements. The y coordinate of $\mathbf{a} - \mathbf{b}$ is $y_a - y_b$, and the y coordinate of $\mathbf{b} - \mathbf{a}$ is $y_b - y_a$. The rules of pre-algebra tell us that

$$y_b - y_a = -(y_a - y_b)$$

Therefore, we know that

$$\mathbf{b} - \mathbf{a} = [-(x_a - x_b), -(y_a - y_b)]$$

By definition, that's the Cartesian negative of $\mathbf{a} - \mathbf{b}$.

4. We are given the two Cartesian vectors

$$\mathbf{a} = (4,5)$$

and

$$\mathbf{b} = (-2,-3)$$

Their Cartesian sum is

$$\mathbf{a} + \mathbf{b} = \{[4 + (-2)],[5 + (-3)]\} = (2,2)$$

The individual Cartesian negatives are

$$-\mathbf{a} = (-4,-5)$$

and

$$-\mathbf{b} = (2,3)$$

These vectors add up to

$$-\mathbf{a} + (-\mathbf{b}) = [(-4 + 2),(-5 + 3)] = (-2,-2)$$

In this case, the sum of the Cartesian negatives is equal to the negative of the Cartesian sum.

5. Let's begin by working out a formula for the negative of a vector sum. Suppose we're given two Cartesian vectors

$$\mathbf{a} = (x_a, y_a)$$

and

$$\mathbf{b} = (x_b, y_b)$$

The sum vector $\mathbf{a} + \mathbf{b}$ is

$$\mathbf{a} + \mathbf{b} = [(x_a + x_b),(y_a + y_b)]$$

The negative of this sum vector is

$$-(\mathbf{a} + \mathbf{b}) = [-(x_a + x_b),-(y_a + y_b)]$$

Using the rules of pre-algebra, we can rewrite the right-hand side of this equation to get

$$-(\mathbf{a} + \mathbf{b}) = [-x_a + (-x_b)],[-y_a + (-y_b)]$$

Now let's go back to the original two vectors. We can state their Cartesian negatives as

$$-\mathbf{a} = (-x_a,-y_a)$$

and

$$-\mathbf{b} = (-x_b,-y_b)$$

When we add these, we obtain

$$-\mathbf{a} + (-\mathbf{b}) = [-x_a + (-x_b)],[-y_a + (-y_b)]$$

That's the same thing we got when we worked out $-(\mathbf{a} + \mathbf{b})$, so we know that

$$-(\mathbf{a} + \mathbf{b}) = -\mathbf{a} + (-\mathbf{b})$$

6. We are given the two polar vectors

$$\mathbf{a} = (\pi/2,4)$$

and

$$\mathbf{b} = (\pi,3)$$

We want to find their polar sum. First, we convert the vectors to Cartesian form. When we do that, we get

$$\mathbf{a} = \{[4\cos(\pi/2)],[4\sin(\pi/2)]\} = [(4 \times 0),(4 \times 1)] = (0,4)$$

and

$$\mathbf{b} = [(3\cos\pi),(3\sin\pi)] = \{[(3 \times (-1)],(3 \times 0)]\} = (-3,0)$$

When we add these, we obtain

$$\mathbf{a} + \mathbf{b} = \{[0 + (-3)],(4 + 0)\} = (-3,4)$$

Let's call this Cartesian sum vector $\mathbf{c} = (x_c, y_c)$, so we have

$$x_c = -3$$

and

$$y_c = 4$$

The point defined by these coordinates lies in the second quadrant of the Cartesian plane. We want to know the polar sum vector $\mathbf{c} = (\theta_c, r_c)$, where θ_c is the direction angle of $\mathbf{c}$ and r_c is the magnitude of $\mathbf{c}$. Using the applicable angle-conversion formula, we get

$$\theta_c = \pi + \text{Arctan} \ [4/(-3)] = \pi + \text{Arctan} \ (-4/3)$$

That's an irrational number. If we want to be exact, we must leave it in this form; there's no way to make it simpler! A calculator set to work in radians can give us approximate values to four decimal places of

$$\text{Arctan} \ (-4/3) \approx -0.9273$$

and

$$\pi \approx 3.1416$$

From this, we can calculate

$$\theta_c \approx 2.2143$$

Using the formula for the polar magnitude, we obtain

$$r_c = (x_c^2 + y_c^2)^{1/2} = [(-3)^2 + 4^2]^{1/2} = (9 + 16)^{1/2} = 25^{1/2} = 5$$

This value is exact. Putting the coordinates into an ordered pair, we derive our exact final answer as

$$\mathbf{c} = \mathbf{a} + \mathbf{b} = (\theta_c, r_c) = \{[\pi + \text{Arctan} \ (-4/3)], 5\}$$

The approximate-angle version is

$$\mathbf{c} = \mathbf{a} + \mathbf{b} \approx (2.2143, 5)$$

Don't get confused here. This ordered pair looks deceptively like the rendition of a vector in the Cartesian plane, but it really defines the vector in the polar coordinate plane. The first coordinate is in radians, and the second coordinate is in linear units.

7. To find the polar negative of the vector we derived in the solution to Problem 6, we reverse the direction but leave the magnitude the same. In this situation, $0 \leq \theta_c < \pi$, so we should add π to the angle to reverse the direction. That gives us the exact answer as

$$-(\mathbf{a} + \mathbf{b}) = \{[\pi + \pi + \text{Arctan } (-4/3)], 5\}$$
$$= \{[2\pi + \text{Arctan } (-4/3)], 5\}$$

If we say that $2\pi \approx 6.2832$, then we can approximate the angle to four decimal places and define the vector as

$$-(\mathbf{a} + \mathbf{b}) \approx (5.3559, 5)$$

8. The original two vectors are

$$\mathbf{a} = (\pi/2, 4)$$

and

$$\mathbf{b} = (\pi, 3)$$

To find the polar negatives, we reverse the directions but leave the magnitudes the same. We want to keep the angles less than 2π without letting either of them become negative. In this case, that means we should add π to θ_a, but we should subtract π from θ_b. When we make these changes, we get

$$-\mathbf{a} = (3\pi/2, 4)$$

and

$$-\mathbf{b} = (0, 3)$$

We must be careful to avoid confusion about what the coordinates of $-\mathbf{b}$ actually mean. The first entry in the ordered pair is an angle, while the second entry is a radius.

9. This time, we want to find the polar sum of the vectors

$$-\mathbf{a} = (3\pi/2, 4)$$

and

$$-\mathbf{b} = (0, 3)$$

Converting them to Cartesian form, we get

$$-\mathbf{a} = \{[4 \cos (3\pi/2)], [4 \sin (3\pi/2)]\} = \{(4 \times 0), [4 \times (-1)]\} = (0, -4)$$

and

$$-\mathbf{b} = [(3 \cos 0),(3 \sin 0)] = [(3 \times 1),(3 \times 0)] = (3,0)$$

Adding, we get the Cartesian vector sum

$$-\mathbf{a} + (-\mathbf{b}) = [(0 + 3),(-4 + 0)] = (3,-4)$$

Let's call this Cartesian sum vector $\mathbf{d} = (x_d, y_d)$. We have

$$x_d = 3$$

and

$$y_d = -4$$

This is in the fourth quadrant of the Cartesian plane. We seek the polar sum vector $\mathbf{d} = (\theta_d, r_d)$, where θ_d is the direction angle of $\mathbf{d}$ and r_d is the magnitude of $\mathbf{d}$. Using the applicable angle-conversion formula, we get

$$\theta_d = 2\pi + \text{Arctan}\ (-4/3)$$

The formula for the polar magnitude tells us that

$$r_d = (x_d^2 + y_d^2)^{1/2} = [3^2 + (-4)^2]^{1/2} = (9 + 16)^{1/2} = 25^{1/2} = 5$$

This gives us the ordered pair

$$\mathbf{d} = -\mathbf{a} + (-\mathbf{b}) = (\theta_d, r_d) = \{[2\pi + \text{Arctan}\ (-4/3)],5\}$$

This is precisely the same vector that we got when we solved Problem 7. Now we know that in the specific polar-vector case where

$$\mathbf{a} = (\pi/2,4)$$

and

$$\mathbf{b} = (\pi,3)$$

the following formula holds:

$$-(\mathbf{a} + \mathbf{b}) = -\mathbf{a} + (-\mathbf{b})$$

Of course, demonstrating this single example doesn't prove the general case. We know it works in general for Cartesian vectors. If you're ambitious and would like some extra credit, go ahead and *rigorously prove* that polar vector negation always distributes through polar vector addition. You're on your own!

10. Here are the original two polar vectors, stated once again for reference:

$$\mathbf{a} = (\pi/2, 4)$$

and

$$\mathbf{b} = (\pi, 3)$$

We want to find their polar differences both ways. Before we do that, we must know the Cartesian forms of the vectors. We worked them out in the solution to Problem 6, getting

$$\mathbf{a} = (0, 4)$$

and

$$\mathbf{b} = (-3, 0)$$

When we subtract **b** from **a**, we get

$$\mathbf{a} - \mathbf{b} = \{[0 - (-3)], (4 - 0)\} = (3, 4)$$

Let's call this Cartesian difference vector $\mathbf{p} = (x_p, y_p)$. The individual coordinates are

$$x_p = 3$$

and

$$y_p = 4$$

This puts us in the first quadrant of the Cartesian plane. We seek the polar sum vector $\mathbf{p} = (\theta_p, r_p)$, where θ_p is the direction angle of **p** and r_p is the magnitude of **p**. Using the appropriate Cartesian-to-polar angle-conversion formula, we come up with

$$\theta_p = \text{Arctan} \ (4/3)$$

A calculator set to work in radians can give us an approximate value to four decimal places of

$$\text{Arctan} \ (4/3) \approx 0.9273$$

Using the formula for the polar magnitude, we obtain the exact result

$$r_p = (x_p^2 + y_p^2)^{1/2} = (3^2 + 4^2)^{1/2} = (9 + 16)^{1/2} = 25^{1/2} = 5$$

Putting the angle and magnitude coordinates into an ordered pair, we derive our exact answer as

$$\mathbf{p} = \mathbf{a} - \mathbf{b} = (\theta_p, r_p) = [\text{Arctan} \ (4/3), 5]$$

The approximate-angle version is

$$\mathbf{p} = \mathbf{a} - \mathbf{b} \approx (0.9273, 5)$$

We must remember that the first coordinate is in radians, and the second is in linear units. Now let's go the other way. When we subtract $\mathbf{a}$ from $\mathbf{b}$, we get

$$\mathbf{b} - \mathbf{a} = [(-3 - 0), (0 - 4)] = (-3, -4)$$

Let's call this Cartesian difference vector $\mathbf{q} = (x_q, y_q)$. We have

$$x_q = -3$$

and

$$y_q = -4$$

This time, we're in the third quadrant. We seek the polar sum vector $\mathbf{q} = (\theta_q, r_q)$, where θ_q is the direction angle of $\mathbf{q}$ and r_q is the magnitude of $\mathbf{q}$. Converting the angle to polar form using the applicable formula, we get

$$\theta_q = \pi + \text{Arctan} \ [-4/(-3)] = \pi + \text{Arctan} \ (4/3)$$

As before, a calculator tells us that

$$\text{Arctan} \ (4/3) \approx 0.9273$$

Using the formula for the polar magnitude yields the exact value

$$r_q = (x_q^2 + y_q^2)^{1/2} = [(-3)^2 + (-4)^2)]^{1/2} = (9 + 16)^{1/2} = 25^{1/2} = 5$$

Our exact final answer is therefore

$$\mathbf{q} = \mathbf{b} - \mathbf{a} = (\theta_q, r_q) = \{[\pi + \text{Arctan} \ (4/3)], 5\}$$

If we let $\pi \approx 3.1416$, the approximate-angle version is

$$\mathbf{q} = \mathbf{b} - \mathbf{a} \approx (4.0689, 5)$$

The first coordinate is in radians, and the second is in linear units.

Chapter 5

1. We've been given the Cartesian vectors

$$\mathbf{a} = (5, -5)$$

and

$$\mathbf{b} = (-5, 5)$$

When we multiply **a** on the left by 4, we get

$$4\mathbf{a} = 4 \times (5,-5) = \{(4 \times 5), [4 \times (-5)]\} = (20,-20)$$

When we multiply **b** on the left by −4, we get

$$-4\mathbf{b} = -4 \times (-5,5) = \{[-4 \times (-5)],(-4 \times 5)\} = (20,-20)$$

2. The Cartesian vector **a** has the coordinates $x_a = 5$ and $y_a = -5$, so it terminates in the fourth quadrant. The direction angle for the polar form of **a** can be found using the conversion formula for a vector in the fourth quadrant, giving us

$$\theta_a = 2\pi + \text{Arctan}\ (-5/5) = 2\pi + \text{Arctan}\ (-1) = 2\pi + (-\pi/4)$$
$$= 7\pi/4$$

The magnitude of **a** is found by the distance formula

$$r_a = [5^2 + (-5)^2]^{1/2} = (25 + 25)^{1/2} = 50^{1/2}$$

Therefore, the polar version of **a** is

$$\mathbf{a} = (7\pi/4,50^{1/2})$$

The Cartesian version of **b** has $x_b = -5$ and $y_b = 5$. It terminates in the second quadrant. Using the conversion formula for the direction angle of a vector in that quadrant, we get

$$\theta_b = \pi + \text{Arctan}\ [5/(-5)] = \pi + \text{Arctan}\ (-1) = \pi + (-\pi/4)$$
$$= 3\pi/4$$

The magnitude of **b** is

$$r_b = [(-5)^2 + 5^2]^{1/2} = (25 + 25)^{1/2} = 50^{1/2}$$

Therefore, the polar version of **b** is

$$\mathbf{b} = (3\pi/4,50^{1/2})$$

When we multiply **a** on the left by 4, we get

$$4\mathbf{a} = 4 \times (7\pi/4,50^{1/2}) = (7\pi/4,800^{1/2})$$

When we multiply **b** on the left by −4, we get

$$-4\mathbf{b} = -4 \times (3\pi/4,50^{1/2}) = (3\pi/4,-800^{1/2}) = (7\pi/4,800^{1/2})$$

In the last step in the equation for $-4\mathbf{b}$, we must take the absolute value of the negative magnitude coordinate, because we can't allow a vector to have negative magnitude. We do this by reversing the direction, in this case by adding π to the angle.

3. We want to prove that positive-scalar multiplication is right-hand distributive over vector subtraction in the Cartesian xy plane. Let's start with

$$(a - b)k_+$$

where $\mathbf{a} = (x_a, y_a)$, $\mathbf{b} = (x_b, y_b)$, and k_+ is a positive real number. Expanding the vectors into their ordered pairs in our initial expression, we get

$$(\mathbf{a} - \mathbf{b})k_+ = [(x_a - x_b), (y_a - y_b)]k_+$$

The definition of right-hand scalar multiplication tells us that we can morph this equation to obtain

$$(\mathbf{a} - \mathbf{b})k_+ = \{[(x_a - x_b)k_+], [(y_a - y_b)]k_+\}$$

The right-hand distributive law for real numbers allows us to transform the equation further, getting

$$(\mathbf{a} - \mathbf{b})k_+ = [(x_a k_+ - x_b k_+), (y_a k_+ - y_b k_+)]$$

Let's put this equation aside for moment. We'll come back to it!

Now, instead of the product of the vector difference and the constant, let's start with the difference between the products

$$\mathbf{a}k_+ - \mathbf{b}k_+$$

We can expand the individual vectors into ordered pairs to get

$$\mathbf{a}k_+ - \mathbf{b}k_+ = (x_a, y_a)k_+ - (x_b, y_b)k_+$$

By the definition of right-hand scalar multiplication, we have

$$\mathbf{a}k_+ - \mathbf{b}k_+ = (x_a k_+, y_a k_+) - (x_b k_+, y_b k_+)$$

When we add the elements of the ordered pairs individually to get a new ordered pair, we obtain

$$\mathbf{a}k_+ - \mathbf{b}k_+ = [(x_a k_+ - x_b k_+), (y_a k_+ - y_b k_+)]$$

The right-hand side of this equation is the same as the right-hand side of the equation we put aside a minute ago. That equation, once again, is

$$(\mathbf{a} - \mathbf{b})k_+ = [(x_a k_+ - x_b k_+), (y_a k_+ - y_b k_+)]$$

Taken together, the above two equations show us that

$$(\mathbf{a} - \mathbf{b})k_+ = \mathbf{a}k_+ - \mathbf{b}k_+$$

4. We've been given the Cartesian vectors

$$\mathbf{a} = (4,4)$$

and

$$\mathbf{b} = (-7,7)$$

We can define the coordinate values as $x_a = 4$, $x_b = -7$, $y_a = 4$, and $y_b = 7$. The Cartesian dot product of $\mathbf{a}$ and $\mathbf{b}$, in that order, is therefore

$$\mathbf{a} \bullet \mathbf{b} = x_a x_b + y_a y_b = 4 \times (-7) + 4 \times 7 = -28 + 28 = 0$$

The Cartesian dot product of $\mathbf{b}$ and $\mathbf{a}$, in that order, is

$$\mathbf{b} \bullet \mathbf{a} = x_b x_a + y_b y_a = -7 \times 4 + 7 \times 4 = -28 + 28 = 0$$

5. The Cartesian vector $\mathbf{a}$ has the coordinates $x_a = 4$ and $y_a = 4$, so it terminates in the first quadrant. The direction angle for the polar form of $\mathbf{a}$ is therefore

$$\theta_a = \text{Arctan}\,(4/4) = \text{Arctan}\,1 = \pi/4$$

The magnitude of $\mathbf{a}$ is

$$r_a = [4^2 + 4^2]^{1/2} = (16 + 16)^{1/2} = 32^{1/2}$$

so the polar form of $\mathbf{a}$ is

$$\mathbf{a} = (\pi/4, 32^{1/2})$$

The Cartesian vector $\mathbf{b}$ has $x_b = -7$ and $y_b = 7$. It terminates in the second quadrant. Using the conversion formula for the direction angle of a vector in the second quadrant, we get

$$\theta_b = \pi + \text{Arctan}\,[7/(-7)] = \pi + \text{Arctan}\,(-1) = \pi + (-\pi/4)$$

$$= 3\pi/4$$

The magnitude of $\mathbf{b}$ is

$$r_b = [(-7)^2 + 7^2]^{1/2} = (49 + 49)^{1/2} = 98^{1/2}$$

Therefore, the polar version of $\mathbf{b}$ is

$$\mathbf{b} = (3\pi/4, 98^{1/2})$$

Let's assign the coordinate values $\theta_a = \pi/4$, $\theta_b = 3\pi/4$, $r_a = 32^{1/2}$, and $r_b = 98^{1/2}$. The Cartesian polar product of **a** and **b**, in that order, is

$$\mathbf{a} \bullet \mathbf{b} = r_a r_b \cos\left(\theta_b - \theta_a\right) = 32^{1/2} \times 98^{1/2} \times \cos\left(3\pi/4 - \pi/4\right)$$
$$= 3{,}136^{1/2} \cos\left(\pi/2\right) = 56 \times 0 = 0$$

The Cartesian dot product of **b** and **a**, in that order, is

$$\mathbf{b} \bullet \mathbf{a} = r_b r_a \cos\left(\theta_a - \theta_b\right) = 98^{1/2} \times 32^{1/2} \times \cos\left(\pi/4 - 3\pi/4\right)$$
$$= 3{,}136^{1/2} \cos\left(-\pi/2\right) = 56 \times 0 = 0$$

6. Consider two standard-form vectors **a** and **b** in Cartesian coordinates, defined by the ordered pairs

$$\mathbf{a} = (x_a, y_a)$$

and

$$\mathbf{b} = (x_b, y_b)$$

By definition, the Cartesian dot product of **a** and **b**, in that order, is

$$\mathbf{a} \bullet \mathbf{b} = x_a x_b + y_a y_b$$

The commutative law for real-number multiplication allows us to reverse the order of both terms in the sum on the right-hand side of this equation, getting

$$\mathbf{a} \bullet \mathbf{b} = x_b x_a + y_b y_a$$

By definition, the right-hand side of the above equation is the Cartesian dot product of **b** and **a**, in that order. Therefore

$$\mathbf{a} \bullet \mathbf{b} = \mathbf{b} \bullet \mathbf{a}$$

for any two standard-form Cartesian-plane vectors **a** and **b**.

7. Suppose we're given two vectors **a** and **b** in the polar plane, defined by

$$\mathbf{a} = (\theta_a, r_a)$$

and

$$\mathbf{b} = (\theta_b, r_b)$$

The polar dot product of **a** and **b**, in that order, is

$$\mathbf{a} \bullet \mathbf{b} = r_a r_b \cos\left(\theta_b - \theta_a\right)$$

The commutative law for real-number multiplication allows us to reverse the order of the multiplication on the right-hand side of this equation to obtain

$$\mathbf{a} \bullet \mathbf{b} = r_b r_a \cos (\theta_b - \theta_a)$$

Now let's look at the difference between the direction angles. From pre-algebra, we recall that when we reverse the order of a difference, we get the negative of that difference. Using this rule, we can modify the angular difference in the above equation to get

$$\mathbf{a} \bullet \mathbf{b} = r_b r_a \cos [-(\theta_a - \theta_b)]$$

Basic trigonometry tells us that the cosine of the negative of an angle is the same as the cosine of the angle itself. Therefore

$$\mathbf{a} \bullet \mathbf{b} = r_b r_a \cos (\theta_a - \theta_b)$$

By definition, the right-hand side of this equation is the polar dot product of **b** and **a**, in that order, telling us that

$$\mathbf{a} \bullet \mathbf{b} = \mathbf{b} \bullet \mathbf{a}$$

for any two vectors **a** and **b** in the polar plane.

8. Let's do the Cartesian proof first. We have a positive scalar k_+ along with two standard-form vectors **a** and **b** in the xy plane. Suppose that the coordinates are

$$\mathbf{a} = (x_a, y_a)$$

and

$$\mathbf{b} = (x_b, y_b)$$

When we multiply these vectors individually on the left by k_+, we get

$$k_+\mathbf{a} = (k_+x_a, k_+y_a)$$

and

$$k_+\mathbf{b} = (k_+x_b, k_+y_b)$$

The Cartesian dot product of these vectors is

$$k_+\mathbf{a} \bullet k_+\mathbf{b} = (k_+x_a k_+x_b + k_+y_a k_+y_b) = (k_+^2 x_a x_b + k_+^2 y_a y_b)$$
$$= k_+^2 (x_a x_b + y_a y_b) = k_+^2 (\mathbf{a} \bullet \mathbf{b})$$

Now let's work through the polar case. Suppose that the coordinates of **a** and **b** are

$$\mathbf{a} = (\theta_a, r_a)$$

and

$$\mathbf{b} = (\theta_b, r_b)$$

When we multiply the individual vectors on the left by the positive scalar k_+ and expand the results into ordered pairs, we get

$$k_+\mathbf{a} = (\theta_a, k_+r_a)$$

and

$$k_+\mathbf{b} = (\theta_b, k_+r_b)$$

Does this step confuse you? If so, remember that because the scalar k_+ is positive, multiplying any polar vector by k_+ doesn't change the vector direction. It only affects the magnitude, making it k_+ times as large. When we take the polar dot product of these new vectors, we get

$$k_+\mathbf{a} \bullet k_+\mathbf{b} = k_+r_a k_+r_b \cos(\theta_b - \theta_a)$$
$$= k_+^2 r_a r_b \cos(\theta_b - \theta_a) = k_+^2(\mathbf{a} \bullet \mathbf{b})$$

9. We want to find the cross product $\mathbf{a} \times \mathbf{b}$ of the polar vectors

$$\mathbf{a} = (\pi/3, 4)$$

and

$$\mathbf{b} = (3\pi/2, 1)$$

The coordinate values are $\theta_a = \pi/3$, $\theta_b = 3\pi/2$, $r_a = 4$, and $r_b = 1$. Before we begin our calculations, we should note that

$$\theta_b - \theta_a = 3\pi/2 - \pi/3 = 7\pi/6$$

That's larger than π, so $\mathbf{a} \times \mathbf{b}$ points straight away from us as we look down on the polar plane. To find the magnitude $r_{a \times b}$, we use the formula for cases where $\pi < \theta_b - \theta_a < 2\pi$. That gives us

$$r_{a \times b} = r_a r_b \sin(2\pi + \theta_a - \theta_b) = 4 \times 1 \times \sin(5\pi/6)$$
$$= 4 \times 1 \times 1/2 = 2$$

so we know that the magnitude of $\mathbf{a} \times \mathbf{b}$ is 2.

10. We've been told to find the cross product of the polar vectors

$$\mathbf{a} = (\pi, 8)$$

and

$$\mathbf{b} = (7\pi/6, 5)$$

The coordinate values are $\theta_a = \pi$, $\theta_b = 7\pi/6$, $r_a = 8$, and $r_b = 5$. In this situation, we have

$$\theta_b - \theta_a = 7\pi/6 - \pi = \pi/6$$

That's smaller than π, so the cross product vector points directly toward us as we look down on the polar plane and imagine going counterclockwise from **a** to **b**. To find the magnitude $r_{a\times b}$, we use the formula for situations in which $0 < \theta_b - \theta_a < \pi$, getting

$$r_{a\times b} = r_a r_b \sin(\theta_b - \theta_a) = 8 \times 5 \times \sin(\pi/6)$$
$$= 8 \times 5 \times 1/2 = 20$$

so we know that the magnitude of $\mathbf{a} \times \mathbf{b}$ is 20.

Chapter 6

1. We know that $j^2 = -1$ by definition, and we derived the fact that $(-j)^2 = -1$ in the chapter text. We might think that

$$-j = j$$

Let's suppose, for the sake of argument, that the above equation is true. Multiplying each side by j gives us

$$-j \times j = j \times j$$

which can be rewritten as

$$-1 \times j \times j = j \times j$$

Because $j \times j = j^2 = -1$ by definition, we can rewrite the above as

$$-1 \times (-1) = -1$$

and finally simplify it to

$$1 = -1$$

This statement is obviously false. By *reductio ad absurdum*, we must conclude that our original assumption, $-j = j$, is also false. Therefore

$$-j \neq j$$

2. The quantity j^{-1} can also be written as $1/j$. It's an unknown, so let's call it x and then set up the simple equation

$$1/j = x$$

We can multiply through by j to get

$$j/j = jx$$

Because any nonzero quantity divided by itself is equal to 1, we can simplify to

$$1 = jx$$

Multiplying through by $-j$ gives us

$$-j = -jjx$$

which can be rewritten as

$$-j = -j^2x$$

We know that $j^2 = -1$, so the above equation becomes

$$-j = -(-1)x$$

which simplifies to

$$-j = x$$

Our unknown quantity is equal to $-j$. We have just demonstrated that

$$j^{-1} = -j$$

The *multiplicative inverse* (reciprocal) of j is the same as its *additive inverse* (negative). No real number behaves like that!

3. First, let's add $-3 + j4$ and $1 + j5$. When we add the real parts, we get

$$-3 + 1 = -2$$

When we add the imaginary parts, we get

$$j4 + j5 = j9$$

The sum can be expressed directly as

$$(-3 + j4) + (1 + j5) = -2 + j9$$

The parentheses are superfluous, but they help to separate the individual complex-number addends on the left-hand side of the equation. Now let's subtract $1 + j5$ from $-3 + j4$. First, we multiply $(1 + j5)$ by -1, getting

$$-1 \times (1 + j5) = -1 - j5$$

Now we add $-3 + j4$ to $-1 - j5$. When we sum the real parts, we get

$$-3 + (-1) = -4$$

Adding the imaginary parts gives us

$$j4 + (-j5) = -j$$

The difference can be expressed directly as

$$(-3 + j4) - (1 + j5) = -4 - j$$

4. We want to find a general formula for the ratio of a complex number to its conjugate. We can do this by evaluating

$$(a + jb)/(a - jb)$$

where a and b are real numbers, and neither a nor b is equal to 0. The general ratio formula is

$$(a + jb)/(c + jd) = [(ac + bd)/(c^2 + d^2)] + j\,[(bc - ad)/(c^2 + d^2)]$$

In this situation, we can let $c = a$ and $d = -b$. Then we can substitute in the ratio formula to get

$$(a + jb)/(a - jb) = \{[aa + b(-b)]/[a^2 + (-b)^2]\} + j\,\{[ba - a(-b)]/[a^2 + (-b)^2]\}$$

$$= [(a^2 - b^2)/(a^2 + b^2)] + j\,[(2ab)/(a^2 + b^2)]$$

The curly braces in the second part, and the square brackets in the third part, are technically unnecessary. But they help to visually set apart the real and imaginary components of the complex quantities.

5. First, let's work out the square of $a + jb$. When we go through the arithmetic, we obtain

$$(a + jb)^2 = (a + jb)(a + jb)$$

$$= a^2 + jab + jba + j^2b^2$$

$$= a^2 + j2ab - b^2$$

$$= (a^2 - b^2) + j2ab$$

Note that in the expression $j2ab$, the numeral 2 is not an exponent! Now let's find the square of the complex number $a - jb$. Paying careful attention to the signs, we get

$$(a - jb)^2 = (a - jb)(a - jb)$$
$$= a^2 + a(-jb) + (-jb)a + (-jb)^2$$
$$= a^2 - jab - jba + (-j)^2 b^2$$
$$= a^2 - j2ab - b^2$$
$$= (a^2 - b^2) - j2ab$$

The two final products we've derived are

$$(a^2 - b^2) + j2ab$$

and

$$(a^2 - b^2) - j2ab$$

which, by definition, are complex conjugates.

6. First, let's find the product of the polar complex vectors $(\pi/4, 2^{1/2})$ and $(3\pi/4, 2^{1/2})$. We must add the direction angles and multiply the magnitudes. The sum of the angles is

$$\pi/4 + 3\pi/4 = \pi$$

The product of the magnitudes is

$$2^{1/2} \times 2^{1/2} = 2$$

Therefore, the product vector is $(\pi, 2)$. The angle θ is equal to π, and the magnitude r is equal to 2. To convert this polar vector $(\theta, r) = (\pi, 2)$ to the complex-number form $a + jb$ where a and b are real-number coefficients, we use the formula for that purpose, getting

$$a + jb = r \cos \theta + j(r \sin \theta) = 2 \cos \pi + j(2 \sin \pi)$$
$$= 2 \times (-1) + j \times 2 \times 0 = -2 + j0 = -2$$

The product of the two original polar complex vectors $(\pi/4, 2^{1/2})$ and $(3\pi/4, 2^{1/2})$ is a vector representing the pure real number -2.

7. Let's convert the polar vector $(\theta, r) = (\pi/4, 2^{1/2})$ to a complex number in the traditional "real-plus-imaginary" form. The conversion formula tells us that

$$r \cos \theta + j(r \sin \theta) = 2^{1/2} \cos (\pi/4) + j[2^{1/2} \sin (\pi/4)]$$
$$= 2^{1/2} \times 2^{1/2}/2 + j(2^{1/2} \times 2^{1/2}/2) = 1 + j$$

Repeating the process with the polar vector $(\theta, r) = (3\pi/4, 2^{1/2})$, we get

$$r \cos \theta + j(r \sin \theta) = 2^{1/2} \cos (3\pi/4) + j[2^{1/2} \sin (3\pi/4)]$$
$$= 2^{1/2} \times (-2^{1/2}/2) + j(2^{1/2} \times 2^{1/2}/2) = -1 + j$$

When we multiply these two complex numbers as binomials, we get

$$(1 + j)(-1 + j) = 1 \times (-1) + 1 \times j + j \times (-1) + j \times j$$
$$= -1 + j + (-j) + (-1) = -2$$

This agrees with the solution to Problem 6. It should! We've been multiplying the same two vectors, representing the same two complex numbers, all along. If we hadn't gotten identical results using the polar method and the Cartesian method, we'd have made a mistake somewhere.

8. Let's convert the polar vector $(\theta, r) = (2\pi/3, 1)$ to the "real-plus-imaginary" complex-number form. The conversion formula tells us that

$$r \cos \theta + j(r \sin \theta) = \cos (2\pi/3) + j \sin (2\pi/3)]$$
$$= -1/2 + j(3^{1/2}/2)$$

If you don't remember why $\sin (2\pi/3) = 3^{1/2}/2$, you might want to verify it for extra credit. (Here's a hint: Use the Pythagorean theorem to solve for the height of a right triangle whose base is 1/2 unit wide and whose hypotenuse is 1 unit long.) Now let's repeat the process with the polar vector $(\theta, r) = (4\pi/3, 1)$. The conversion formula gives us

$$r \cos \theta + j(r \sin \theta) = \cos (4\pi/3) + j \sin (4\pi/3)]$$
$$= -1/2 + j(-3^{1/2}/2)$$
$$= -1/2 - j(3^{1/2}/2)$$

9. Figure A-6 is a graph of the three cube roots of 1 as polar complex vectors. Each radial division represents 1/5 unit.

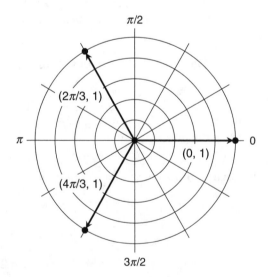

Figure A-6   Illustration for the solution to Problem 9 in Chap. 6. Each radial division represents 1/5 unit.

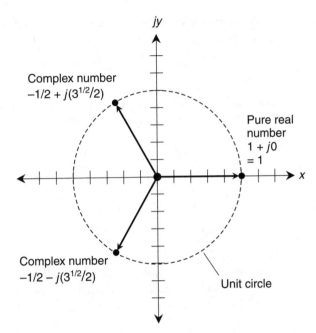

Figure A-7   Illustration for the solution to Problem 10 in Chap. 6. Each axis division represents 1/5 unit.

10. Figure A-7 is a graph of three cube roots of 1 as Cartesian complex vectors. Each radial division represents 1/5 unit. All three vectors terminate on the unit circle.

Chapter 7

1. In Fig. 7-7, the x axis goes from left to right, the y axis goes from bottom to top, and the z axis goes from far to near. According to the following rules:

   - The origin has $x = 0$, $y = 0$, and $z = 0$
   - The point P has $x = 3$, $y = -3$, and $z = 4$
   - The point Q has $x = -5$, $y = 4$, and $z = 0$
   - The point R has $x = 0$, $y = 0$, and $z = 6$

2. We have $P = (3,-3,4)$. Let's call the coordinates $x_p = 3$, $y_p = -3$, and $z_p = 4$. When we plug these values into the formula for the distance c of a point from the origin, we get

$$c = (x_p^2 + y_p^2 + z_p^2)^{1/2} = [3^2 + (-3)^2 + 4^2]^{1/2}$$
$$= (9 + 9 + 16)^{1/2} = 34^{1/2}$$

That's an irrational number. When we use a calculator to approximate its value to three decimal places, we get

$$c \approx 5.831$$

3. In this case, $Q = (-5,4,0)$, so we can say that $x_q = -5$, $y_q = 4$, and $z_q = 0$. When we plug these values into the distance-from-the-origin formula, we get

$$c = (x_q^2 + y_q^2 + z_q^2)^{1/2} = [(-5)^2 + 4^2 + 0^2]^{1/2}$$
$$= (25 + 16 + 0)^{1/2} = 41^{1/2}$$

A calculator approximates this irrational number to

$$c \approx 6.403$$

4. This distance can be read straightaway from the graph if we use the z axis as a measuring stick. If we want to go through the mathematics, we have $R = (0,0,6)$, so we can assign $x_r = 0$, $y_r = 0$, and $z_r = 6$. The distance formula yields

$$c = (x_r^2 + y_r^2 + z_r^2)^{1/2} = (0^2 + 0^2 + 6^2)^{1/2} = 36^{1/2} = 6$$

This value is exact.

5. Line segment L connects points Q and R, where

$$Q = (x_q, y_q, z_q) = (-5,4,0)$$

and

$$R = (x_r, y_r, z_r) = (0,0,6)$$

Plugging the coordinates into the formula for the distance d between two points in Cartesian three-space, we get

$$d = [(x_r - x_q)^2 + (y_r - y_q)^2 + (z_r - z_q)^2]^{1/2}$$
$$= \{[0 - (-5)]^2 + (0 - 4)^2 + (6 - 0)^2\}^{1/2}$$
$$= [5^2 + (-4)^2 + 6^2]^{1/2} = (25 + 16 + 36)^{1/2} = 77^{1/2}$$

When we use a calculator to round this irrational number off to three decimal places, we get

$$d \approx 8.775$$

6. Line segment M connects points P and R, where

$$P = (x_p, y_p, z_p) = (3,-3,4)$$

and

$$R = (x_r, y_r, z_r) = (0,0,6)$$

Plugging the coordinates into the formula for the distance d between P and R gives us

$$d = [(x_r - x_p)^2 + (y_r - y_p)^2 + (z_r - z_p)^2]^{1/2}$$
$$= \{(0 - 3)^2 + [0 - (-3)]^2 + (6 - 4)^2\}^{1/2}$$
$$= [(-3)^2 + 3^2 + 2^2]^{1/2} = (9 + 9 + 4)^{1/2} = 22^{1/2}$$

A calculator rounds this value to three decimal places as

$$d \approx 4.690$$

7. Line segment N connects points P and Q, where

$$P = (x_p, y_p, z_p) = (3, -3, 4)$$

and

$$Q = (x_q, y_q, z_q) = (-5, 4, 0)$$

The distance d between these points is

$$d = [(x_q - x_p)^2 + (y_q - y_p)^2 + (z_q - z_p)^2]^{1/2}$$
$$= \{(-5 - 3)^2 + [4 - (-3)]^2 + (0 - 4)^2\}^{1/2}$$
$$= [(-8)^2 + 7^2 + (-4)^2]^{1/2} = (64 + 49 + 16)^{1/2} = 129^{1/2}$$

A calculator rounds this to three decimal places as

$$d \approx 11.358$$

8. We want to find the midpoint of the line segment L connecting the points

$$Q = (x_q, y_q, z_q) = (-5, 4, 0)$$

and

$$R = (x_r, y_r, z_r) = (0, 0, 6)$$

Let's call the midpoint A (for "average") in this situation, because we're already using M as the name of a line segment. The midpoint formula tells us that the coordinates of A are

$$(x_a, y_a, z_a) = [(x_q + x_r)/2, (y_q + y_r)/2, (z_q + z_r)/2]$$
$$= \{(-5 + 0)/2, (4 + 0)/2, (0 + 6)/2\}$$
$$= (-5/2, 4/2, 6/2) = (-5/2, 2, 3)$$

9. We want to find the midpoint of the line segment M connecting the points

$$P = (x_p, y_p, z_p) = (3, -3, 4)$$

and

$$R = (x_r, y_r, z_r) = (0, 0, 6)$$

If we again call the midpoint A, our formula tells us that the coordinates of A are

$$(x_a, y_a, z_a) = [(x_p + x_r)/2, (y_p + y_r)/2, (z_p + z_r)/2]$$
$$= [(3 + 0)/2, (-3 + 0)/2, (4 + 6)/2\}$$
$$= (3/2, -3/2, 10/2) = (3/2, -3/2, 5)$$

10. We want to identify the midpoint of the line segment N connecting the points

$$P = (x_p, y_p, z_p) = (3, -3, 4)$$

and

$$Q = (x_q, y_q, z_q) = (-5, 4, 0)$$

Let's call the midpoint A once more. Plugging the values into the formula, we obtain the coordinates

$$(x_a, y_a, z_a) = [(x_p + x_q)/2, (y_p + y_q)/2, (z_p + z_q)/2]$$
$$= \{[(3 + (-5)]/2, (-3 + 4)/2, (4 + 0)/2\}$$
$$= (-2/2, 1/2, 4/2) = (-1, 1/2, 2)$$

Chapter 8

1. We want to find the magnitude r_a of the standard-form vector

$$\mathbf{a} = (8, -1, -6)$$

Let's call the coordinates $x_a = 8$, $y_a = -1$, and $z_a = -6$. Using the formula for vector magnitude, we obtain

$$r_a = (x_a^2 + y_a^2 + z_a^2)^{1/2} = [8^2 + (-1)^2 + (-6)^2]^{1/2}$$
$$= (64 + 1 + 36)^{1/2} = 101^{1/2}$$

When we round this irrational number to three decimal places, we get

$$r_a \approx 10.050$$

2. The vector $\mathbf{a'}$ originates at $(-2,0,4)$ and terminates at $(0,0,0)$ in *xyz* space. To find the standard form of this vector (let's call it $\mathbf{a}$), we subtract the originating coordinates from the terminating ones. According to that rule, the *x* coordinate of $\mathbf{a}$ is

$$x_a = 0 - (-2) = 2$$

The *y* coordinate of $\mathbf{a}$ is

$$y_a = 0 - 0 = 0$$

The *z* coordinate of $\mathbf{a}$ is

$$z_a = 0 - 4 = -4$$

Putting these together, we get

$$\mathbf{a} = (x_a, y_a, z_a) = (2,0,-4)$$

3. The vector $\mathbf{b'}$ originates at $(2,3,4)$ and terminates at $(6,7,8)$. Let $\mathbf{b}$ be the standard form of this vector. We find the terminating coordinates of $\mathbf{b}$ by subtracting the starting coordinates of $\mathbf{b'}$ from its ending coordinates. For the *x* value of $\mathbf{b}$, we get

$$x_b = 6 - 2 = 4$$

For the *y* value of $\mathbf{b}$, we get

$$y_b = 7 - 3 = 4$$

For the *z* value of $\mathbf{b}$, we get

$$z_b = 8 - 4 = 4$$

Assembling these coordinates into an ordered triple, we have

$$\mathbf{b} = (x_b, y_b, z_b) = (4,4,4)$$

When we multiply $\mathbf{b}$ on the left by the scalar 4, we obtain

$$4\mathbf{b} = 4 \, (x_b, y_b, z_b) = (4x_b, 4y_b, 4z_b)$$
$$= [(4 \times 4), (4 \times 4), (4 \times 4)] = (16,16,16)$$

That's the standard form of $4\mathbf{b}$, so it must also be the standard form of $4\mathbf{b'}$.

4. We have the two standard-form vectors

$$\mathbf{a} = (-7,-10,0)$$

and

$$\mathbf{b} = (8,-1,-6)$$

Let's assign the coordinates and pair them off as follows:

$$x_a = -7 \quad \text{and} \quad x_b = 8$$
$$y_a = -10 \quad \text{and} \quad y_b = -1$$
$$z_a = 0 \quad \text{and} \quad z_b = -6$$

When we plug these values into the formula for the dot product of two vectors in Cartesian xyz space, we get

$$\mathbf{a} \cdot \mathbf{b} = x_a x_b + y_a y_b + z_a z_b$$
$$= -7 \times 8 + (-10) \times (-1) + 0 \times (-6)$$
$$= -56 + 10 + 0$$
$$= -46$$

5. We have the two standard-form vectors

$$\mathbf{a} = (2,6,0)$$

and

$$\mathbf{b} = (7,4,3)$$

Let's call the coordinates $x_a = 2$, $y_a = 6$, $z_a = 0$, $x_b = 7$, $y_b = 4$, and $z_b = 3$. We can use the formula for the cross product of two vectors in xyz space to get

$$\mathbf{a} \times \mathbf{b} = [(y_a z_b - z_a y_b),(z_a x_b - x_a z_b),(x_a y_b - y_a x_b)]$$
$$= [(6 \times 3 - 0 \times 4),(0 \times 7 - 2 \times 3),(2 \times 4 - 6 \times 7)]$$
$$= [(18 - 0),(0 - 6),(8 - 42)] = (18,-6,-34)$$

6. We can use the formula for the dot product of two vectors, based on their magnitudes and the angle between them. In this case, $r_f = 4$ and $r_g = 7$, and the angle between them, expressible as θ_{fg}, is 0 because the vectors point in the same direction. Therefore

$$\mathbf{f} \cdot \mathbf{g} = r_f r_g \cos \theta_{fg} = 4 \times 7 \times \cos 0$$
$$= 4 \times 7 \times 1 = 28$$

7. As in the previous solution, we can use the formula for the dot product of two vectors, based on their magnitudes and the angle between them. Here, $r_f = 4$ and $r_g = 7$, and the angle between them, θ_{fg}, is π because the vectors point in opposite directions. Therefore

$$\mathbf{f} \cdot \mathbf{g} = r_f r_g \cos \theta_{fg} = 4 \times 7 \times \cos \pi$$
$$= 4 \times 7 \times (-1) = -28$$

8. Once again, let's use the formula for the dot product of two vectors, based on their magnitudes and the angle between them. The magnitudes are $r_f = 4$ and $r_g = 7$. We see the angle θ_{fg}, going counterclockwise from $\mathbf{f}$ to $\mathbf{g}$, as $\pi/2$. Then

$$\mathbf{f} \cdot \mathbf{g} = r_f r_g \cos \theta_{fg} = 4 \times 7 \times \cos (\pi/2)$$
$$= 4 \times 7 \times 0 = 0$$

When we go clockwise from **g** to **f**, we negate the angle between them because we've reversed our direction. The angle θ_{gf}, going clockwise from **g** to **f**, is $-\pi/2$. In this case we have

$$\mathbf{g} \bullet \mathbf{f} = r_g r_f \cos \theta_{gf} = 7 \times 4 \times \cos (-\pi/2)$$

$$= 7 \times 4 \times 0 = 0$$

9. Let's use the formula for the magnitude of the cross product of two vectors, based on their individual magnitudes and the angle between them. We have $r_f = 4$ and $r_g = 7$. The angle θ_{fg}, going counterclockwise from **f** to **g**, is $\pi/2$. The magnitude $r_{f \times g}$ of the cross product vector **f** × **g** is therefore

$$r_{f \times g} = r_f r_g \sin \theta_{fg} = 4 \times 7 \times \sin (\pi/2)$$

$$= 4 \times 7 \times 1 = 28$$

Because we see the rotation going counterclockwise from **f** to **g**, the vector **f** × **g** points toward us. According to the reverse-directional commutative law for cross products, we know that **g** × **f** has the same magnitude as **f** × **g**, but **g** × **f** points in the opposite direction (that is, away from us).

10. Our vector **b** starts out (2,0,0), so **b** points in the same direction as **a**. This fact tells us that the cross product **a** × **b** is the zero vector. As **b** starts rotating counterclockwise, the vector **a** × **b** "sprouts and grows" toward us along the +z axis. When **b** = (0,2,0), having gone through $\pi/2$ radians of rotation in the xy plane, the magnitude $r_{a \times b}$ of the cross product vector is $2 \times 2 = 4$, and **a** × **b** points toward us. As **b** keeps rotating, **a** × **b** starts to "shrink" while continuing to point toward us along the +z axis. When **b** = (−2,0,0), having gone through π radians of rotation, vector **b** points in the opposite direction from vector **a**, so **a** × **b** is the zero vector again. As the rotation continues, **a** × **b** "sprouts and grows" directly away from us along the −z axis. When **b** = (0,−2,0), having turned through $3\pi/2$ radians of rotation, $r_{a \times b} = 2 \times 2 = 4$, and **a** × **b** points away from us. After that, **a** × **b** "shrinks" again while continuing to point away from us along the −z axis, vanishing to the zero vector when **b** = (2,0,0), having passed through a full rotation in the xy plane. If **b** keeps rotating indefinitely, the cross product vector **a** × **b** oscillates alternately toward and away from us along the z axis, attaining peak magnitudes of 4 at the instants when **b** lies along the y axis, and vanishing to the zero vector at the instants when **b** lies along the x axis.

Chapter 9

1. In cylindrical coordinates, the graph of the equation $\theta = 0$ appears as a vertical plane perpendicular to the reference plane, and passing through the reference axis. In xyz space, this would be the xz plane. The graph of $r = 0$ is a vertical straight line that coincides with the h axis. In xyz space, this would be the z axis. The graph of $h = 0$ is a horizontal plane that coincides with the reference plane. In xyz space, this would be the xy plane.

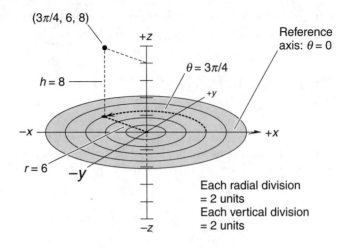

Figure A-8   Illustration for the solution to Problem 2 in Chap. 9. Each radial division represents 2 units. Each vertical division also represents 2 units.

2. Figure A-8 is a plot of the point $(\theta, r, h) = (3\pi/4, 6, 8)$ in cylindrical coordinates.

3. We want to find the (x, y, z) representation of the point $(\theta, r, h) = (\pi/4, 0, 1)$. Let's use the conversion formulas we learned. Here are the formulas again:

$$x = r \cos \theta$$
$$y = r \sin \theta$$
$$z = h$$

Plugging in the values, we get

$$x = 0 \cos \pi/4 = 0$$
$$y = 0 \sin \pi/4 = 0$$
$$z = h = 1$$

Therefore, the Cartesian equivalent point is

$$(x, y, z) = (0, 0, 1)$$

4. We want to convert the xyz space point $(-4, 1, 0)$ to cylindrical coordinates. In this situation, we have $x = -4$ and $y = 1$. To find the angle, we should use the formula

$$\theta = \pi + \text{Arctan}\ (y/x)$$

because $x < 0$ and $y > 0$. When we plug in the values for x and y, we get

$$\theta = \pi + \text{Arctan}\ [1/(-4)] = \pi + \text{Arctan}\ (-1/4)$$

When we input the values for x and y to the formula for r, we get

$$r = [(-4)^2 + 1^2]^{1/2} = (16 + 1)^{1/2} = 17^{1/2}$$

Because $z = 0$, we know that

$$h = z = 0$$

The exact cylindrical equivalent point is

$$(\theta, r, h) = \{[\pi + \text{Arctan}\ (-1/4)], 17^{1/2}, 0\}$$

Using a calculator set for radians to approximate the angle coordinate to four decimal places, we get

$$\theta = \pi + \text{Arctan}\ (-1/4) \approx 3.1416 + (-0.2450) \approx 2.8966$$

When we approximate the radius coordinate to four decimal places, we obtain

$$r = 17^{1/2} \approx 4.1231$$

The approximate ordered triple representing our point in cylindrical coordinates is therefore

$$(\theta, r, h) \approx (2.8966, 4.1231, 0)$$

The first coordinate represents an angle in radians. The second and third coordinates represent linear displacements in space.

5. We want to find the (x,y,z) equivalent of $(\theta, r, h) = (\pi/4, 2^{1/2}, 1)$. First, let's find x. Using the cylindrical-to-Cartesian conversion equation, we get

$$x = r \cos \theta = 2^{1/2} \cos (\pi/4) = 2^{1/2} \times 2^{1/2}/2 = 2/2 = 1$$

The cylindrical-to-Cartesian conversion equation for y tells us that

$$y = r \sin \theta = 2^{1/2} \sin (\pi/4) = 2^{1/2} \times 2^{1/2}/2 = 2/2 = 1$$

Finding z involves no conversion at all. We have simply

$$z = h = 1$$

Therefore, the Cartesian equivalent point of the cylindrical $(\theta, r, h) = (\pi/4, 2^{1/2}, 1)$ is

$$(x, y, z) = (1, 1, 1)$$

That's what we began with when we worked out the example in the chapter text.

6. In spherical coordinates, the graph of the equation $\theta = 0$ appears as a vertical plane containing the reference axis. In *xyz* space, this would be the *xz* plane. It's exactly the same situation as we had in the cylindrical coordinate system when we solved Problem 1, because the horizontal direction angles are identical in both systems. The graph of $\phi = 0$ in spherical coordinates is a vertical straight line that coincides with the Cartesian *z* axis. The graph of $r = 0$ in spherical coordinates is the origin point. In *xyz* space, it's (0,0,0).

7. Figure A-9 is a plot of the point $(\theta,\phi,r) = (3\pi/4,\pi/4,8)$ in spherical coordinates.

8. We have a point in spherical three-space whose coordinates are given by

$$P = (\theta,\phi,r) = (\pi/4,0,1)$$

The formula for *x* is

$$x = r \sin \phi \cos \theta$$

When we plug in the spherical values, we get

$$x = 1 \sin 0 \cos (\pi/4) = 1 \times 0 \times 2^{1/2}/2 = 0$$

The formula for *y* is

$$y = r \sin \phi \sin \theta$$

Plugging in the spherical values, we get

$$y = 1 \sin 0 \sin (\pi/4) = 1 \times 0 \times 2^{1/2}/2 = 0$$

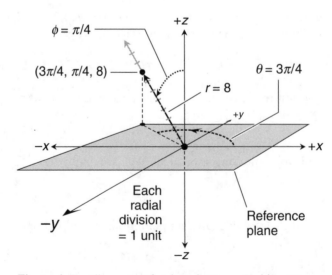

Figure A-9 Illustration for the solution to Problem 7 in Chap. 9. Each radial division represents 1 unit.

The formula for z is

$$z = r \cos \phi$$

Plugging in the spherical values, we get

$$z = 1 \cos 0 = 1 \times 1 = 1$$

Therefore, the coordinates in *xyz* space are

$$P = (0,0,1)$$

9. We want to convert the *xyz* space point $(-4,1,0)$ to spherical coordinates. To find the radius, we use the formula

$$r = (x^2 + y^2 + z^2)^{1/2}$$

Plugging in the values, we get

$$r = [(-4)^2 + 1^2 + 0^2]^{1/2} = (16 + 1 + 0)^{1/2} = 17^{1/2}$$

To find the horizontal angle, we use the formula

$$\theta = \pi + \text{Arctan}\,(y/x)$$

because $x < 0$ and $y > 0$. When we plug in the values for x and y, we get

$$\theta = \pi + \text{Arctan}\,[1/(-4)] = \pi + \text{Arctan}\,(-1/4)$$

To find the vertical angle, we can use the formula

$$\phi = \text{Arccos}\,(z/r)$$

We already know that $r = 17^{1/2}$, so

$$\phi = \text{Arccos}\,(0/17^{1/2}) = \text{Arccos}\,0 = \pi/2$$

Our spherical ordered triple, listing the coordinates in the order $P = (\theta,\phi,r)$, is

$$P = \{[\pi + \text{Arctan}\,(-1/4)], \pi/2, 17^{1/2}\}$$

Using a calculator set for radians and rounding the irrational values to four decimal places, we get

$$\theta = \pi + \text{Arctan}\,(-1/4) \approx 3.1416 + (-0.2450) \approx 2.8966$$
$$\phi = \pi/2 \approx 3.1416 / 2 \approx 1.5708$$
$$r = 17^{1/2} \approx 4.1231$$

Therefore, we can approximate the spherical coordinates as

$$P \approx (2.8966, 1.5708, 4.1231\}$$

10. We're given the point P in cylindrical three-space as

$$P = (\theta, r, h) = [3\pi/4, 6^{1/2}/2, 6^{1/2}/2]$$

Our first task is to find the equivalent coordinates in xyz space. Here are the conversion formulas once again, for reference:

$$x = r \cos \theta$$
$$y = r \sin \theta$$
$$z = h$$

Plugging in the numbers to these formulas gives us

$$x = 6^{1/2}/2 \cos (3\pi/4) = 6^{1/2}/2 \times (-2^{1/2}/2) = -3^{1/2}/2$$
$$y = 6^{1/2}/2 \sin (3\pi/4) = 6^{1/2}/2 \times 2^{1/2}/2 = 3^{1/2}/2$$
$$z = h = 6^{1/2}/2$$

Therefore, we have the Cartesian equivalent point

$$P = (x, y, z) = (-3^{1/2}/2, 3^{1/2}/2, 6^{1/2}/2)$$

When we check this against the intermediate result we got as we solved the last challenge in the chapter text, we see that the two agree. So far, we're doing okay! Now let's convert this Cartesian ordered triple to spherical coordinates. To find the spherical radius, we use the formula

$$r = (x^2 + y^2 + z^2)^{1/2}$$

Plugging in the values, we get

$$r = [(-3^{1/2}/2)^2 + (3^{1/2}/2)^2 + (6^{1/2}/2)^2]^{1/2} = (3/4 + 3/4 + 6/4)^{1/2}$$
$$= (12/4)^{1/2} = 3^{1/2}$$

To find the horizontal angle, we use the formula

$$\theta = \pi + \text{Arctan} \, (y/x)$$

because $x < 0$ and $y > 0$. When we plug in the values for x and y, we get

$$\theta = \pi + \text{Arctan} \, [3^{1/2}/(-3^{1/2})] = \pi + \text{Arctan} \, (-1) = \pi + (-\pi/4) = 3\pi/4$$

To find the vertical angle, we can use the formula

$$\phi = \text{Arccos} \, (z/r)$$

We already know that $r = 3^{1/2}$, so

$$\phi = \text{Arccos } [(6^{1/2}/2)/3^{1/2}] = \text{Arccos } (2^{1/2}/2) = \pi/4$$

Our spherical ordered triple, listing the coordinates in the order $P = (\theta, \phi, r)$, is therefore

$$P = (3\pi/4, \pi/4, 3^{1/2})$$

This is the original spherical angle in the challenge from the chapter text. We've worked the problem out in both directions without running into any trouble, so we can be confident that we didn't make any errors either way.

Worked-Out Solutions to Exercises: Chapter 11-19

These solutions do not necessarily represent the only ways the chapter-end problems can be figured out. If you think you can solve a particular problem in a quicker or better way than you see here, by all means go ahead! But always check your work to be sure your alternative answer is correct.

Chapter 11

1. The domain of the relation shown in Fig. 11-10 is set X. We've been told that the relation never maps any element of set X into more than one element of set Y. Set Y contains no elements outside the co-domain. Therefore, the relation is an injection. The illustration shows that the relation maps elements X completely onto set Y, so the relation is a surjection. Because the relation is both an injection and a surjection, it's a bijection by definition. In this example, the range happens to be the same as the co-domain. That's not true of all relations. This relation is a function, because no element in the domain maps to more than one element in the range.

2. Every positive integer y in set Y (the range) has infinitely many rational numbers x from set X (the domain) assigned to it. For example, if we take the integer $y = 5$ in set Y, it can correspond to any rational x in set X such that $4 < x \le 5$. The relation is clearly not one-to-one, so it's not an injection. For any positive integer y in set Y, we can find at least one positive rational x in set X that maps to it, so the relation is a surjection. The relation is not a bijection; it would have to be both an injection and a surjection to "qualify" for that status. If we take any positive rational number x in the domain X, we can never map it to more than one positive integer y in the range Y. Therefore, our relation is a function of x.

3. This relation, like the one described in Problem 2, is not one-to-one, so it isn't an injection. For any positive rational number y in set Y, we can find a positive integer x in set X that maps to it, so we have a surjection. The relation is not a bijection, because it isn't both an injection and a surjection. If we take any positive integer x in the domain X,

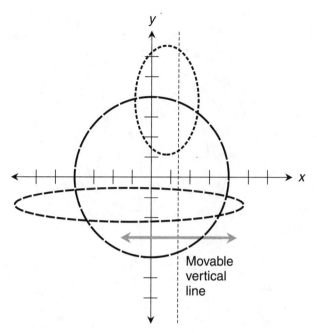

Figure B-1   Illustration for the solution to Problem 4 in Chap. 11.

we can map it to infinitely many positive rationals y in the range Y. Therefore, this relation is not a function of x.

4. A relation whose graph is a circle or ellipse in the Cartesian xy plane can never be a function of x, because such a graph always fails the vertical-line test. Figure B-1 shows several examples.

5. A relation whose graph is a circle or ellipse in the polar θr plane is a function of θr if the origin is inside the circle or ellipse. Figure B-2A shows a simple example in which a circle is centered at the origin in the polar plane. When we graph this relation the "Cartesian way" as shown in Fig. B-2B, we get a straight, horizontal line that passes the vertical-line test.

6. We've been given the functions

$$f(x) = x + 2$$

and

$$g(x) = 3$$

Their sums are

$$(f + g)(x) = f(x) + g(x) = (x + 2) + 3 = x + 5$$

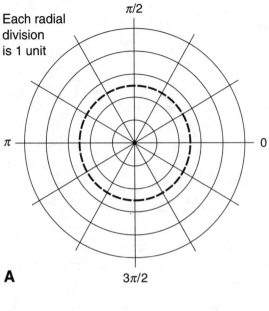

A

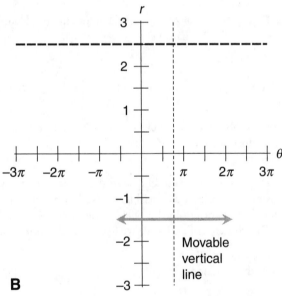

B

Figure B-2 Illustration for the solution to Problem 5 in Chap. 11. At A, each radial division represents 1 unit. At B, the divisions are as labeled.

and

$$(g+f)(x) = g(x) + f(x) = 3 + (x + 2) = x + 5$$

Their differences are

$$(f-g)(x) = f(x) - g(x) = (x + 2) - 3 = x - 1$$

and

$$(g-f)(x) = g(x) - f(x) = 3 - (x + 2) = -x + 1$$

Their products are

$$(f \times g)(x) = f(x) \times g(x) = (x + 2) \times 3 = 3x + 6$$

and

$$(g \times f)(x) = g(x) \times f(x) = 3 \times (x + 2) = 3x + 6$$

Their ratios are

$$(f/g)(x) = f(x)/g(x) = (x + 2)/3 = x/3 + 2/3$$

and

$$(g/f)(x) = g(x)/f(x) = 3/(x + 2)$$

7. We've been given the functions

$$f(x) = x + 1$$

and

$$g(x) = x - 1$$

Their sums are

$$(f+g)(x) = f(x) + g(x) = (x + 1) + (x - 1) = 2x$$

and

$$(g+f)(x) = g(x) + f(x) = (x - 1) + (x + 1) = 2x$$

Their differences are

$$(f-g)(x) = f(x) - g(x) = (x + 1) - (x - 1) = 2$$

and

$$(g-f)(x) = g(x) - f(x) = (x-1) - (x+1) = -2$$

Their products are

$$(f \times g)(x) = f(x) \times g(x) = (x+1)(x-1) = x^2 - 1$$

and

$$(g \times f)(x) = g(x) \times f(x) = (x-1)(x+1) = x^2 - 1$$

Their ratios are

$$(f/g)(x) = f(x)/g(x) = (x+1)/(x-1)$$

and

$$(g/f)(x) = g(x)/f(x) = (x-1)/(x+1)$$

8. We've been given the functions

$$f(x) = x^{-1}$$

and

$$g(x) = x^{-2}$$

Their sums are

$$(f+g)(x) = f(x) + g(x) = x^{-1} + x^{-2}$$

and

$$(g+f)(x) = g(x) + f(x) = x^{-2} + x^{-1} = x^{-1} + x^{-2}$$

Their differences are

$$(f-g)(x) = f(x) - g(x) = x^{-1} - x^{-2}$$

and

$$(g-f)(x) = g(x) - f(x) = x^{-2} - x^{-1} = -x^{-1} + x^{-2}$$

Their products are

$$(f \times g)(x) = f(x) \times g(x) = x^{-1}x^{-2} = x^{-3}$$

and

$$(g \times f)(x) = g(x) \times f(x) = x^{-2}x^{-1} = x^{-3}$$

Their ratios are

$$(f/g)(x) = f(x)/g(x) = (x^{-1})/(x^{-2}) = x^{-1}x^{2} = x$$

and

$$(g/f)(x) = g(x)/f(x) = (x^{-2})/(x^{-1}) = x^{-2}x = x^{-1}$$

9. We've been given the functions

$$f(x) = \sin^{2}\theta$$

and

$$g(x) = \cos^{2}\theta$$

Their sums are

$$(f+g)(x) = f(x) + g(x) = \sin^{2}\theta + \cos^{2}\theta = 1$$

and

$$(g+f)(x) = g(x) + f(x) = \cos^{2}\theta + \sin^{2}\theta = 1$$

Their differences are

$$(f-g)(x) = f(x) - g(x) = \sin^{2}\theta - \cos^{2}\theta$$

and

$$(g-f)(x) = g(x) - f(x) = \cos^{2}\theta - \sin^{2}\theta$$

Their products are

$$(f \times g)(x) = f(x) \times g(x) = \sin^{2}\theta \cos^{2}\theta$$

and

$$(g \times f)(x) = g(x) \times f(x) = \cos^{2}\theta \sin^{2}\theta = \sin^{2}\theta \cos^{2}\theta$$

Their ratios are

$$(f/g)(x) = f(x)/g(x) = (\sin^{2}\theta)/(\cos^{2}\theta) = \tan^{2}\theta$$

and

$$(g/f)(x) = g(x)/f(x) = (\cos^{2}\theta)/(\sin^{2}\theta) = \cot^{2}\theta$$

10. We must remember that the domain of a sum, difference, product, or ratio function is the intersection between the domains of the constituent functions. Also, in the case of a ratio, the domain can't include any value where the denominator becomes 0. With these facts in mind, let's go through the solutions we derived for Problems 6 through 9 and evaluate all the possible real-number domains:

Problem 6 and its solutions: The domains of both f and g encompass all real numbers. Therefore, the domains of all the sum, difference, and product functions are the entire set of reals. The domain of f/g is the entire set of reals, because g never attains the value 0. The domain of g/f is the set of all reals except -2, because

$$f(-2) = -2 + 2 = 0$$

Problem 7 and its solution: The domains of both f and g encompass all reals, so the domains of all the sum, difference, and product functions are the entire set of reals. The domain of f/g is the entire set of reals except 1, because

$$g(1) = 1 - 1 = 0$$

The domain of g/f is the entire set of reals except -1, because

$$f(-1) = -1 + 1 = 0$$

Problem 8 and its solutions: The domains of both f and g include all reals except 0, so the domains of all the sum, difference, and product functions encompass all reals except 0. The same holds for both ratio functions. There are no additional restrictions, because neither denominator can ever become 0.

Problem 9 and its solutions: The domains of both f and g are the entire set of reals, so the domains of all the sum, difference, and product functions encompass all reals. The domain of f/g is the entire set of reals except all odd-integer multiples of $\pi/2$, because the square of the cosine of any odd-integer multiple of $\pi/2$ is equal to 0. The domain of g/f is the entire set of reals except all integer multiples of π, because the square of the sine of any integer multiple of π is equal to 0.

Chapter 12

1. We want to find the inverse of the relation

$$f(x) = 2x + 4$$

If we call the dependent variable y, then

$$y = 2x + 4$$

Swapping the names of the variables, we get

$$x = 2y + 4$$

which can be manipulated with algebra to obtain

$$y = x/2 - 2$$

If we replace the new variable y by the relation notation $f^{-1}(x)$, we get

$$f^{-1}(x) = x/2 - 2$$

2. We want to find the inverse of the relation

$$g(x) = x^2 - 4x + 4$$

Calling the dependent variable y, we can write this as

$$y = x^2 - 4x + 4$$

which factors to

$$y = (x - 2)^2$$

When we swap the names of the variables, we get

$$x = (y - 2)^2$$

Taking the complete square root of both sides produces

$$\pm x^{1/2} = y - 2$$

which can be manipulated with algebra to obtain

$$y = 2 \pm x^{1/2}$$

Replacing y by $g^{-1}(x)$, we get

$$g^{-1}(x) = 2 \pm x^{1/2}$$

3. We want to find the inverse of the relation

$$h(x) = x^3 - 5$$

Calling the dependent variable y, we have

$$y = x^3 - 5$$

Swapping the names of the variables yields

$$x = y^3 - 5$$

which can be morphed algebraically into

$$y = (x + 5)^{1/3}$$

Replacing y by $h^{-1}(x)$, we get

$$h^{-1}(x) = (x + 5)^{1/3}$$

4. Let's examine each situation by imagining what happens when we try to input real numbers to the original relations, and working from there:

Problem 1 and its solution: The relation and its inverse are

$$f(x) = 2x + 4$$

and

$$f^{-1}(x) = x/2 - 2$$

We can input any real number into either of these relations, and we always get a meaningful result. Therefore, the domains of f and f^{-1} both span the entire set of real numbers. The range of f^{-1} is the same as the domain of f, and the range of f is the same as the domain of f^{-1}. Therefore, the ranges of f and f^{-1} both include all real numbers.

Problem 2 and its solution: The relation and its inverse are

$$g(x) = x^2 - 4x + 4$$

and

$$g^{-1}(x) = 2 \pm x^{1/2}$$

We can input any real number we choose into g, and we always get a meaningful result. However, that result is always nonnegative, because it's the square of the real-number quantity $(x - 2)$. It's possible for the output of g to equal 0; that happens when the input is 2. The range of g is therefore the set of all nonnegative reals. The domain of g^{-1} is identical to the range of g, so the domain of g^{-1} spans the set of all nonnegative reals. The range of g^{-1} is the same as the domain of g, which is the set of all reals.

Problem 3 and its solution: The relation and its inverse are

$$h(x) = x^3 - 5$$

and

$$h^{-1}(x) = (x + 5)^{1/3}$$

We can input any real number into h, and we always get a real-number output value. Therefore, the domain of h spans the entire set of reals. The same is true of the domain of h^{-1}; the relation is defined for all possible real-number input values. The range of h^{-1} is the same as the domain of h, and the range of h is the same as the domain of h^{-1}. Therefore, the ranges of h and h^{-1} both span the set of all reals.

5. The equation of our relation, as given in the problem, is

$$x^2/4 - y^2/9 = 1$$

Let's use algebra to morph this equation so it's stated with y all by itself on the left-hand side and an expression containing only the variable x on the right-hand side. If we multiply through by 36, we get

$$9x^2 - 4y^2 = 36$$

which can be rewritten as

$$-4y^2 = 36 - 9x^2$$

Multiplying through by -1, we get

$$4y^2 = 9x^2 - 36$$

When we divide through by 4, we obtain

$$y^2 = 9x^2/4 - 9$$

Taking the complete square root of both sides yields

$$y = \pm(9x^2/4 - 9)^{1/2}$$

Using relation notation to express this equation and name the relation f, we get

$$f(x) = \pm(9x^2/4 - 9)^{1/2}$$

That's the relation! Now let's find its inverse. We begin by restating the relation with y on the left instead of $f(x)$, so we have

$$y = \pm(9x^2/4 - 9)^{1/2}$$

Swapping the names of the variables gives us

$$x = \pm(9y^2/4 - 9)^{1/2}$$

When we square both sides, we obtain

$$x^2 = 9y^2/4 - 9$$

Adding 9 to each side, we get

$$x^2 + 9 = 9y^2/4$$

Multiplying through by 4, and then transposing the left- and right-hand sides of the equation, we come up with

$$9y^2 = 4x^2 + 36$$

Dividing through by 9 yields

$$y^2 = 4x^2/9 + 4$$

After we take the complete square root of both sides, we have

$$y = \pm(4x^2/9 + 4)^{1/2}$$

Replacing the variable y by the relation notation $f^{-1}(x)$, we conclude that

$$f^{-1}(x) = \pm(4x^2/9 + 4)^{1/2}$$

6. With the information stated in Problem 5, along with the graph of the relation shown in Fig. 12-2, we can see that the real-number domain of f is the set of all reals greater than or equal to 2, or smaller than or equal to -2. Another way of stating this is to say that the domain of f is the set of all reals except those in the open interval $(-2,2)$. The range of f is clearly the set of all reals.

7. Figure B-3 shows the graph of the original relation

$$f(x) = \pm(9x^2/4 - 9)^{1/2}$$

as a pair of gray curves, and the graph of the inverse relation

$$f^{-1}(x) = \pm(4x^2/9 + 4)^{1/2}$$

as a pair of black curves. The "point reflector" is the dashed line. The real-number domain of f^{-1} is the same as the real-number range of f; that's the set of all reals. The real-number range of f^{-1} is identical to the real-number domain of f; that's the set of all reals except those in the open interval $(-2,2)$.

8. Figure B-4 shows the graph of the original relation with its domain restricted to the reals greater than or equal to 2 (gray curve) and its inverse (black curve). The curve for the inverse relation is a half-hyperbola that opens upward and intersects the y axis at $(0,2)$. A movable vertical line never intersects the black curve at more than one point. Therefore, the inverse relation

$$f^{-1}(x) = (4x^2/9 + 4)^{1/2}$$

is a true function of x.

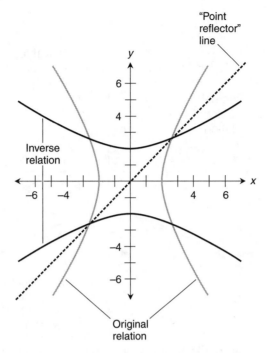

Figure B-3   Illustration for the solution to
Problem 7 in Chap. 12.

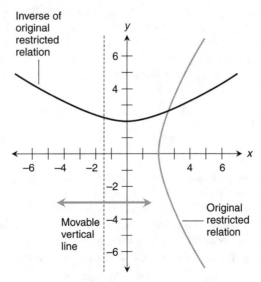

Figure B-4   Illustration for the solution to
Problem 8 in Chap. 12.

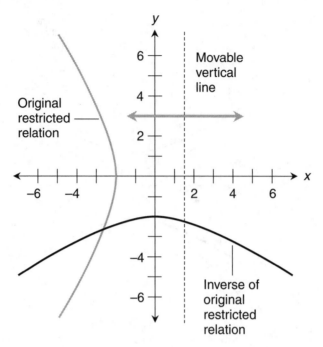

Figure B-5   Illustration for the solution to Problem 9 in Chap. 12.

9. Figure B-5 shows the graph of the original relation with its domain restricted to the reals smaller than or equal to −2 (gray curve) and its inverse (black curve). The curve for the inverse relation is a half-hyperbola that opens downward and intersects the *y* axis at (0,−2). A movable vertical line never intersects the black curve at more than one point. Therefore, the inverse relation

$$f^{-1}(x) = -(4x^2/9 + 4)^{1/2}$$

is a true function of *x*.

10. Figure B-6 shows the graph of the original relation with its range restricted to the nonnegative reals (pair of gray curves) and the graph of its inverse (pair of black curves). A movable vertical line never intersects the graph of the original restricted relation at more than one point, so it's a true function. Whenever the movable vertical line intersects the graph of the inverse, that line crosses both curves. On this basis, we know that the inverse of the original restricted relation is not a true function.

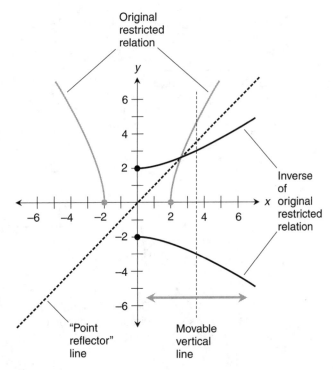

Figure B-6 Illustration for the solution to Problem 10 in Chap. 12.

Chapter 13

1. If the plane passes through the point where the apexes of the two cones meet, the intersection is a point, a straight line, or a pair of lines that intersect. Here's how the situations break down:

   - If the plane is slanted so that we'd get a circle or ellipse if the plane didn't pass through the apexes, then we get a single point (figs. B-7A and B).
   - If the plane slants so that we'd get a parabola if the plane didn't pass through the apexes, then we get a straight line (fig. B-7C).
   - If the plane slants so that we'd get a hyperbola if the plane didn't pass through the apexes, we get a pair of lines that cross (fig. B-7D).

2. When we examine the graph of the ellipse, we see that the lower focus is at (0,0) and the lower vertex is at (0,−2). Therefore, the distance between the lower focus and the lower vertex is 2 units, telling us that the focal length is

$$f = 2$$

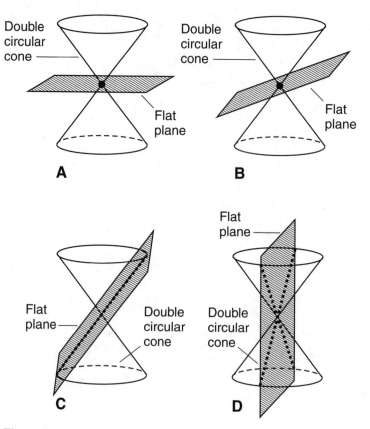

Figure B-7   Illustrations for the solution to Problem 1 in Chap. 13.

We can also see that the distance between the lower vertex and the directrix is 4 units, because the lower vertex is at $(0,-2)$ and the directrix passes through $(0,-6)$. Therefore

$$f/e = 4$$

Combining the above two equations and solving, we get

$$e = 1/2$$

That's the eccentricity of the ellipse.

3. When we solved Problem 2, we determined that $f = 2$ and $e = 1/2$. Plugging these values into the formulas relating the parameters of the ellipse, we obtain

$$u = 2 + 2/(1/2) + y$$

and

$$x^2 + y^2 = [(1/2)\ u]^2$$

These simplify to

$$u = 6 + y$$

and

$$x^2 + y^2 = u^2/4$$

Let's substitute the quantity $(6 + y)$ for u in the second equation above. That gives us

$$x^2 + y^2 = (6 + y)^2/4$$

This is an equation for our ellipse, although it's not in standard form.

4. We've been told that both foci and both vertices lie on the y axis, so $x = 0$ at the lower vertex and also at the upper vertex. If we plug $x = 0$ into the equation we got in the solution to Problem 3, we'll be left with an equation that tells us the y values at both vertices of the ellipse. Here it is

$$y^2 = (6 + y)^2/4$$

Multiplying through by 4, we obtain

$$4y^2 = (6 + y)^2$$

When we multiply out the squared binomial on the right-hand side, we get

$$4y^2 = 36 + 12y + y^2$$

This is a quadratic equation. Let's morph it with algebra into the standard form for a quadratic. That gives us

$$3y^2 - 12y - 36 = 0$$

which factors into

$$(3y + 6)(y - 6) = 0$$

The solutions are therefore

$$y = -2$$

or

$$y = 6$$

The first solution corresponds to the point $(0,-2)$ on the ellipse. The second solution corresponds to $(0,6)$ on the ellipse. We already know that the coordinates of the lower vertex are $(0,-2)$, so the coordinates of the upper vertex must be $(0,6)$.

Now let's find the coordinates of the upper focus. We already know that the lower focus is at (0,0). We also know that the lower vertex is at (0,−2), which is 2 units below the lower focus on the *y* axis. Because all ellipses are symmetrical with respect to the foci, the upper focus must be 2 units below the upper vertex on the *y* axis. Therefore, the coordinates of the upper focus must be (0,4).

5. We can find the coordinates of our ellipse's center by averaging the coordinates of the foci. The lower focus is at (0,0), while the upper focus is at (0,4). Therefore, the center must be at (0,2).

The length of the vertical semi-axis (let's call it *b*) is the distance between the center and the lower vertex, as shown in Fig. B-8. The center is at (0,2) and the lower vertex is at (0,−2), so the vertical semi-axis is 4 units long. (We get the same result if we use the center and the upper vertex.)

To find the length of the horizontal semi-axis, we must know the coordinates of either the left-most point or the right-most point on the ellipse. These are the points where the horizontal line *y* = 2, which passes through the center of the ellipse, intersects the curve. To find these points, we can plug *y* = 2 into the equation we got for the ellipse when we solved Problem 3. When we make the substitutions, we obtain

$$x^2 + 2^2 = (6 + 2)^2/4$$

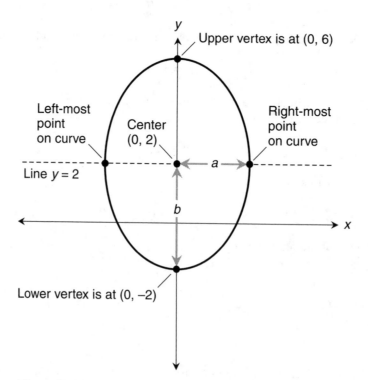

Figure B-8 Illustration for the solution to Problem 5 in Chap. 13.

which simplifies to

$$x^2 = 12$$

This equation solves easily to

$$x = 12^{1/2}$$

or

$$x = -12^{1/2}$$

telling us that the left-most point on the curve is at $(-12^{1/2},2)$, and the right-most point on the curve is at $(12^{1/2},2)$. The horizontal semi-axis (call it *a*) is therefore $12^{1/2}$ units long.

In the generalized standard form for the equation of an ellipse, the ordered pair (x_0,y_0) represents the coordinates of the center point. We know that this point is $(0,2)$. We've now figured out these four values:

$$x_0 = 0$$
$$y_0 = 2$$
$$a = 12^{1/2}$$
$$b = 4$$

As you can guess, we chose the names of these parameters so that they'd fit neatly into the generalized standard form for the equation of an ellipse. That form is

$$(x - x_0)^2/a^2 + (y - y_0)^2/b^2 = 1$$

Plugging in the numbers straightaway, we obtain

$$(x - 0)^2/(12^{1/2})^2 + (y - 2)^2/4^2 = 1$$

which simplifies to

$$x^2/12 + (y - 2)^2/16 = 1$$

At last, we've found the standard-form equation for the ellipse graphed in Fig. 13-12! If you're skeptical and ambitious, you'll demand proof that this equation is equivalent to the one we got in the solution to Problem 3. Why not demonstrate that fact as an extra-credit exercise? Start with

$$x^2/12 + (y - 2)^2/16 = 1$$

and morph it into

$$x^2 + y^2 = (6 + y)^2/4$$

6. Here's the original equation again, for reference:

$$x^2 + 9y^2 = 9$$

We can divide this equation through by 9 to obtain

$$x^2/9 + y^2 = 1$$

This equation is in the standard form for an ellipse. We recall that the general version is

$$(x - x_0)^2/a^2 + (y - y_0)^2/b^2 = 1$$

where x_0 and y_0 are the coordinates of the center, a is the length of the horizontal semi-axis, and b is the length of the vertical semi-axis. In this case, we have

$$x_0 = 0$$
$$y_0 = 0$$
$$a = 9^{1/2} = 3$$
$$b = 1^{1/2} = 1$$

The center is at (0,0). The horizontal semi-axis measures 3 units. The vertical semi-axis measures 1 unit. We can now sketch the graph as shown in Fig. B-9.

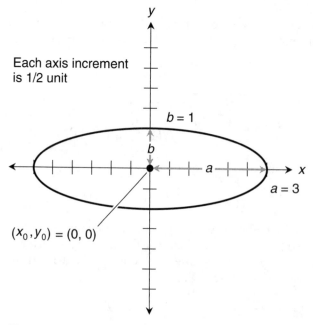

Figure B-9   Illustration for the solution to Problem 6 in Chap. 13. Each axis division represents 1/2 unit.

7. Here's the original equation again, for reference:

$$x^2 + y^2 + 2x - 2y + 2 = 4$$

It takes some mathematical intuition to see how this can be morphed into the standard form for a conic section. We can split it into the sum of two trinomials as

$$(x^2 + 2x + 1) + (y^2 - 2y + 1) = 4$$

The parentheses, while not technically necessary, clarify the identities of the trinomials. Both of these trinomials happen to be perfect squares. They can be factored to get

$$(x + 1)^2 + (y - 1)^2 = 4$$

which is in the standard form for a circle. We remember that the general equation for a circle is

$$(x - x_0)^2 + (y - y_0)^2 = r^2$$

where (x_0, y_0) are the coordinates of the center, and r the radius. In this particular example, we have

$$x_0 = -1$$
$$y_0 = 1$$
$$r = 4^{1/2} = 2$$

The center is at $(-1, 1)$. The radius is 2 units. We can now sketch the graph as shown in Fig. B-10.

Figure B-10   Illustration for the solution to Problem 7 in Chap. 13. Each axis division represents 1/2 unit.

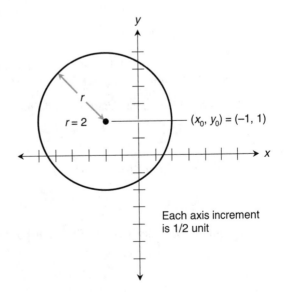

Each axis increment is 1/2 unit

8. Here's the original equation again, for reference:

$$x^2 - y^2 + 2x + 2y = 4$$

As in Problem 7, we must rely on our algebra experience to see how this can be transformed into an equation that's in the standard form for a conic section. We can split it into a difference between two trinomials as

$$(x^2 + 2x + 1) - (y^2 - 2y + 1) = 4$$

In this equation, the parentheses *are* necessary! The trinomials can be factored exactly as they were in the solution to Problem 7; when we do that, we obtain

$$(x + 1)^2 - (y - 1)^2 = 4$$

We can divide through by 4, getting

$$(x + 1)^2/4 - (y - 1)^2/4 = 1$$

This equation is in the standard form for a hyperbola. The general version, as we've learned, is

$$(x - x_0)^2/a^2 - (y - y_0)^2/b^2 = 1$$

where (x_0, y_0) are the coordinates of the center, a is the length of the horizontal semi-axis, and b is the length of the vertical semi-axis. In this case, we have

$$x_0 = -1$$
$$y_0 = 1$$
$$a = 4^{1/2} = 2$$
$$b = 4^{1/2} = 2$$

The center is at $(-1, 1)$. The horizontal and vertical semi-axes are both 2 units long. With this information, we can sketch the graph as shown in Fig. B-11.

9. Here's the original equation again, for reference:

$$x^2 - 3x - y + 3 = 1$$

We can subtract 1 from each side to get

$$x^2 - 3x - y + 2 = 0$$

Adding y to each side and then transposing the sides left-to-right yields

$$y = x^2 - 3x + 2$$

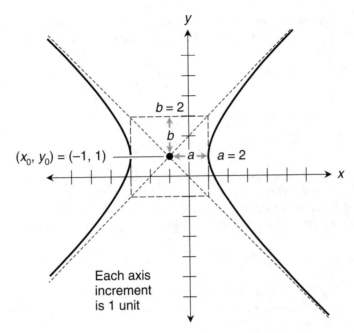

Figure B-11   Illustration for the solution to Problem 8 in
Chap. 13. Each axis division represents 1 unit.

This equation is in the standard form for a parabola. The coefficient of x^2 is positive, so we know that the parabola opens upward. When we divide the negative of the coefficient for x by twice the coefficient for x^2, we get the x value of the vertex point, which we can call x_0. We have

$$x_0 = -(-3)/(2 \times 1) = 3/2$$

To find the y value of the vertex point (let's call it y_0), we plug 3/2 into the function for x and grind out the arithmetic:

$$y_0 = (3/2)^2 - 3 \times 3/2 + 2 = 9/4 - 9/2 + 2$$
$$= 9/4 - 18/4 + 8/4 = (9 - 18 + 8)/4$$
$$= -1/4$$

Now we know that the coordinates of the vertex are $(3/2, -1/4)$. Because the parabola opens upward, we know that this vertex is the absolute minimum point.

The right-hand side of the standard-form equation factors into a product of two clean-cut binomials

$$y = (x - 1)(x - 2)$$

If we set $y = 0$ and then solve the resulting quadratic equation, we get two roots that tell us the x-intercepts of the curve. The roots here can be read straight from the factors as

$$x = 1$$

or

$$x = 2$$

revealing that (1,0) and (2,0) lie on the parabola. These two points are close together, and they're also close to the vertex. It's difficult to draw a good image of the parabola based on these three points alone. But we can find the y-intercept, which is "farther out," to help us draw the curve. When we plug in 0 for x, we get

$$y = 0^2 - 3 \times 0 + 2 = 0 - 0 + 2 = 2$$

indicating that (0,2) is on the parabola. Now that we know the coordinates of four points that lie on the parabola, we can do some "mental curve-fitting" and sketch the graph as shown in Fig. B-12.

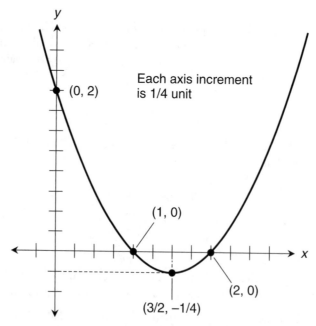

Figure B-12 Illustration for the solution to Problem 9 in Chap. 13. Each axis division represents 1/4 unit.

10. The standard-form general equation for a hyperbola in the xy plane is

$$(x - x_0)^2 / a^2 - (y - y_0)^2 / b^2 = 1$$

where (x_0, y_0) are the coordinates of the center, a is the length of the horizontal semi-axis, and b is the length of the vertical semi-axis. The specific equation of our hyperbola is

$$(x - 1)^2 / 4 - (y + 2)^2 / 9 = 1$$

From this equation, we can see that the constants are

$$x_0 = 1$$
$$y_0 = -2$$
$$a = 4^{1/2} = 2$$
$$b = 9^{1/2} = 3$$

This information tells us that the hyperbola's center is at $(1, -2)$. The horizontal semi-axis is 2 units wide. The vertical semi-axis is 3 units tall. We can therefore sketch the graph as shown in Fig. B-13.

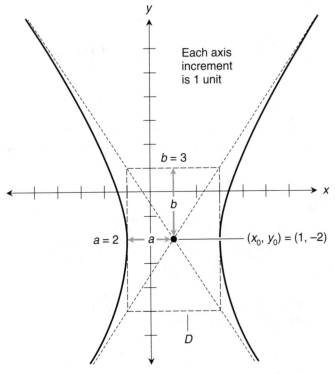

Figure B-13   Illustration for the solution to Problem 10 in Chap. 13. Each axis division represents 1 unit.

To find the equations of the asymptotes, we must know the coordinates of a point on each of them, and we must also know their slopes. The asymptotes intersect at the center of the hyperbola, which is at

$$(x_0, y_0) = (1, -2)$$

We know that this point lies on both asymptote lines. Let's construct a rectangle D whose width is twice the length of the horizontal semi-axis, and whose height is twice the length of the vertical semi-axis. The asymptotes pass through the corners of this rectangle, as shown in Fig. B-13. When we go from the center to the upper right-hand corner of D, the "rise over run" (an informal term for slope) is

$$b/a = 3/2$$

so $m = 3/2$. We can therefore write the equation of the "up-ramping" asymptote in point-slope form as

$$y + 2 = (3/2)(x - 1)$$

When we travel from the center to the lower right-hand corner of D, the "rise over run" is

$$b/a = -3/2$$

so $m = -3/2$. We can therefore write the equation of the "down-ramping" asymptote in point-slope form as

$$y + 2 = (-3/2)(x - 1)$$

If you insist on having the equations of these lines appear in the standard form for a linear equation in two-space, feel free to convert them to that form!

Chapter 14

1. Suppose that there's a real number x that satisfies the equation

$$e^x = 0$$

We know that our mystery number x can't be equal to 0, because we can plug $x = 0$ straightaway into the equation and get

$$e^0 = 1$$

based on the fact that any positive real number (including e) raised to the zeroth power is equal to 1. We've assumed that x is real, and we've discovered that $x \neq 0$, so we can be certain that $1/x$ is a nonzero real number. Therefore, it's okay for us to take the $(1/x)$th power of both sides of the original equation to obtain

$$(e^x)^{(1/x)} = 0^{(1/x)}$$

which we can rewrite using the algebraic rules for exponents as

$$e^{[x(1/x)]} = 0^{(1/x)}$$

When 0 is raised to any nonzero real power, the result is 0. That fact, along with another "dose" of the algebraic rules for exponents, allows us to streamline the above equation, getting

$$e^{(x/x)} = 0$$

which simplifies further to

$$e^1 = 0$$

and finally to

$$e = 0$$

This statement is patently untrue. According to *reductio ad absurdum*, it follows that our original assumption must be false. We must conclude that no real-number power of e is equal to 0.

2. The dashed gray curves in Fig. B-14 are the graphs of

$$y = e^x$$

Figure B-14   Illustration for the solution to Problem 2 in Chap. 14.

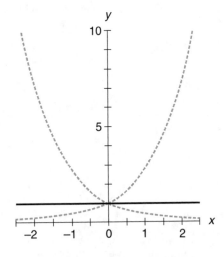

and

$$y = e^{-x}$$

taken directly from Figs. 14-1A and 14-2A. Now let's consider the function

$$y = e^x e^{-x}$$

Using the algebraic rules for exponents, we can rewrite this as

$$y = e^{[x+(-x)]}$$

which simplifies to

$$y = e^0$$

and further to

$$y = 1$$

The domain of this constant function encompasses all real numbers. The range is the set containing the single element 1. The graph is a solid black horizontal line passing through the point (0,1) in Fig. B-14.

3. The dashed gray curves in Fig. B-15 are the graphs of

$$y = 10^x$$

and

$$y = 10^{-x}$$

Figure B-15   Illustration for the solution to Problem 3 in Chap. 14.

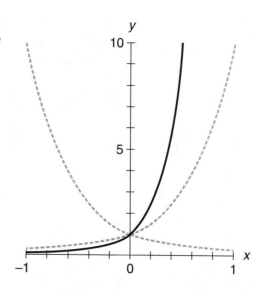

taken directly from Figs. 14-1B and 14-2B. Now let's look at the ratio function

$$y = 10^x/10^{-x}$$

Using the algebraic rules for exponents, we can rewrite this as

$$y = 10^{[x-(-x)]}$$

which simplifies to

$$y = 10^{2x}$$

The domain encompasses all real numbers. The range is the set of all positive real numbers. The graph is the solid black curve in Fig. B-15.

4. Figure B-16 shows the same graphs as Fig. B-15. However, in this illustration, the y axis is logarithmic, spanning the three orders of magnitude from 0.1 to 100. The dashed gray lines are the graphs of

$$y = 10^x$$

and

$$y = 10^{-x}$$

Figure B-16 Illustration for the solution to Problem 4 in Chap. 14.

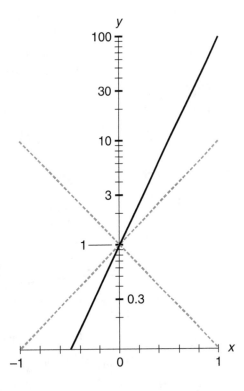

The solid black line is the graph of

$$y = 10^{2x}$$

5. Figure B-17 is the graph of the function

$$y = 10^{(1/x)}$$

for values of x ranging from -10 to 10. When we input $x = 0$, we get $10^{1/0}$, which is undefined. For any other real value of x, the output value y is a positive real, so the domain is the set of all nonzero reals. No matter how large we want y to be when $y > 1$, we can always find some value of x that will produce it. No matter how small we want y to be when $0 < y < 1$, we can always find some value of x that will produce it. However, we can't find any value for x that will give us $y = 1$. For that to happen, we must raise 10 to the zeroth power, meaning that we must find some x such that $1/x = 0$. No such x exists, so the range of the function is the set of all positive reals except 1. The graph has a horizontal asymptote whose equation is $y = 1$, and a vertical asymptote corresponding to the y axis. The open circle at $(0,0)$ tells us that this point is not part of the graph.

6. Suppose that there's a real number x that satisfies the equation

$$\ln 0 = x$$

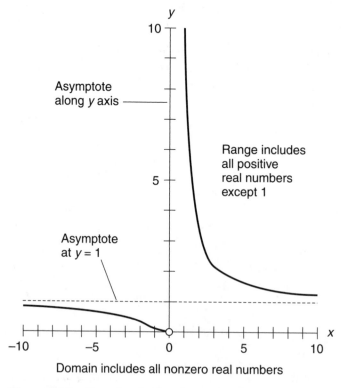

Figure B-17   Illustration for the solution to Problem 5 in Chap. 14.

Let's take the natural exponential of each side of this equation. That gives us

$$e^{(\ln 0)} = e^x$$

We know that the natural log function and the natural exponential function are inverses of each other. They "undo" each other's work, as long as we stay within the domain of the natural log function. We've assumed that ln 0 is a real number, and therefore that it's in the domain of the natural log function. Based on that assumption, we can rewrite the above equation as

$$0 = e^x$$

In the solution to Problem 1, we proved that no real number x can satisfy this equation. That result contradicts our original assumption here. By *reductio ad absurdum*, we are forced to conclude that the natural log of 0 is not a real number.

7. The dashed gray curves in Fig. B-18 are the graphs of

$$y = \ln x$$

and

$$y = \log_{10} x$$

We want to graph the sum function

$$y = \ln x + \log_{10} x$$

When we input several values, use a calculator to obtain the outputs, plot the points, and then connect the points by curve fitting, we get the solid black curve. The domain of this sum function is the set of positive reals, and the range is the set of all reals.

Figure B-18 Illustration for the solution to Problem 7 in Chap. 14.

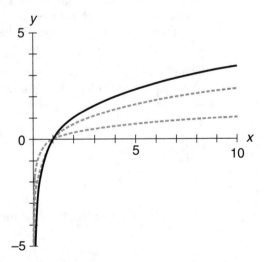

Figure B-19   Illustration for the solution to Problem 8 in Chap. 14.

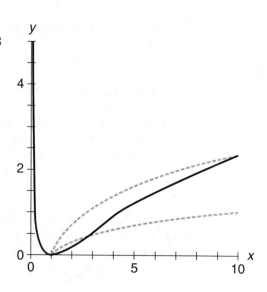

8. The dashed gray curves in Fig. B-19 are the graphs of

$$y = \ln x$$

and

$$y = \log_{10} x$$

The portions of these curves below the x axis (that is, where $y < 0$) are cut off here, because we haven't included any of the negative y axis. But the function values are still there, of course! We want to graph the product function

$$y = (\ln x)(\log_{10} x)$$

When we input several values for x, use a calculator to obtain the outputs, plot the points, and then connect the points by curve fitting, we get the solid black curve. The domain of this product function is the set of all positive reals, and the range is the set of all nonnegative reals.

9. Figure B-20 shows the same graphs as Fig. B-18. Here, the x axis is logarithmic, spanning the two orders of magnitude from 0.1 to 10. The dashed gray lines are the graphs of

$$y = \ln x$$

and

$$y = \log_{10} x$$

The solid black line is the graph of the sum function

$$y = \ln x + \log_{10} x$$

Figure B-20 Illustration for the solution to Problem 9 in Chap. 14.

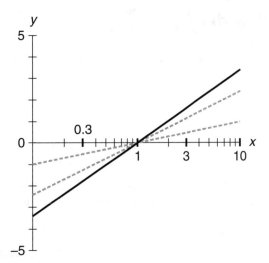

10. Figure B-21 shows the same graphs as Fig. B-19. In this coordinate system, the x axis is logarithmic, spanning the single order of magnitude from 1 to 10. The y axis is linear, spanning the values from 0 to 2.5. The dashed gray lines are the graphs of

$$y = \ln x$$

and

$$y = \log_{10} x$$

The solid black curve is the graph of the product function

$$y = (\ln x)(\log_{10} x)$$

Figure B-21 Illustration for the solution to Problem 10 in Chap. 14.

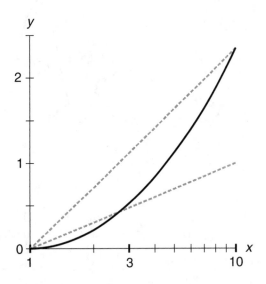

Chapter 15

1. Figure B-22 shows superimposed graphs of the sine and cosine functions (dashed gray curves) along with a graph of their difference function

$$h(\theta) = \sin\theta - \cos\theta$$

shown as a solid black curve. The domain of h includes all real numbers. The range of h is the set of all reals in the closed interval $[-2^{1/2}, 2^{1/2}]$.

2. The solid black curve in Fig. B-23 is a graph of the difference between the squares of the sine and the cosine functions, which are shown as superimposed dashed gray curves. We have

$$h(\theta) = \sin^2\theta - \cos^2\theta$$

The domain of h includes all of the real numbers. The range is the set of all real numbers in the closed interval $[-1,1]$.

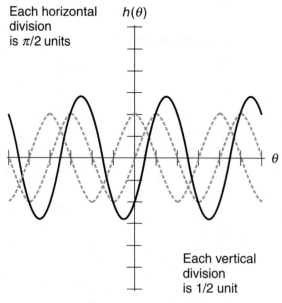

Each horizontal division is $\pi/2$ units

$h(\theta)$

Each vertical division is 1/2 unit

Figure B-22   Illustration for the solution to Problem 1 in Chap. 15. Each horizontal division represents $\pi/2$ units. Each vertical division represents 1/2 unit.

Figure B-23 Illustration for the solution to Problem 2 in Chap. 15. Each horizontal division represents $\pi/2$ units. Each vertical division represents 1/4 unit.

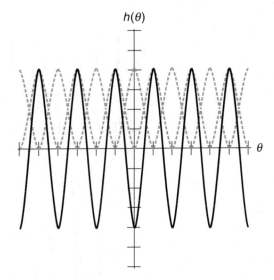

Each horizontal division is $\pi/2$ units
Each vertical division is 1/4 unit

3. The solid black complex of curves in Fig. B-24 is the graph of the ratio of the square of the cosine to the square of the sine. If we call the function h, then

$$h\,(\theta) = (\cos^2\,\theta)/(\sin^2\,\theta)$$

The superimposed gray curves are graphs of the original sine-squared and cosine-squared functions. The domain of h includes all real numbers except the integer multiples of π. The range of h spans the set of all nonnegative reals.

Figure B-24 Illustration for the solution to Problem 3 in Chap. 15. Each horizontal division represents $\pi/2$ units. Each vertical division represents 1/2 unit. The vertical dashed lines are asymptotes of h. The positive dependent-variable axis is also an asymptote of h.

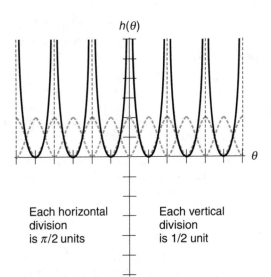

Each horizontal division is $\pi/2$ units

Each vertical division is 1/2 unit

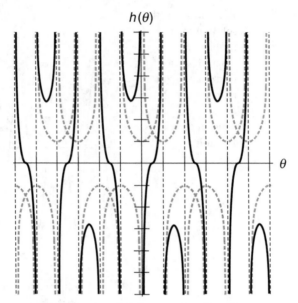

Each horizontal division is π/2 units
Each vertical division is 1 unit

Figure B-25   Illustration for the solution to Problem 4 in Chap. 15. Each horizontal division represents $\pi/2$ units. Each vertical division represents 1 unit. The vertical dashed lines are asymptotes of h. The dependent-variable axis is also an asymptote of h.

4. The dashed gray curves in Fig. B-25 are the superimposed graphs of the secant and cosecant functions. The complex of solid black curves is a graph of the difference function

$$h\,(\theta) = \sec\,\theta - \csc\,\theta$$

The domain of h includes all real numbers except the integer multiples of $\pi/2$. The range of h spans the set of all real numbers.

5. The dashed gray curves in Fig. B-26 are the superimposed graphs of the secant-squared and cosecant-squared functions. The complex of solid black curves is a graph of the difference function

$$h\,(\theta) = \sec^2\,\theta - \csc^2\,\theta$$

The domain of h includes all real numbers except the integer multiples of $\pi/2$. The range of h includes all real numbers.

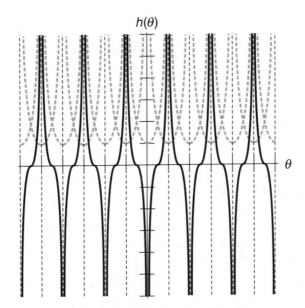

$h(\theta)$

θ

Each horizontal division is $\pi/2$ units
Each vertical division is 1 unit

Figure B-26   Illustration for the solution to
Problem 5 in Chap. 15. Each
horizontal division represents
$\pi/2$ units. Each vertical division
represents 1 unit. The vertical
dashed lines are asymptotes of h.
The dependent-variable axis is also
an asymptote of h.

6. We want to find a graph of the ratio function

$$h(\theta) = (\csc^2 \theta)/(\sec^2 \theta)$$

We can simplify this function with some algebra, along with our knowledge of
trigonometry. The secant is the reciprocal of the cosine, so the converse is also true.
We have

$$1/(\sec \theta) = \cos \theta$$

Squaring both sides, we get

$$1/(\sec^2 \theta) = \cos^2 \theta$$

Substituting in the equation for our original ratio function, we get

$$h(\theta) = (\csc^2 \theta)(\cos^2 \theta)$$

The cosecant is the reciprocal of the sine, so

$$\csc \theta = 1/(\sin \theta)$$

Squaring both sides gives us

$$\csc^2 \theta = 1/(\sin^2 \theta)$$

Substituting in the modified equation for our original function, we obtain

$$f(\theta) = [1/(\sin^2 \theta)] \, (\cos^2 \theta) = [(\cos \theta)/(\sin \theta)]^2$$

The cosine divided by the sine is the cotangent, so we can substitute again to conclude that our original function is

$$h(\theta) = \cot^2 \theta$$

with the restriction that we can't define h for any input value where either the secant or the cosecant become singular.

The solid black curves in Fig. B-27 show the result of squaring all the values of the cotangent function, noting the undefined values as asymptotes or open circles. The domain includes all real numbers except integer multiples of $\pi/2$, where one or the other of the original squared functions is singular. The range is the set of all positive reals.

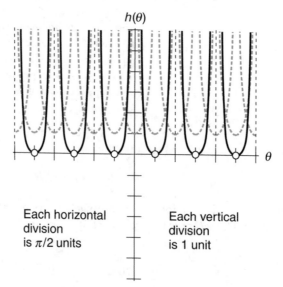

Figure B-27   Illustration for the solution to Problem 6 in Chap. 15. Each horizontal division represents $\pi/2$ units. Each vertical division represents 1 unit. The vertical dashed lines are asymptotes of h. The positive dependent-variable axis is also an asymptote of h.

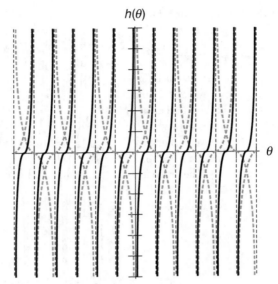

$h(\theta)$

Each horizontal division is $\pi/2$ units
Each vertical division is 1 unit

Figure B-28   Illustration for the solution to Problem 7 in Chap. 15. Each horizontal division represents $\pi/2$ units. Each vertical division represents 1 unit. The vertical dashed lines are asymptotes of h. The dependent-variable axis is also an asymptote of h.

7. In Fig. B-28, the dashed gray curves are graphs of the tangent and cotangent functions. The solid black curves compose the graph of

$$h\,(\theta) = \tan\,\theta - \cot\,\theta$$

The domain of h is the set of all reals except the integer multiples of $\pi/2$. The range is the set of all real numbers.

8. In Fig. B-29, the dashed gray curves are graphs of the tangent-squared and cotangent-squared functions. The solid black curves compose the graph of

$$h\,(\theta) = \tan^2\,\theta - \cot^2\,\theta$$

The domain of h includes all reals except the integer multiples of $\pi/2$. The range is the set of all real numbers.

9. In Fig. B-30, the dashed gray curves represent the squares of the secant and tangent functions. The solid black line with holes is a graph of

$$f(\theta) = \sec^2\,\theta - \tan^2\,\theta$$

Figure B-29 Illustration for the solution to Problem 8 in Chap. 15. Each horizontal division represents $\pi/2$ units. Each vertical division represents 1 unit. The vertical dashed lines are asymptotes of h. The dependent-variable axis is also an asymptote of h.

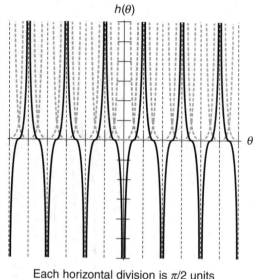

$h(\theta)$

θ

Each horizontal division is $\pi/2$ units
Each vertical division is 1 unit

The domain of f is the set of all reals except the odd-integer multiples of $\pi/2$. The range of f is the set containing the number 1.

10. In Fig. B-31, the dashed gray curves represent the squares of the cosecant and cotangent functions. The solid black line with "holes" is a graph of

$$f(\theta) = \csc^2 \theta - \cot^2 \theta$$

The domain of f is the set of all reals except the integer multiples of π. The range of f is the set containing the number 1.

Figure B-30 Illustration for the solution to Problem 9 in Chap. 15. Each horizontal division represents $\pi/2$ units. Each vertical division represents 1 unit.

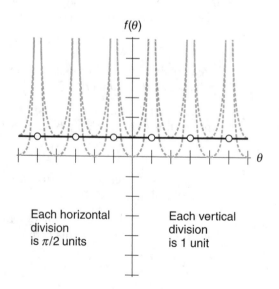

$f(\theta)$

θ

Each horizontal division is $\pi/2$ units

Each vertical division is 1 unit

Figure B-31   Illustration for the solution to Problem 10 in Chap. 15. Each horizontal division represents $\pi/2$ units. Each vertical division represents 1 unit.

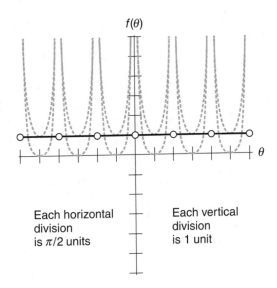

Each horizontal division is $\pi/2$ units

Each vertical division is 1 unit

Chapter 16

1. For reference, the parametric equations are

$$x = t^2$$

and

$$y = t^3$$

We can take the 1/3 power of both sides of the second equation to get

$$y^{1/3} = t$$

Substituting $y^{1/3}$ for t in the first equation yields

$$x = (y^{1/3})^2$$

which can be simplified to

$$x = y^{2/3}$$

This equation contains the variables x and y only, without the parameter t. There's another way to approach this problem. We can take the positive-or-negative 1/2 power of both sides of the first original equation, getting

$$\pm x^{1/2} = t$$

Then we can substitute $\pm x^{1/2}$ for t in the second original equation to obtain

$$y = (\pm x^{1/2})^3 = \pm x^{3/2}$$

which contains the variables x and y only, without the parameter t.

2. The second answer to Problem 1 is an expression of the relation in which x is the independent variable and y is the dependent variable. This relation is not a function of x. We can see this by applying the vertical-line test to the graph of Fig. 16-4. The graph fails the test because, for all positive values of x, there are two values of y.

3. For reference, the parametric equations are

$$\theta = t^{-1}$$

and

$$r = \ln t$$

We can take the natural exponential of both sides of the second equation to obtain

$$e^r = e^{(\ln t)}$$

which simplifies to

$$e^r = t$$

provided that $t > 0$, so we're sure that $\ln t$ is defined. Substituting e^r for t in the first original parametric equation yields

$$\theta = (e^r)^{-1}$$

which simplifies to

$$\theta = e^{-r}$$

This equation contains the variables θ and r only, without the parameter t. We can approach this problem another way. If we take the reciprocal of both sides of the first original parametric equation, we get

$$\theta^{-1} = (t^{-1})^{-1}$$

as long as $t \neq 0$, so we're sure that t^{-1} is defined. This simplifies to

$$\theta^{-1} = t$$

When we substitute θ^{-1} for t in the second parametric equation, we get

$$r = \ln (\theta^{-1}) = -\ln \theta$$

which, again, contains the variables θ and r only, without the parameter t. This relation is defined only for positive real-number values of θ.

4. The second answer to Problem 3 is an expression of the relation in which θ is the independent variable and r is the dependent variable. We can't apply a line test *directly* to Fig. 16-7, but we can graph the relation in a Cartesian coordinate plane with θ on the horizontal axis and r on the vertical axis. When we do that, we get the curve shown in Fig. 16-6, with θ in place of x and r in place of y. This graph passes the vertical-line test, indicating that r is a function of θ.

5. For reference, the parametric equations are

$$x = a \cos t$$

and

$$y = a \sin t$$

We've been assured that $a \neq 0$, so we can divide the equations both through by a to obtain

$$x/a = \cos t$$

and

$$y/a = \sin t$$

Squaring both sides of both equations gives us

$$(x/a)^2 = \cos^2 t$$

and

$$(y/a)^2 = \sin^2 t$$

When we add these two equations, left-to-left and right-to-right, we have

$$(x/a)^2 + (y/a)^2 = \cos^2 t + \sin^2 t$$

The Pythagorean trigonometric identity for the sine and the cosine tells us that

$$\cos^2 t + \sin^2 t = 1$$

for all real numbers t. Therefore, the preceding equation can be rewritten as

$$(x/a)^2 + (y/a)^2 = 1$$

Expanding the squared ratios on the left-hand side gives us

$$x^2/a^2 + y^2/a^2 = 1$$

Multiplying through by a^2, we get

$$x^2 + y^2 = a^2$$

which is the equation of a circle of radius a, centered at the origin.

6. When we subtract x^2 from both sides of the solution to Problem 5, we obtain

$$y^2 = a^2 - x^2$$

Taking the positive-or-negative square root of both sides gives us

$$y = \pm(a^2 - x^2)^{1/2}$$

This relation is not a function of x, as we can see when we apply the vertical-line test to the graph of Fig. 16-10. Whenever we input any value of the independent variable x that lies within the open interval $(-a,a)$, our relation produces two values of the dependent variable y.

7. For reference, the parametric equations are

$$x = \sec t$$

and

$$y = \tan t$$

When we square both sides of both equations, we obtain

$$x^2 = \sec^2 t$$

and

$$y^2 = \tan^2 t$$

Subtracting the second equation from the first, left-to-left and right-to-right, we get

$$x^2 - y^2 = \sec^2 t - \tan^2 t$$

From trigonometry, the Pythagorean identity for the secant and the tangent tells us that

$$\sec^2 t - \tan^2 t = 1$$

for all real numbers t except odd-integer multiples of $\pi/2$. The preceding equation can therefore be rewritten as

$$x^2 - y^2 = 1$$

which represents the unit hyperbola in the Cartesian xy plane.

8. For reference, the parametric equations are

$$x = a \csc t$$

and

$$y = b \cot t$$

We've been assured that $a \neq 0$, so we can divide the first equation through by a to obtain

$$x/a = \csc t$$

We've also been told that $b \neq 0$, so we can divide the second equation through by b, getting

$$y/b = \cot t$$

Squaring both sides of both equations gives us

$$(x/a)^2 = \csc^2 t$$

and

$$(y/b)^2 = \cot^2 t$$

When we subtract the second equation, left-to-left and right-to-right, from the first one, we obtain

$$(x/a)^2 - (y/b)^2 = \csc^2 t - \cot^2 t$$

The Pythagorean identity for the cosecant and the cotangent tells us that

$$\csc^2 t - \cot^2 t = 1$$

for all real numbers t except integer multiples of π. Knowing this, we can rewrite the preceding equation as

$$(x/a)^2 - (y/b)^2 = 1$$

Expanding the squared ratios on the left-hand side gives us

$$x^2/a^2 - y^2/b^2 = 1$$

which represents a hyperbola centered at the origin in the Cartesian xy plane. The width of the horizontal (x-coordinate) semi-axis is a units, and the height of the vertical (y-coordinate) semi-axis is b units.

9. We start with the relation

$$x = \sin (\cos y)$$

Suppose we assign our parameter t such that

$$\cos y = t$$

Taking the arccosine of both sides, we get

$$\arccos (\cos y) = \arccos t$$

which simplifies to

$$y = \arccos t$$

That's one of our parametric equations. We can substitute t for $\cos y$ in the original equation to get

$$x = \sin t$$

That's the other parametric equation.

10. As in Problem 9, we start with the relation

$$x = \sin (\cos y)$$

Taking the arcsine of both sides gives us

$$\arcsin x = \arcsin [\sin (\cos y)]$$

which simplifies to

$$\arcsin x = \cos y$$

When we take the arccosine of both sides, we obtain

$$\arccos (\arcsin x) = \arccos (\cos y)$$

which simplifies to

$$\arccos (\arcsin x) = y$$

Transposing the left- and right-hand sides, we get the sought-after equation

$$y = \arccos (\arcsin x)$$

Now let's derive this equation from the parametric equations in the solution to Problem 9. Those equations are

$$x = \sin t$$

and

$$y = \arccos t$$

We can take the arcsine of both sides of the first equation, getting

$$\arcsin x = \arcsin (\sin t)$$

This simplifies to

$$\arcsin x = t$$

Substituting arcsin x for t in the right-hand side of the second parametric equation, we obtain

$$y = \arccos (\arcsin x)$$

which is the same equation we got from the original relation in terms of x and y without the parameter t.

Chapter 17

1. The standard-form vector

$$-4\mathbf{i} + 4\mathbf{j} - 4\mathbf{k}$$

can be described by the ordered triple

$$(a,b,c) = (-4,4,-4)$$

We've been told that one of the points in our plane is

$$(x_0,y_0,z_0) = (0,0,0)$$

The general formula for a plane in Cartesian xyz space is

$$a(x - x_0) + b(y - y_0) + c(z - z_0) = 0$$

Plugging in the known values, we get

$$-4(x - 0) + 4[y - 0] + (-4)(z - 0) = 0$$

which simplifies to

$$-4x + 4y - 4z = 0$$

2. The standard-form vector

$$-2\mathbf{i} + 0\mathbf{j} + 0\mathbf{k}$$

can be described by the ordered triple

$$(a,b,c) = (-2,0,0)$$

One of the points in the plane is

$$(x_0,y_0,z_0) = (4,5,6)$$

The general formula for a plane in Cartesian xyz space is

$$a(x - x_0) + b(y - y_0) + c(z - z_0) = 0$$

Plugging in the known values, we get

$$-2(x - 4) + 0[y - 5] + 0(z - 6) = 0$$

which simplifies to

$$-2x + 8 = 0$$

3. We've been told that the equation of a certain sphere is

$$x^2 + 2x + 1 + y^2 - 2y + 1 + z^2 + 8z + 16 = 64$$

Grouping the addends by threes, we get

$$(x^2 + 2x + 1) + (y^2 - 2y + 1) + (z^2 + 8z + 16) = 64$$

Factoring each of the trinomials enclosed by parentheses, we obtain

$$(x + 1)^2 + (y - 1)^2 + (z + 4)^2 = 64$$

The general equation for a sphere in Cartesian xyz space is

$$(x - x_0)^2 + (y - y_0)^2 + (z - z_0)^2 = r^2$$

where (x_0,y_0,z_0) are the coordinates of the center, and r is the radius. Based on this information, we can deduce that the coordinates of this sphere's center are

$$(x_0,y_0,z_0) = (-1,1,-4)$$

and the radius r is the positive square root of 64, which is 8.

4. We've been told that the coordinates of the center of a certain sphere are

$$(x_0, y_0, z_0) = (5, 7, -3)$$

and the radius is

$$r = 23^{1/2}$$

Once again, the general equation for a sphere in Cartesian *xyz* space is

$$(x - x_0)^2 + (y - y_0)^2 + (z - z_0)^2 = r^2$$

where (x_0, y_0, z_0) are the coordinates of the center, and r is the radius. Plugging in the known values directly, we conclude that the equation for this particular sphere is

$$(x - 5)^2 + (y - 7)^2 + (z + 3)^2 = 23$$

5. Stated again for reference, the equation of our object is

$$8(x - 1)^2 + 8(y + 2)^2 + 6(z + 7)^2 = 24$$

Dividing through by 24, we obtain

$$(x - 1)^2/3 + (y + 2)^2/3 + (z + 7)^2/4 = 1$$

This is the equation for a distorted sphere centered at

$$(x_0, y_0, z_0) = (1, -2, -7)$$

The length of the axial radius in the *x* direction is the positive square root of 3. The length of the axial radius in the *y* direction is also the positive square root of 3. The length of the axial radius in the *z* direction is the positive square root of 4, or 2, which is a little longer than the other two axes. Therefore, our object is an ellipsoid.

6. Stated again for reference, the equation for the object under scrutiny is

$$400(x + 2)^2 + 225(y - 4)^2 + 144z^2 - 3600 = 0$$

When we add 3600 to each side, we get

$$400(x + 2)^2 + 225(y - 4)^2 + 144z^2 = 3600$$

Dividing through by 3600 yields

$$(x + 2)^2/9 + (y - 4)^2/16 + z^2/25 = 1$$

This is the equation for a distorted sphere centered at

$$(x_0, y_0, z_0) = (-2, 4, 0)$$

The length of the axial radius in the x direction is $9^{1/2}$, which is 3. The length of the axial radius in the y direction is $16^{1/2}$, which is 4. The length of the axial radius in the z direction is $25^{1/2}$, which is 5. Because no two of the axial radii are the same, our object is an oblate ellipsoid.

7. We've been told that the equation of a certain object is

$$x^2 + 2x + 1 + y^2 - 2y + 1 - z^2 + 6z - 9 = 36$$

Grouping the terms by threes, we get

$$(x^2 + 2x + 1) + (y^2 - 2y + 1) + (-z^2 + 6z - 9) = 36$$

which can be rewritten as

$$(x^2 + 2x + 1) + (y^2 - 2y + 1) - (z^2 - 6z + 9) = 36$$

Factoring each of the trinomials enclosed by parentheses, we obtain

$$(x + 1)^2 + (y - 1)^2 - (z - 3)^2 = 36$$

Dividing through by 36 gives us

$$(x + 1)^2/36 + (y - 1)^2/36 - (z - 3)^2/36 = 1$$

This equation represents a hyperboloid of one sheet whose center is at

$$(x_0, y_0, z_0) = (-1, 1, 3)$$

and whose axis is a line parallel to the coordinate z axis.

8. We've been told that the coordinates of the vertex of an elliptic cone are $(-2, 3, 4)$, and that the cone's axis is parallel to the coordinate y axis. The equation must therefore be of the form

$$(x + 2)^2/a^2 - (y - 3)^2/b^2 + (z - 4)^2/c^2 = 0$$

where a, b, and c determine the eccentricity and orientation of the cross-sectional ellipses that we get when we slice through the cone with planes perpendicular to its axis. On the basis of the information given, all we can say about these constants is that they're positive real numbers.

9. Here's the generalized equation for the elliptic cone described in Problem 8:

$$(x + 2)^2/a^2 - (y - 3)^2/b^2 + (z - 4)^2/c^2 = 0$$

This cone intersects the xz plane in a curve where the y value is always equal to 0. If we set $y = 0$ in the above equation, we get

$$(x + 2)^2/a^2 - (0 - 3)^2/b^2 + (z - 4)^2/c^2 = 0$$

which simplifies to

$$(x+2)^2/a^2 - 9/b^2 + (z-4)^2/c^2 = 0$$

If we add the quantity $9/b^2$ to both sides, we obtain

$$(x+2)^2/a^2 + (z-4)^2/c^2 = 9/b^2$$

Now we can divide through by the quantity $9/b^2$, getting

$$(x+2)^2/(9a^2/b^2) + (z-4)^2/(9c^2/b^2) = 1$$

which can also be expressed as

$$(x+2)^2/(3a/b)^2 + (z-4)^2/(3c/b)^2 = 1$$

This is a generalized equation for an ellipse in the Cartesian plane, where the variables are x and z. The center of the ellipse has the coordinates

$$(x_0, z_0) = (-2, 4)$$

The semi-axes have lengths $3a/b$ and $3c/b$.

10. Here, once again, is the generalized equation that we derived for the elliptic cone described in Problem 8:

$$(x+2)^2/a^2 - (y-3)^2/b^2 + (z-4)^2/c^2 = 0$$

This cone intersects the xy plane in a curve where $z = 0$ at every point. If we set $z = 0$ in the above equation, we get

$$(x+2)^2/a^2 - (y-3)^2/b^2 + (0-4)^2/c^2 = 0$$

We can simplify this equation to

$$(x+2)^2/a^2 - (y-3)^2/b^2 + 16/c^2 = 0$$

If we subtract the quantity $16/c^2$ from both sides, we obtain

$$(x+2)^2/a^2 - (y-3)^2/b^2 = -16/c^2$$

Dividing through by the quantity $16/c^2$, we obtain

$$(x+2)^2/(16a^2/c^2) - (y-3)^2/(16b^2/c^2) = -1$$

which can be rewritten as

$$(x+2)^2/(4a/c)^2 - (y-3)^2/(4b/c)^2 = -1$$

Finally, we can multiply the entire equation through by -1 to obtain

$$(y - 3)^2 / (4b/c)^2 - (x + 2)^2 / (4a/c)^2 = 1$$

This is a generalized equation for a hyperbola in the Cartesian xy plane. It's oriented differently than the hyperbolas in Chap. 13, however. Instead of opening in the positive and negative x directions as the hyperbolas in Chap. 13 do, this pair of curves opens in the positive and negative y directions. The coordinates of the center are

$$(x_0, y_0) = (-2, 3)$$

The lengths of the semi-axes are $4a/c$ and $4b/c$.

Chapter 18

1. We have a three-way equation for a straight line. We want to find the preferred (lowest) direction numbers, and then find the direction vector based on those numbers. We also want to know the coordinates of a specific point on the line; any point will suffice. Stated again for reference, our three-way equation is

$$x - 1 = y - 2 = z - 4$$

This equation is in the standard symmetric form for a straight line in Cartesian xyz space, so we don't have to manipulate anything. The generalized standard-form symmetric equation is

$$(x - x_0)/a = (y - y_0)/b = (z - z_0)/c$$

where (x_0, y_0, z_0) are the coordinates of a specific point on the line, and a, b, and c are the direction numbers. Comparing the symmetric equation we've been given with the generalized form, we can see that

$$x_0 = 1$$
$$y_0 = 2$$
$$z_0 = 4$$

This tells us that $(1, 2, 4)$ is a point on the line. We can also see that

$$a = 1$$
$$b = 1$$
$$c = 1$$

so the direction numbers are $(1, 1, 1)$. This set of numbers is in lowest form, because there's no common divisor other than 1. We can write down a standard-form direction vector **m** from these values as

$$\mathbf{m} = \mathbf{i} + \mathbf{j} + \mathbf{k}$$

2. Stated again for reference, the three-way equation for our line is

$$4x = 5y = 6z$$

Dividing the entire equation through by 60, we obtain

$$x/15 = y/12 = z/10$$

The generalized standard-form symmetric equation is

$$(x - x_0)/a = (y - y_0)/b = (z - z_0)/c$$

where (x_0, y_0, z_0) are the coordinates of a specific point on the line, and a, b, and c are the direction numbers. In this case, we have

$$x_0 = 0$$
$$y_0 = 0$$
$$z_0 = 0$$

This tells us that the origin $(0,0,0)$ is on the line. We also have

$$a = 15$$
$$b = 12$$
$$c = 10$$

so the line's direction numbers are $(15,12,10)$. This ordered triple is in lowest form, so the direction vector **m** is

$$\mathbf{m} = 15\mathbf{i} + 12\mathbf{j} + 10\mathbf{k}$$

3. We've been given the symmetric equation

$$(x - 2)/3 = (4y - 8)/4 = (z + 5)/(-2)$$

We can divide out the middle portion to get

$$(x - 2)/3 = y - 2 = (z + 5)/(-2)$$

This equation is in the standard symmetric form. Once again, the generalized standard-form symmetric equation is

$$(x - x_0)/a = (y - y_0)/b = (z - z_0)/c$$

In this situation, we have

$$x_0 = 2$$
$$y_0 = 2$$
$$z_0 = -5$$

so we know that the point (2,2,−5) is on the line. We can also see that

$$a = 3$$
$$b = 1$$
$$c = -2$$

so the line's direction numbers are (3,1,−2). This ordered triple is in lowest form. (We could divide it through by −1, but then we'd get two negative elements instead of only one.) The direction vector **m** is therefore

$$\mathbf{m} = 3\mathbf{i} + \mathbf{j} - 2\mathbf{k}$$

4. Stated again for convenience, the parametric equations for our object are

$$x = -4$$
$$y = t$$
$$z = -t^2 - 1$$

The first equation tells us that the object lies in the plane $x = -4$, which is perpendicular to the x axis, parallel to the yz plane, and 4 units distant from the yz plane on the $-x$ side. We can draw projections of the coordinate y and z axes into the plane $x = -4$, obtaining a Cartesian yz grid. In that system, our object is a parabola defined by

$$y = t$$

and

$$z = -t^2 - 1$$

Substituting y for t in the second equation gives us

$$z = -y^2 - 1$$

Figure B-32 is a graph of this curve as it looks when we see the plane $x = -4$ broadside from a point on the positive x axis at a considerable distance from the origin.

Figure B-32 Illustration for the solution to Problem 4 in Chap. 18.

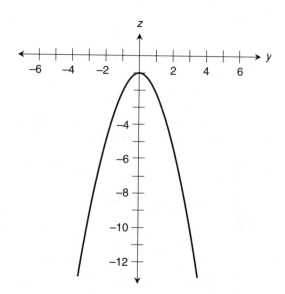

5. Stated again for convenience, the parametric equations are

$$x = t^2 + 2t$$

$$y = t$$

$$z = 0$$

According to the last equation, the whole object lies in the plane $z = 0$, which coincides with the *xy* plane. In that system, the object is a parabola defined by

$$x = t^2 + 2t$$

and

$$y = t$$

Substituting *y* for *t* in the first equation, we obtain

$$x = y^2 + 2y$$

Figure B-33 is a graph of this curve as it looks when we observe the *xy* plane broadside from a point on the +*z* axis at a considerable distance from the origin.

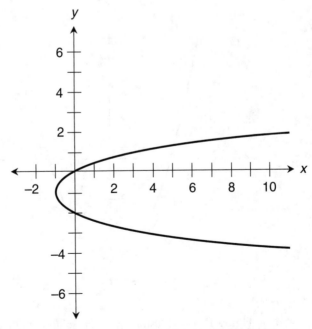

Figure B-33   Illustration for the solution to Problem 5 in Chap. 18.

6. We have been given the parametric equations

$$x = t$$
$$y = -7$$
$$z = t^2/2 - 5$$

According to the second equation, our object lies entirely in the plane $y = -7$. This plane is perpendicular to the y axis, parallel to the xz plane, and 7 units distant from the xz plane on the $-y$ side. When we draw projections of the three-space x and z axes onto the plane $y = -7$, we create a coordinate grid for a parabola defined by

$$x = t$$

and

$$z = t^2/2 - 5$$

Let's substitute x directly into the second of these equations to obtain

$$z = x^2/2 - 5$$

Figure B-34 is a graph of this equation as it appears when seen from a point of view broadside to the plane $y = -7$. We're looking in the $+y$ direction from somewhere along the $-y$ axis, but quite a lot farther away from the origin than the point where $y = -7$.

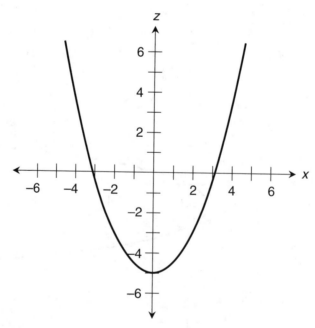

Figure B-34   Illustration for the solution to Problem 6 in Chap. 18.

7. Repeated for convenience, the parametric equations are

$$x = 4 \cos t$$

$$y = 4 \sin t$$

$$z = 1$$

According to the last equation, the object is contained entirely in the plane $z = 1$, which is parallel to the xy plane and 1 unit away from it on the $+z$ side. Within this plane, the parametric equations of the object reduce to

$$x = 4 \cos t$$

and

$$y = 4 \sin t$$

The graph is a circle in the plane $z = 1$, centered on the point $(0,0,1)$ and having a radius of 4 units. Figure B-35 is a graph of this object as we would see it looking broadside at the plane $z = 1$, from a point fairly far from the origin on the $+z$ axis.

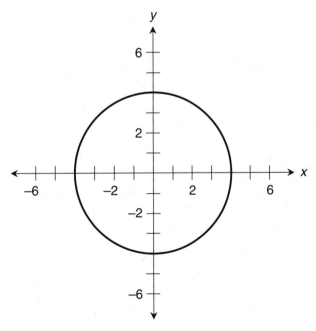

Figure B-35 Illustration for the solution to Problem 7 in Chap. 18.

8. Repeated for convenience, the parametric equations are

$$x = 5 \cos t$$

$$y = 0$$

$$z = 5 \sin t$$

According to the middle equation, the entire object lies in the plane $y = 0$, which is the xz plane. Within the Cartesian xz system, the equations describing the object are

$$x = 5 \cos t$$

and

$$z = 5 \sin t$$

The graph is a circle in the xz plane, centered at the origin and having a radius of 5 units. Figure B-36 illustrates this circle as seen from somewhere along the $-y$ axis. We're fairly far from the origin, and we're looking in the $+y$ direction.

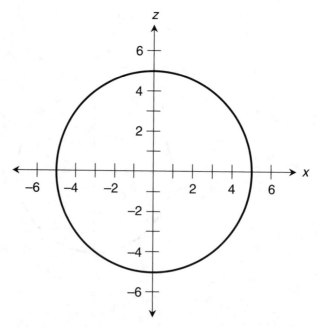

Figure B-36   Illustration for the solution to Problem 8 in Chap. 18.

9. Repeated for convenience, the parametric equations are

$$x = 5 \cos t$$
$$y = 3 \sin t$$
$$z = \pi$$

According to the last equation, the object is contained entirely in the plane $z = \pi$, which is parallel to the xy plane and π units away from it on the +z side. Within that plane, the parametric equations are

$$x = 5 \cos t$$

and

$$y = 3 \sin t$$

The graph is an ellipse in the plane $z = \pi$ and centered on $(0,0,\pi)$. The major semi-axis is parallel to the x axis, and measures 5 units wide. The minor semi-axis is parallel to the y axis, and measures 3 units high. Figure B-37 is a graph of this ellipse as we gaze broadside at the plane $z = \pi$, from some location on the +z axis that's considerably farther from the origin than $(0,0,\pi)$.

10. Stated again for reference, the parametric equations are

$$x = 2 \cos t$$
$$y = t/(2\pi)$$
$$z = 2 \sin t$$

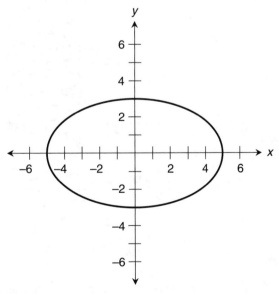

Figure B-37 Illustration for the solution to Problem 9 in Chap. 18.

The object described by these equations is a circular helix having a radius of 2 units, and centered on the *y* axis. Here are some values of *x*, *y*, and *z* that we can calculate as *t* varies, causing a point on the helix to complete a single revolution in a moving plane perpendicular to the *y* axis:

- When $t = 0$, we have $x = 2$, $y = 0$, and $z = 0$.
- When $t = \pi/2$, we have $x = 0$, $y = 1/4$, and $z = 2$.
- When $t = \pi$, we have $x = -2$, $y = 1/2$, and $z = 0$.
- When $t = 3\pi/2$, we have $x = 0$, $y = 3/4$, and $z = -2$.
- When $t = 2\pi$, we have $x = 2$, $y = 1$, and $z = 0$.

Every time *t* increases by 2π, our point makes exactly one revolution in a moving plane that's always perpendicular to the *y* axis. In addition, we can see that every time *t* increases by 2π, our point gets 1 unit more distant from the *xz* plane in the +*y* direction. The pitch of the helix is therefore 1 linear unit per revolution.

The sense of rotation is rather tricky to describe. Suppose that we're somewhere on the *y* axis, and we direct our gaze in the −*y* direction. As the value of the parameter *t* increases, causing a point on the helix to move generally toward us, the point appears to revolve counterclockwise around the *y* axis. The helix's sense of rotation is therefore counterclockwise in the +*y* direction. Figure B-38 is a perspective drawing of the graph of this object in Cartesian *xyz* space.

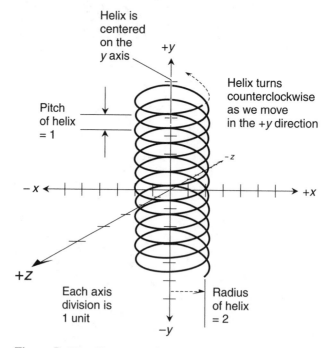

Figure B-38 Illustration for the solution to Problem 10 in Chap. 18. Each axis division represents 1 unit.

Chapter 19

1. Here's the sequence S again, for reference:

$$S = 5, 3, 1, -1, -3, -5, \ldots$$

We can read the value of the first term directly as $s_0 = 5$. After that, the terms proceed as follows:

- The second term is 3, which is 2 less than the first term.
- The third term is 1, which is 2 less than the second term.
- The fourth term is -1, which is 2 less than the third term.
- The fifth term is -3, which is 2 less than the fourth term.
- The sixth term is -5, which is 2 less than the fifth term.

We're told that S is an arithmetic sequence, so we can be sure that the terms always differ by the same constant c. The general form is

$$S = s_0, (s_0 + c), (s_0 + 2c), (s_0 + 3c), \ldots$$

where s_0 is the initial term and c is the constant. In this case, $c = -2$, giving us

$$
\begin{aligned}
S &= [5], [5 + (-2)], [5 + 2 \times (-2)], [5 + 3 \times (-2)], \ldots \\
&= 5, (5 - 2), (5 - 4), (5 - 6), (5 - 8), (5 - 10), \ldots \\
&= 5, 3, 1, -1, -3, -5, \ldots
\end{aligned}
$$

The brackets in the first line and the parentheses in the second line aren't technically necessary, but they serve to visually isolate the terms. In general, the nth term of the series is what we get when we multiply the constant c by $n - 1$ and then add it to s_0. The hundredth term in S is therefore

$$
\begin{aligned}
s_0 + (n - 1)c &= 5 + (100 - 1) \times (-2) \\
&= 5 + 99 \times (-2) \\
&= 5 + (-198) \\
&= 5 - 198 \\
&= -193
\end{aligned}
$$

2. The values of the terms in the sequence S become large negatively without bound. We can choose an integer that's as large negatively as we want, and we'll always be able to find an element in S that's still more negative. Therefore, S does not converge.

3. We're told that $t_0 = 2$ and $k = -4$. Let's plug these values into the general formula for an infinite geometric sequence

$$T = t_0, t_0 k, t_0 k^2, t_0 k^3, t_0 k^4, \ldots$$

When we do that for the first seven terms, we get

$$t_0 = 2$$
$$t_1 = 2 \times (-4) = -8$$
$$t_2 = 2 \times (-4)^2 = 2 \times 16 = 32$$
$$t_3 = 2 \times (-4)^3 = 2 \times (-64) = -128$$
$$t_4 = 2 \times (-4)^4 = 2 \times 256 = 512$$
$$t_5 = 2 \times (-4)^5 = 2 \times (-1024) = -2048$$
$$t_6 = 2 \times (-4)^6 = 2 \times 4096 = 8192$$

This series does not converge. The terms' *actual values* alternate between positive and negative. The *absolute values* increase by a factor of 4 with each succeeding term. In an intuitive sense, T "approaches both positive infinity and negative infinity."

4. This time, we're given the values $t_0 = 2$ and $k = -1/4$. We can plug these values into the general formula for an infinite geometric sequence

$$T = t_0, t_0k, t_0k^2, t_0k^3, t_0k^4, \ldots$$

Calculating the first seven terms, we obtain

$$t_0 = 2$$
$$t_1 = 2 \times (-1/4) = -1/2$$
$$t_2 = 2 \times (-1/4)^2 = 2 \times 1/16 = 1/8$$
$$t_3 = 2 \times (-4)^3 = 2 \times (-1/64) = -1/32$$
$$t_4 = 2 \times (-4)^4 = 2 \times 1/256 = 1/128$$
$$t_5 = 2 \times (-4)^5 = 2 \times (-1/1024) = -1/512$$
$$t_6 = 2 \times (-4)^6 = 2 \times 1/4096 = 1/2048$$

This series converges on 0. The terms' *actual values* alternate between positive and negative. The *absolute values* decrease by a factor of 4 with each succeeding term. In an intuitive sense, T "approaches 0 from both sides."

5. Here's the sequence B again, for reference:

$$B = 0/1, 1/2, 2/3, 3/4, 4/5, \ldots, (n-1)/n, \ldots$$

When we add the elements of B, we obtain the series

$$B_+ = 0/1 + 1/2 + 2/3 + 3/4 + 4/5 + \cdots + (n-1)/n + \cdots$$

In summation notation, we can write

$$B_+ = \sum_{n=1}^{\infty} (n-1)/n$$

The first five terms of the sequence B_*, listing the partial sums of B_+, are

$$B_* = 0/1, \ 1/2, \ 7/6, \ 23/12, \ 163/60, \ ...$$

As we continue to calculate partial sums, we keep adding values that get closer and closer to 1. The terms in B_* grow at an ever-increasing rate. If we choose any positive real number, no matter how large, we can eventually generate an element of B_* that exceeds it. Therefore, the sequence B_* of partial sums does not converge.

6. Here's the series again, expressed in summation notation:

$$S_+ = \sum_{i=1}^{n} 1/10^i$$

When we write out the first five terms of S_+ as fractions, we get the sum

$$S_+ = 1/10 + 1/100 + 1/1000 + 1/10,000 + 1/100,000 + \cdots$$

Expressing the terms as powers of 10, we have

$$S_+ = 10^{-1} + 10^{-2} + 10^{-3} + 10^{-4} + 10^{-5} + \cdots$$

Expressing the terms as decimal quantities, we have

$$S_+ = \ 0.1 + 0.01 + 0.001 + 0.0001 + 0.00001 + \cdots$$

The first five terms in the sequence of partial sums S_* are

$$S_* = 0.1, \ 0.11, \ 0.111, \ 0.1111, \ 0.11111, \ ...$$

7. We want to find the limit

$$\underset{n \to \infty}{Lim} \sum_{i=1}^{n} 1/10^i$$

if it exists. From the results of Problem 6, we see that it's the limit of the sequence of partial sums

$$S_* = 0.1, \ 0.11, \ 0.111, \ 0.1111, \ 0.11111, \ ...$$

so we know that

$$\underset{n \to \infty}{Lim} \sum_{i=1}^{n} 1/10^i = 0.11111...$$

We learned in our algebra courses that $0.11111... = 1/9$. Therefore

$$\underset{n \to \infty}{Lim} \sum_{i=1}^{n} 1/10^i = 1/9$$

8. When $n = 2$, the partial sum is

$$(1 + 2)/2^2 = 0.75$$

When $n = 6$, the partial sum is approximately

$$(1 + 2 + 3 + 4 + 5 + 6)/6^2 = 0.58333$$

When $n = 10$, the partial sum is

$$(1 + 2 + 3 + \cdots + 8 + 9 + 10)/10^2 = 0.55$$

When $n = 20$, the partial sum is

$$(1 + 2 + 3 + \cdots + 18 + 19 + 20)/20^2 = 0.525$$

We're approaching 1/2 from the right (the positive side). If you're a computer expert, try programming your machine to work out the partial sums for much larger values of n, and see more clearly that 1/2 (or 0.5) is indeed the limit of this sequence of partial sums.

9. Refer to Table B-1 for squares of integers from 1 to 20. When $n = 2$, the partial sum is

$$(1^2 + 2^2)/2^3 = 0.625$$

Table B-1. Squares and cubes of positive integers from 1 to 20.

n	n^2	n^3
1	1	1
2	4	8
3	9	27
4	16	64
5	25	125
6	36	216
7	49	343
8	64	512
9	81	729
10	100	1000
11	121	1331
12	144	1728
13	169	2197
14	196	2744
15	225	3375
16	256	4096
17	289	4913
18	324	5832
19	361	6859
20	400	8000

When $n = 6$, the partial sum is approximately

$$(1^2 + 2^2 + 3^2 + 4^2 + 5^2 + 6^2)/6^3 \approx 0.42130$$

Remember that the "wavy" equals sign means "is approximately equal to." When $n = 10$, the partial sum is

$$(1^2 + 2^2 + 3^2 + \cdots + 8^2 + 9^2 + 10^2)/10^3 = 0.385$$

When $n = 20$, the partial sum is

$$(1^2 + 2^2 + 3^2 + \cdots + 18^2 + 19^2 + 20^2)/20^3 = 0.35875$$

We're approaching 1/3 from the right. If you're a computer expert, try programming your machine to work out the partial sums for much larger values of n, and see more clearly that 1/3 (or 0.33333...) is indeed the limit of this sequence of partial sums.

10. Refer to Table B-1 for cubes of integers from 1 to 20. When $n = 2$, the partial sum is

$$(1^3 + 2^3)/2^4 = 0.5625$$

When $n = 6$, the partial sum is approximately

$$(1^3 + 2^3 + 3^3 + 4^3 + 5^3 + 6^3)/6^4 \approx 0.34028$$

When $n = 10$, the partial sum is

$$(1^3 + 2^3 + 3^3 + \cdots + 8^3 + 9^3 + 10^3)/10^4 = 0.3025$$

When $n = 20$, the partial sum is

$$(1^3 + 2^3 + 3^3 + \cdots + 18^3 + 19^3 + 20^3)/20^4 = 0.27563$$

We're approaching 1/4 from the right. If you're a computer expert, try programming your machine to work out the partial sums for much larger values of n, and see more clearly that 1/4 (or 0.25) is indeed the limit of this sequence of partial sums.

Answers to Final Exam Questions

1. d	2. a	3. b	4. d	5. c
6. e	7. e	8. d	9. e	10. c
11. c	12. b	13. e	14. c	15. e
16. c	17. e	18. d	19. b	20. d
21. a	22. a	23. e	24. b	25. c
26. b	27. a	28. d	29. e	30. d
31. a	32. b	33. c	34. d	35. e
36. a	37. a	38. e	39. b	40. a
41. e	42. c	43. a	44. a	45. d
46. c	47. e	48. d	49. a	50. c
51. e	52. e	53. c	54. b	55. a
56. b	57. c	58. a	59. d	60. e
61. e	62. c	63. a	64. a	65. d
66. c	67. a	68. d	69. b	70. b
71. b	72. d	73. a	74. a	75. e
76. a	77. c	78. b	79. d	80. c
81. d	82. b	83. b	84. c	85. e
86. a	87. e	88. e	89. a	90. e
91. c	92. d	93. e	94. c	95. c
96. e	97. a	98. a	99. a	100. d

Special Characters in Order of Appearance

Symbol	First use	Meaning
$\in$	Chapter 1	Set symbol meaning "is an element of"
(x,y)	Chapter 1	Ordered pair (x,y)
		Open interval where $x < y$, such that neither x nor y is included
$[x,y)$	Chapter 1	Half-open interval where $x < y$, such that x is included but y is not
$(x,y]$	Chapter 1	Half-open interval where $x < y$, such that y is included but x is not
$[x,y]$	Chapter 1	Closed interval where $x < y$, such that x and y are both included
Δ	Chapter 1	Uppercase Greek letter delta, symbolizing difference in coordinate values
$\approx$	Chapter 1 (App. A)	"Squiggly" or "wavy" equals sign, symbolizing approximate equality
θ	Chapter 2	Lowercase Greek letter theta, symbolizing an angular variable
ϕ	Chapter 2	Lowercase Greek letter phi, symbolizing an angular variable
$\pm$	Chapter 2	Positive or negative value
∞	Chapter 2	Lemniscate, denoting infinity
$\bullet$	Chapter 5	Boldface dot, symbolizing the dot product of vectors
$\times$	Chapter 5	Boldface multiplication symbol, denoting the cross product of vectors
i	Chapter 6	Positive square root of -1
j	Chapter 6	Positive square root of -1, also called j operator
$\mathbf{i}$	Chapter 8	Standard unit vector $(1,0,0)$ in Cartesian xyz space
$\mathbf{j}$	Chapter 8	Standard unit vector $(0,1,0)$ in Cartesian xyz space
$\mathbf{k}$	Chapter 8	Standard unit vector $(0,0,1)$ in Cartesian xyz space
$\Leftrightarrow$	Chapter 10	Logical equivalence symbol, meaning "if and only if"

f^{-1}	Chapter 12	Inverse of relation or function f
$\log_e$	Chapter 14	Natural (base-e) logarithm
$\ln$		
$\log_{10}$	Chapter 14	Common (base-10) logarithm
Lim	Chapter 19	Limit of a sequence, series, or function
Σ	Chapter 19	Uppercase Greek letter sigma, symbolizing summation of a sequence

Suggested Additional Reading

Bluman, A., *Math Word Problems Demystified*, New York: McGraw-Hill, 2005.

Bluman, A., *Pre-Algebra Demystified*, New York: McGraw-Hill, 2004.

Gibilisco, S., *Algebra Know-It-ALL*, New York: McGraw-Hill, 2008.

Gibilisco, S., *Calculus Know-It-ALL*, New York: McGraw-Hill, 2009.

Gibilisco, S., *Mastering Technical Mathematics,* 3d ed., New York: McGraw-Hill, 2008.

Gibilisco, S., *Technical Math Demystified*, New York: McGraw-Hill, 2006.

Huettenmueller, R., *Algebra Demystified*, New York: McGraw-Hill, 2003.

Huettenmueller, R., *College Algebra Demystified*, New York: McGraw-Hill, 2004.

Huettenmueller, R., *Pre-Calculus Demystified*, New York: McGraw-Hill, 2005.

Krantz, S., *Calculus Demystified*, New York: McGraw-Hill, 2003.

Krantz, S., *Differential Equations Demystified*, New York: McGraw-Hill, 2005.

Olive, J., *Maths: A Student's Survival Guide,* 2d ed., Cambridge, England: Cambridge University Press, 2003.

Prindle, A., *Math the Easy Way,* 3d ed., Hauppauge, NY: Barron's Educational Series, 1996.

Shankar, R., *Basic Training in Mathematics: A Fitness Program for Science Students*, New York: Plenum Publishing Corporation, 1995.

Index